2nd International Conference on Computational Mathematics and Engineering Sciences (CMES 2017)

ITM Web of Conferences Volume 13 (2017)

Istanbul, Turkey
20 - 22 May 2017

Editors:

Abdon Atangana
Hasan Bulut
Zakia Hammouch
Haci Mehmet Baskonus
Toufik Mekkaoui
Fethi Bin Muhammad Belgacem

ISBN: 978-1-5108-4884-9

Printed from e-media with permission by:

Curran Associates, Inc.
57 Morehouse Lane
Red Hook, NY 12571

CURRAN ASSOCIATES INC.
proceedings
.com

Some format issues inherent in the e-media version may also appear in this print version.

Printed by Curran Associates, Inc. (2017)

For additional information, please contact EDP Sciences – Web of Conferences at the address below.

EDP Sciences – Web of Conferences
17, Avenue du Hoggar
Parc d'Activité de Courtabœuf
BP 112
F-91944 Les Ulis Cedex A
France

Phone: +33 (0) 1 69 18 75 75
Fax: +33 (0) 1 69 28 84 91

contact-edps@webofconferences.org

Additional copies of this publication are available from:

Curran Associates, Inc.
57 Morehouse Lane
Red Hook, NY 12571 USA
Phone: 845-758-0400
Fax: 845-758-2633
Email: curran@proceedings.com
Web: www.proceedings.com

TABLE OF CONTENTS

A PRACTICAL METHOD FOR ANALYTICAL EVALUATION OF APPROXIMATE SOLUTIONS OF FISHER'S EQUATIONS 1
Bildik Necdet, Deniz Sinan

PISA QUESTION AND REASONING SKILL 14
Ersoy Esen, Bal-Incebacak Belgin

DEVELOPING AN ACHIEVEMENT TEST FOR FRACTION TEACHING: VALIDITY AND RELIABILITY ANALYSIS 39
Bal-Incebacak Belgin, Ersoy Esen

ASYMPTOTICALLY STATISTICAL EQUIVALENT OF ORDER α IN AMENABLE SEMIGROUPS 53
Altin Yavuz, Et Mikail, Altinok Hifsi

A NOTE ON STURM-LIOUVILLE PROBLEM FOR DIFFERENCE EQUATIONS 58
Bas Erdal, Ozarslan Ramazan

TRIGONOMETRIC FUNCTION SOLUTIONS OF FRACTIONAL DRINFELD'S SOKOLOV - WILSON SYSTEM 64
Koçak Zeynep Fidan, Yel Gülnur

GENERALIZED CESÀRO SUMMABLE DIFFERENCE SEQUENCE SPACES 68
Et Mikail

THE NUMERICAL SOLUTION OF FRACTIONAL BRATU-TYPE DIFFERENTIAL EQUATIONS 73
Demir Duygu Dönmez, Zeybek Aylin

CRYPTANALYSIS OF APPLICATION OF LAPLACE TRANSFORM FOR CRYPTOGRAPHY 84
Gençoglu Muharrem Tuncay

COMBINING CRYPTOGRAPHY WITH STEGANOGRAPHY 89
Gençoglu Muharrem Tuncay

THE CREATIVITY DIMENSION OF INSTRUCTIONAL MATERIALS DESIGNED BY PROSPECTIVE TEACHERS: THE COMPARISON ACROSS DOMAINS 97
Ersoy Esen, Sendurur Emine, Çetin Ismail

THE EVALUATION OF THE PROBLEM SOLVING IN MATHEMATICS COURSE ACCORDING TO STUDENT VIEWS 108
Ersoy Esen, Bal-Incebacak Belgin

MODELING ROUTING CONSTRUCTS TO REPRESENT DISTRIBUTED WORKFLOW PROCESSES USING EXTENDED PETRI NETS 126
Karay Mehmet

SOME WAVE SIMULATION PROPERTIES OF THE (2+1) DIMENSIONAL BREAKING SOLITON EQUATION 139
Aksan Emine Nesligül, Bulut Hasan, Kayhan Miraç

NOVEL BEHAVIORS TO THE NONLINEAR EVOLUTION EQUATION DESCRIBING THE DYNAMICS OF IONIC CURRENTS ALONG MICROTUBULES 144
Baskonus Haci Mehmet, Erdogan Fevzi, Ozkul Arif, Asmouh Ilham

SOME PROTOTYPE RESULTS OF THE SYMMETRIC REGULARIZED LONG WAVE EQUATION ARISING IN NONLINEAR ION ACOUSTIC WAVES 149
Isik Erdem, Bulut Hasan, Atas Sibel Sehriban

SOME NOVEL EXPONENTIAL AND COMPLEX STRUCTURAL PROPERTIES OF THE FISHER EQUATION ARISING IN MATHEMATICAL BIOSCIENCE 157
Bulut Hasan, Atas Sibel Sehriban, Baskonus Haci Mehmet

REGARDING ON THE NOVEL FORMS OF THE (3+1) - DIMENSIONAL KADOMSTEV-PETVIASHVILI EQUATION 165
Bulut Hasan, Demirdag Betul, Baskonus Haci Mehmet

ON SOME COMPLEX ASPECTS OF THE (2+1)-DIMENSIONAL BROER-KAUP-KUPERSHMIDT SYSTEM 171
Bulut Hasan, Isik Hilal Arslanoglu, Sulaiman Tukur Abdulkadir

A THREE FACTOR REMOTE USER AUTHENTICATION SCHEME USING COLLISION RESIST FUZZY EXTRACTOR IN SINGLE SERVER ENVIRONMENT 178
Giri Debasis, Maitra Tanmoy

CLASSIFICATIONS ON THE TRAVELLING WAVE SOLUTIONS TO THE (3+1)-DIMENSIONAL GENERALIZED KP AND JIMBO-MIWA EQUATIONS 192
Isik Ozlem, Degirmenci Ozge Irmak, Bulut Hasan
SEQUENCES WITH RANDOM INDICE IN CLASSICAL BANACH SPACE 196
Gürcan Mehmet, Çalik Sinan, Güral Yunus
USEFUL INEQUALITIES FOR UPPER AND LOWER BOUNDS FOR THE REPAIRABLE AGING COLD STANDBY SYSTEM 200
Gökdere Gökhan
ON ASYMPTOTICALLY LACUNARY STATISTICAL EQUIVALENT SEQUENCES OF ORDER α IN PROBABILITY 207
Isik Mahmut, Akbas Kübra Elif
NUMERICAL INVESTIGATION ON MHD FLOW AND HEAT TRANSFER OVER AN EXPONENTIALLY STRETCHING SHEET WITH VISCOUS DISSIPATION AND RADIATION EFFECTS 212
Çilingir Süngü Inci
CONTINUOUS ULTRASONIC THICKNESS MEASUREMENT TECHNOLOGY 225
Pronin Vitaliy, Shkatov Petr, Sandulyak Anna, Baskonus Haci Mehmet
DEPTH MEASUREMENT OF CLOSELY SPACED SURFACE CRACKS BY THE ELECTRIC POTENTIAL METHOD 229
Shkatov Petr, Lisitsyna Irina, Sandulyak Anna, Bulut Hasan
BENEFITS OF REVERSE ENGINEERING TECHNOLOGIES IN SOFTWARE DEVELOPMENT MAKERSPACE 232
Aabidi M.H., El Mahi Bouziane, Baidada Chafik, Jakimi Abdeslam, Ammar Hany
TWO-EXPONENTIAL MODEL OF FERROIMPURITIES EXTRACTION DURING MAGNETIC INSPECTION OF THESE IMPURITIES CONTENT IN VARIOUS MEDIA 239
Sandulyak Anna, Sandulyak Alexander, Polismakova Maria, Sandulyak Darya
PREPARATION OF IMPROVED TURKISH DATASET FOR SENTIMENT ANALYSIS IN SOCIAL MEDIA 244
Makinist Semiha, Hallaç Ibrahim Riza, Ay Karakus Betül, Aydin Galip
A STUDY ON THE IMPROVED TAN(Φ (ε) /2) -EXPANSION METHOD 250
Karaagac Berat, Yagmurlu N. Murat, Esen Alaattin
ON NUMERICAL SOLUTION OF FRACTIONAL ORDER BOUNDARY VALUE PROBLEM WITH SHOOTING METHOD 255
Demir Hüseyin, Baltürk Yücel
ANALYSIS OF OUTPUT VOLTAGE HARMONICS OF VOLTAGE SOURCE INVERTER USED PI AND PID CONTROLLERS OPTIMIZED WITH ITAE PERFORMANCE CRITERIA 260
Gül Ozan, Tan Nusret
MATHEMATICAL MODELLING OF THE SYNTHESIS OF NEW MATERIALS 264
Antonov Valery, Davidov Roman, Kalinin Nikolay, Ryakhovskiy Alexey
ON INVERSE STURM-LIOUVILLE PROBLEMS WITH SYMMETRIC POTENTIALS 272
Tuz Münevver
ON SOME SPECTRAL PROBLEMS FOR DIFFUSION OPERATOR 281
Babaoglu Mine, Panakhov Etibar S.
Author Index

Statement of Peer review

In submitting conference proceedings to ITM Web of Conferences, I certify to the Publisher that I adhere to the **Policy on Publishing Integrity** of the journal in order to safeguard good scientific practice in publishing.

1. All articles have been subjected to peer review administered by the proceedings editors.
2. Reviews have been conducted by expert referees, who have been requested to provide unbiased and constructive comments aimed, whenever possible, at improving the work.
3. Proceedings editors have taken all reasonable steps to ensure the quality of the materials they publish and their decision to accept or reject a paper for publication has been based only on the merits of the work and the relevance to the journal.

Title, date and place of the conference

2 nd International Conference on Computational Mathematics and Engineering Sciences (CMES2017), May 20-22 2017, Istanbul, Turkey.

Proceedings editor(s):

Abdon Atangana,
Hasan Bulut,
Zakia Hammouch,
Haci Mehmet Baskonus,
Toufik Mekkaoui,
Fethi Bin Muhammad Belgacem

Date and editor's signature

19.09.2017 Prof. Dr. Hasan BULUT

ITM Web of Conferences **13**, 01001 (2017) DOI: 10.1051/itmconf/20171301001
CMES2017

A Practical Method for Analytical Evaluation of Approximate Solutions of Fisher's Equations

Necdet BİLDİK[1,*] and *Sinan* DENİZ[1,**]

[1] *Department of Mathematics, Faculty of Art and Sciences, Manisa Celal Bayar University, 45140 Manisa, Turkey.*

Abstract. In this article, a framework is developed to get more approximate solutions to nonlinear partial differential equations by applying perturbation iteration technique. This technique is modified and improved to solve nonlinear diffusion equations of the Fisher type. Some problems are investigated to illustrate the efficiency of the method. Comparisons between the new results and the solutions obtained by other techniques prove that this technique is highly effective and accurate in solving nonlinear problems. Convergence analysis and error estimate are also provided by using some related theorems. The basic ideas indicated in this work are anticipated to be further developed to handle nonlinear models.

1 Introduction

In the past few decades many analytical and numerical methods have been suggested to examine the solutions of partial differential equations (PDEs) such as Adomian decomposition method [1, 2], variational iteration method[3, 9], homotopy perturbation method [10–12], homotopy analysis method [13–16] and others [4–8]. However, even these methods may not always be sufficient to reach satisfactory results. For this reason, much attention has been recently devoted to the newly developed techniques for obtaining more accurate and efficient solutions. Two of these analytical methods are the optimal homotopy asymptotic method and optimal perturbation iteration method which are recently developed by Marinca et al and Deniz et al respectively. Many authors have applied these methods to solve nonlinear differential equations arising in scientific problems [17–22].

In this study, we consider the generalized Fisher's equation

$$u_t = au_{xx} + bu\left(1 - u^{\alpha}\right) \tag{1}$$

where a,b and α are real numbers. It is used as a model equation for the evolution of a neutron population in a nuclear reactor and a prototype model for a propagating flame. It also represents the propagation of a viral mutant in an infinitely long habitat [23, 24].
A traveling wave solution $u(x,t) = u(\xi = x - ct)$, spreading with a speed c, must be positive and bounded [25]. Therefore the boundary conditions for the traveling wave solution are usually

$$u(\xi \to -\infty) \to 1, u(\xi \to \infty) \to 0 \tag{2}$$

* e-mail: n.bildik@cbu.edu.tr
** e-mail: sinandeniz01@gmail.com

or

$$u(x,0) = v(x), \lim_{|x|\to 1} u(x,t) = 0, |x| < 1, t > 0 \tag{3}$$

where $|x| < 1, t > 0$ and $v(x)$ is a given function.

In this study, we modify perturbation iteration method (PIM), which is recently constructed by Pakdemirli et al. [26, 27], to apply it to solve PDEs, particularly for the Eq. (1). We also provide the convergence analysis and error estimate of PIM.

2 Method of Solution

To give the basic concepts of PIM, let us consider the Fisher's equation in closed form:

$$F(u_{xx}, u_t, u) = u_t + bu^{\alpha+1} - au_{xx} - bu = 0 \tag{4}$$

where $u = u(x,t)$. Introducing the perturbation parameter $\varepsilon = 1$ with nonlinear term yields

$$F(u_{xx}, u_t, u, \varepsilon) = u_t + \varepsilon\left(-au_{xx} - bu + bu^{\alpha+1}\right) = 0. \tag{5}$$

F can also be divided as

$$F = S + R \tag{6}$$

where

$$S = u_t$$

is the linear simpler part, which is easy to deal with and

$$R(u_{xx}, u, \varepsilon) = \varepsilon\left(-au_{xx} - bu + bu^{\alpha+1}\right)$$

is the remaining part. One correction term (u_c) is taken from the perturbation expansion as:

$$u_{n+1} = u_n + \varepsilon (u_c)_n \tag{7}$$

where $n \in \mathrm{N} \cup \{0\}$. Perturbation iteration algorithms will be classified with respect to the degrees of derivatives in the Taylor expansions (m). Briefly, this process will be represented as PIA-m.

PIA-1:

Substituting (7) into the remaining part R in (6), then expanding in a Taylor series only with first derivative gives PIA-1:

$$R + R_{u_{xx}}((u_c)_n)_{xx}\varepsilon + R_u (u_c)_n \varepsilon + R_\varepsilon \varepsilon = -S \tag{8}$$

where $R_u = \frac{\partial R}{\partial u}, R_\varepsilon = \frac{\partial R}{\partial \varepsilon}, R_{u_{xx}} = \frac{\partial R}{\partial u_{xx}}$. Evaluating all derivatives and functions at $\varepsilon = 0$ yields

$$((u_c)_n)_t = a(u_n)_{xx} + b(u_n) - (u_n)_t - b(u_n)^{\alpha+1}. \tag{9}$$

We can easily obtain first correction term $(u_c)_0$ from (9) by determining a trial function u_0 which satisfies initial condition(s). Then, u_1 is determined by using $(u_c)_0$, (7) and initial condition(s). One can obtain satisfactory results by repeating iteration process with the aid of (7) and (9).

PIA-2:

Following the same procedure as in the previous algorithm with second derivative gives PIA-2:

$$\begin{aligned} &R + R_{u_{xx}} \left((u_c)_n\right)_{xx} \varepsilon + R_u (u_c)_n \varepsilon + R_\varepsilon \varepsilon + \frac{\varepsilon^2 R_{\varepsilon\varepsilon}}{2} + \frac{\varepsilon^2 R_{u_{xx}u_{xx}}}{2} \left((u_c)_n\right)_{xx}^2 \\ &+ \frac{\varepsilon^2 R_{uu}}{2} \left((u_c)_n\right)^2 + \varepsilon^2 R_{u_{xx}u} \left((u_c)_n\right)_{xx} (u_c)_n + \varepsilon^2 R_{u_{xx}\varepsilon} \left((u_c)_n\right)_{xx} + \varepsilon^2 R_{u\varepsilon} (u_c)_n = -S \end{aligned} \quad (10)$$

By computing all derivatives and functions at $\varepsilon = 0$, we have

$$\left((u_c)_n\right)_t + (\alpha + 1)\, b\, (u_n)^\alpha (u_c)_n = a\, (u_n)_{xx} + b\, (u_n) - (u_n)_t - b\, (u_n)^{\alpha+1} \quad (11)$$

for the Fisher's equation (5). By means of (7) and (11), iterative scheme is developed.
After obtaining first correction term $(u_c)_0$, first approximate solutions are defined by using prescribed initial conditions. Suppose now that

$$\begin{aligned} u_0 &= C_0 \\ (u_c)_n &= C_{n+1}. \end{aligned} \quad (12)$$

Then we have

$$\begin{aligned} u_0 &= C_0 \\ u_1 &= u_0 + (u_c)_0 = C_0 + C_1 \\ u_2 &= u_1 + (u_c)_1 = C_0 + C_1 + C_2 \\ u_3 &= u_2 + (u_c)_2 = C_0 + C_1 + C_2 + C_3 \\ &\vdots \\ u_{n+1} &= u_n + (u_c)_n = C_0 + C_1 + C_2 + \cdots + C_{n+1} = \textstyle\sum_{i=0}^{n+1} C_i \end{aligned} \quad (13)$$

Consequently, one can represent the solution as

$$u(x,t) = \lim_{n\to\infty} u_{n+1}(x,t) = \sum_{i=0}^{\infty} C_i. \quad (14)$$

3 Convergence Analysis and Error Estimate

In this section, we investigate the convergence of the PIM by giving some theorems.

Theorem 1:(Banach's fixed point theorem). *Suppose that B be a Banach space and*

$$A : B \to B \quad (15)$$

is a nonlinear mapping, and assume that

$$\|A[u] - A[\bar{u}]\| \le \beta \|u - \bar{u}\|, u, \bar{u} \in B, \quad (16)$$

for some constant $\beta < 1$ *. Then A has a unique fixed point. Also, the sequence*

$$u_{n+1} = A[u_n], \quad (17)$$

with an arbitrary choice of $u_0 \in B$ *, converges to the fixed point of A and*

$$\|u_r - u_s\| \le \|u_1 - u_0\| \sum_{j=s-1}^{r-2} \beta^j. \quad (18)$$

The following theorem, which is needed for our analysis, can be deduced from the Banach fixed point theorem [28].

Theorem 2. *Let B be a Banach space denoted with a suitable norm $\|.\|$ over which the series $\sum_{i=0}^{\infty} C_i$ is defined and assume that the initial guess $u_0 = C_0$ remains inside the ball of the solution u. The series solution $\sum_{i=0}^{\infty} C_i$ converges if there exists β such that*

$$\|C_{n+1}\| \leq \beta \|C_n\| . \tag{19}$$

Proof: We first define a sequence as:

$$\begin{aligned} A_0 &= C_0 \\ A_1 &= C_0 + C_1 \\ A_2 &= C_0 + C_1 + C_2 \\ &\vdots \\ A_n &= C_0 + C_1 + C_2 + \cdots + C_n \end{aligned} \tag{20}$$

Now, we must show that $\{A_n\}_{n=0}^{\infty}$ is a Cauchy sequence in B. To do this, consider that

$$\|A_{n+1} - A_n\| = \|C_{n+1}\| \leq \beta \|C_n\| \leq \beta^2 \|C_{n-1}\| \leq \cdots \leq \beta^{n+1} \|C_0\| . \tag{21}$$

For every $n, k \in \mathrm{N}, n \geq k$:

$$\begin{aligned} &\|A_n - A_k\| = \|(A_n - A_{n-1}) + (A_{n-1} - A_{n-2}) + \cdots + (A_{k+1} - A_k)\| \\ &\leq \|A_n - A_{n-1}\| + \|A_{n-1} - A_{n-2}\| + \cdots + \|A_{k+1} - A_k\| \\ &\leq \beta^n \|C_0\| + \beta^{n-1} \|C_0\| + \cdots + \beta^{k+1} \|C_0\| = \tfrac{1-\beta^{n-k}}{1-\beta} \beta^{k+1} \|C_0\| \end{aligned} \tag{22}$$

Since we also have $0 < \beta < 1$, we can obtain from (22)

$$\lim_{n,k \to \infty} \|A_n - A_k\| = 0. \tag{23}$$

Thus, $\{A_n\}_{n=0}^{\infty}$ is a Cauchy sequence in B and this implies that the series solution (14) is convergent.

Theorem 3. *If the initial guess $u_0 = C_0$ remains inside the ball of the solution u then $A_n = \sum_{i=0}^{n} C_i$ also remains inside the ball of the solution.*
Proof: Suppose that

$$C_0 \in B_r(u) \tag{24}$$

where

$$B_r(u) = \{C \in A | \|u - C\| < r\} . \tag{25}$$

is the ball of the solution u. By hypothesis $u = \lim_{n \to \infty} A_n = \sum_{i=0}^{\infty} C_i$ and from Theorem 2, we have

$$\|u - A_n\| \leq \beta^{n+1} \|C_0\| < \|C_0\| < r \tag{26}$$

where $\beta \in (0, 1)$*and* $n \in \mathrm{N}$. This completes the proof.

Theorem 4. *Assume that the obtained solution $\sum_{i=0}^{\infty} C_i$ is convergent to the solution u. If the truncated series $\sum_{i=0}^{k} C_i$ is used as an approximation to the solution u of problem (4), then the maximum error is given as,*

$$E_k \le \frac{\beta^{k+1}}{1-\beta} \|C_0\| . \tag{27}$$

Proof: From the Eq. (22), we have

$$\|A_n - A_k\| \le \frac{1-\beta^{n-k}}{1-\beta} \beta^{k+1} \|C_0\| \tag{28}$$

for $n \ge k$. By knowing

$$u = \lim_{n\to\infty} A_n = \sum_{i=0}^{\infty} C_i \tag{29}$$

it is obvious that

$$\left\| u - \sum_{i=0}^{k} C_i \right\| \le \frac{1-\beta^{n-k}}{1-\beta} \beta^{k+1} \|C_0\| \tag{30}$$

and also we can write

$$E_k = \left\| u - \sum_{i=0}^{k} C_i \right\| \le \frac{\beta^{k+1}}{1-\beta} \|C_0\| \tag{31}$$

since $1-\beta^{n-k} < 1$. Here β is selected as $\beta = \max\{\beta_i, i = 0, 1, \ldots, n\}$ where

$$\beta_i = \frac{\|C_{n+1}\|}{\|C_n\|}. \tag{32}$$

4 Applications

In this section, we will take a few examples with a known exact and approximate solutions obtained via other methods in order to show the effectiveness of both perturbation iteration algorithms.

Example 1: Consider the nonlinear Fisher's problem [29, 30]

$$u_t = u_{xx} + u(1-u) \tag{33}$$

subject to a constant initial condition

$$u(x,0) = \lambda. \tag{34}$$

An artificial perturbation parameter $\varepsilon = 1$ is substituted into (33) as:

$$F(u_{xx}, u_t, u, \varepsilon) = S + R = u_t + \varepsilon\left(-u_{xx} - u + u^2\right) = 0 \tag{35}$$

where $S = u_t$ and $R = \varepsilon\left(-u_{xx} - u + u^2\right)$.

PIA-1 solution:
For the Eq. (35), we have

$$a = b = \alpha = 1. \tag{36}$$

By substituting (36) into Eq. (9), PIA-1 turns into

$$((u_c)_n)_t = (u_n)_{xx} + u_n - (u_n)_t - u_n^2. \tag{37}$$

One may choose the initial function as:

$$u_0 = C_0 = \lambda \tag{38}$$

which satisfies the initial condition. Using the Eqs. (13), (37), (38) with initial condition, the iterations are reached as:

$$u_1 = \lambda + \lambda t - \lambda^2 t \tag{39}$$

$$u_2 = \lambda + \lambda t - \lambda^2 t - \frac{1}{6} t^2 (-1 + \lambda)\lambda\left(3 - 6\lambda - 2t\lambda + 2t\lambda^2\right) \tag{40}$$

$$\begin{aligned} u_3 = {} & \lambda + \lambda t - \lambda^2 t - \tfrac{1}{6} t^2 (-1 + \lambda)\lambda\left(3 - 6\lambda - 2t\lambda + 2t\lambda^2\right) \\ & - \tfrac{(-1+\lambda)\lambda}{1260} \left(\begin{array}{l} 210t^3 - 840t^3\lambda - 420t^4\lambda - 63t^5\lambda + 840t^3\lambda^2 + 1260t^4\lambda^2 \\ +483t^5\lambda^2 + 70t^6\lambda^2 - 840t^4\lambda^3 - 840t^5\lambda^3 - 280t^6\lambda^3 - 20t^7\lambda^3 \\ +420t^5\lambda^4 + 350t^6\lambda^4 + 60t^7\lambda^4 - 140t^6\lambda^5 - 60t^7\lambda^5 + 20t^7\lambda^6 \end{array} \right) \end{aligned} \tag{41}$$

and so on. In [30] , the exact solution is given by

$$u(x,t) = \frac{\lambda e^t}{1 - \lambda + \lambda e^t}. \tag{42}$$

It is worth mentioning that the coefficients of higher order terms of PIA-1 solutions deviate from the Taylor expansion of (42) after the second term u_1.

PIA-2 solution:

Proceeding as in PIA-1, Eq. (11) reduces to

$$((u_c)_n)_t + 2u_n (u_c)_n = (u_n)_{xx} + (u_n) - (u_n)_t - (u_n)^2 . \tag{43}$$

for the Eq. (35). Using the same starting function $u_0 = C_0 = \lambda$ and applying the initial condition gives

$$u_1 = \lambda - \frac{1}{2} e^{-2t\lambda}(-1 + e^{2t\lambda})(-1 + \lambda) \tag{44}$$

For the other iterations, the term u_n in the left hand side of equation (43) can be approximated as $u_0 = \lambda$ for simplicity. By considering this, we have

$$u_2 = \lambda - \frac{1}{2} e^{-2t\lambda}(-1 + e^{2t\lambda})(-1 + \lambda) - \frac{e^{-4t\lambda}(-1 + \lambda)(1 - 2e^{2t\lambda} + e^{4t\lambda} - \lambda + e^{4t\lambda}\lambda - 4e^{2t\lambda}t\lambda^2)}{8\lambda}. \tag{45}$$

$$u_3 = u_2 + \tfrac{e^{-8\lambda t}}{384\lambda^3} \times \left[\begin{array}{l} 1 - 6e^{2t\lambda} + 18e^{4t\lambda} - 10e^{6t\lambda} - 3e^{8t\lambda} - 4\lambda + 3e^{2t\lambda}\lambda + 36e^{4t\lambda}\lambda \\ -96e^{4t\lambda}t^2\lambda^5 - 35e^{6t\lambda}\lambda + 24e^{6t\lambda}t\lambda + 6\lambda^2 + 27e^{2t\lambda}\lambda^2 - 84e^{4t\lambda}\lambda^2 + \\ 21e^{6t\lambda}\lambda^2 + 30e^{8t\lambda}\lambda^2 - 12e^{2t\lambda}t\lambda^2 + 48e^{4t\lambda}t\lambda^2 + 24e^{6t\lambda}t\lambda^2 - \\ 120e^{6t\lambda}t^2\lambda^5 + 48e^{4t\lambda}t^2\lambda^6 + 120e^{6t\lambda}t^2\lambda^6 + \cdots \end{array} \right] \tag{46}$$

and so on. One can proceed to reach higher iterations by using a symbolic computer program. This problem has been studied by several authors via different methods. Table 1 and Table 2 display the 3- terms approximation obtained by Adomian decomposition method (ADM) [29], variational iteration method (VIM) [31] , homotopy perturbation method (HPM) [32], PIA-1 with 2-terms and 3-terms approximation of PIA-2 for different λ's. It is clear that PIA-2 gives better results than other techniques even for *n*=2. and different values of λ.

Example 2: Consider the nonlinear Fisher's problem [29, 32]

$$u_t = u_{xx} + 6u(1 - u) \tag{47}$$

Table 1: The approximate solutions of Fisher equation (33) using different methods for $\lambda = 2$.

t	ADM,VIM,HPM	PIA-1	PIA-2,n=2	PIA-2,n=3	Exact Solution
0.1	1.82867	1.82604	1.82657	1.82622	1.82621
0.2	1.70933	1.6909	1.69581	1.69327	1.69309
0.3	1.634	1.57935	1.59677	1.58924	1.58833
0.4	1.59467	1.48076	1.52199	1.5068	1.50412
0.5	1.58333	1.3876	1.4659	1.44119	1.43527
0.6	1.592	1.29459	1.42414	1.38902	1.37818
0.7	1.61267	1.19822	1.3933	1.34773	1.3303
0.8	1.63733	1.09665	1.37069	1.31526	1.28976
0.9	1.658	0.989778	1.35422	1.28993	1.25515
1.	1.66667	0.879365	1.34228	1.27032	1.2254

Table 2: The approximate solutions of Fisher equation (33) using different methods for $\lambda = 4$.

t	ADM,VIM,HPM	PIA-1	PIA-2,n=2	PIA-2,n=3	Exact Solution
0.1	3.172	3.10241	3.11587	3.11188	3.11166
0.2	2.896	2.50177	2.62569	2.59558	2.591
0.3	2.884	1.96033	2.34994	2.2723	2.25029
0.4	2.848	1.41888	2.19833	2.06834	2.01102
0.5	2.5	0.953571	2.1172	1.9425	1.83452
0.6	1.552	0.718907	2.07478	1.86721	1.69955
0.7	-0.284	0.710974	2.05303	1.82349	1.59347
0.8	-3.296	0.185132	2.04203	1.79877	1.50829
0.9	-7.772	-3.43782	2.03654	1.7851	1.4387
1.	-14.	-16.3429	2.03383	1.77768	1.38104

subject to the initial condition

$$u(x,0) = (1+e^{x})^{-2}. \tag{48}$$

(47) can be written as

$$F(u_{xx}, u_t, u, \varepsilon) = S + R = u_t + \varepsilon\left(-u_{xx} - 6u + 6u^2\right) = 0 \tag{49}$$

where $R = \varepsilon\left(-u_{xx} - 6u + 6u^2\right)$ and $S = u_t$.

PIA-1 solution:

For the Eq. (47), we have

$$a = \alpha = 1, b = 6. \tag{50}$$

By considering (50) and setting $\varepsilon = 1$, Eq. (9) becomes

$$((u_c)_n)_t = (u_n)_{xx} + 6u_n - (u_n)_t - 6u_n^2. \tag{51}$$

Selecting the Eq.(48) as a starting function and using the formula, the iterations in (13) are reached as:

$$u_0 = C_0 = (1+e^{x})^{-2} \tag{52}$$

$$u_1 = \frac{1}{(1+e^x)^2} + \frac{10e^x t}{(1+e^x)^3} \tag{53}$$

$$u_2 = \frac{1}{(1+e^x)^2} + \frac{10e^x t}{(1+e^x)^3} + \frac{25e^x t^2\left(-1+3e^{2x}+2e^{3x}-8e^x t\right)}{(1+e^x)^6} \tag{54}$$

$$\begin{aligned} u_3 = u_2 &+ \frac{25t^3e^x\left(-6e^x-15e^{2x}+20e^{3x}+5\right)}{3\left(e^x+1\right)^6} - \frac{50t^4e^{2x}\left(5e^x+52e^{2x}-17\right)}{\left(e^x+1\right)^8} \\ &- \frac{150t^5e^{2x}\left(-47e^x+20e^{3x}+5\right)}{\left(e^x+1\right)^9} + \frac{10000t^6e^{3x}\left(2e^x-1\right)}{\left(e^x+1\right)^{10}} - \frac{240000t^7e^{4x}}{7\left(e^x+1\right)^{12}} \end{aligned} \tag{55}$$

PIA-2 solution:
In the light of previous example, one can construct PIA-2 as:

$$((u_c)_n)_t + 12\,(u_n)\,(u_c)_n = (u_n)_{xx} + 6\,(u_n) - (u_n)_t - 6\,(u_n)^2 . \tag{56}$$

For the sake of simplicity and easier calculating, the term u_n in the left hand side of equation (56) is taken as 0.25 which comes from

$$u_0 = (1+e^x)^{-2} = 0.25 - 0.25x + 0.0625x^2 + O\,[x]^3 . \tag{57}$$

Thus, we have

$$((u_c)_n)_t + 3\,(u_c)_n = (u_n)_{xx} + 6\,(u_n) - (u_n)_t - 6\,(u_n)^2 \tag{58}$$

for (49). Using the equations (13), (52) and (58) with initial condition yields

$$u_1 = \frac{1}{(1+e^x)^2} - \frac{10e^{-3t+x}\left(-1+e^{3t}\right)}{3\,(1+e^x)^3} \tag{59}$$

$$u_2 = u_1 + \frac{10e^{-6t+x}\left(\begin{array}{l} -2e^{6t}+20e^x-11e^{6t+x}+24e^{6t+2x} \\ -13e^{3(t+x)}\,(1+3t) - 24e^{3t+2x}\,(1+3t) \\ +e^{3t}\,(2+6t) + e^{3t+x}\,(-9+93t) + 13e^{6t+3x} \end{array}\right)}{9\,(1+e^x)^6} \tag{60}$$

$$u_3 = u_2 + \frac{5e^{-12t}}{243\,(e^x+1)^{12}}\left(\begin{array}{l} 16000e^{4x} - 72e^{9t+x} + 72e^{12t+x} - 840e^{6t+2x} - \\ 138e^{9t+2x} + 978e^{12t+2x} + 16320e^{6t+3x} - 24792e^{9t+3x} \\ +8472e^{12t+3x} - 6000e^{3t+4x} + 79200e^{6t+4x} - 99808e^{9t+4x} \\ +10608e^{12t+4x} + \cdots \end{array}\right) \tag{61}$$

and so on. The solution in a closed form by Wazwaz [29] given by

$$u\,(x,t) = \left(1+e^{x-5t}\right)^{-2} . \tag{62}$$

The problem is already solved with ADM [29], HPM [32]. However, table (3) reveals that PIA-2 solutions gives more effective results than ADM [29], HPM [32] and PIA-1 solutions for *n=3*. Figure 1 also shows the accuracy of the fourth order PIA-2 solutions.

Example 3: Consider the nonlinear generalized Fisher's problem [29, 32]

$$u_t = u_{xx} + u\left(1-u^6\right) \tag{63}$$

Table 3: The four terms approximate solutions of Fisher equation (47) using different methods.

(x, t)	ADM and VIM	PIA-1	PIA-2	Exact Solution
(0.1,0.1)	0.359069	0.358299	0.358301	0.358426
(0.2,0.2)	0.485955	0.473909	0.474520	0.476064
(0.3,0.3)	0.637904	0.578701	0.579214	0.590630
(0.4,0.4)	0.831367	0.651047	1.521991	0.692254
(0.5,0.5)	1.089006	0.668100	0.741226	0.775803
(0.6,0.6)	1.435358	0.608198	0.801921	0.840572
(0.7,0.7)	1.892066	0.453814	0.852762	0.888637
(0.8,0.8)	2.473569	0.195085	0.896315	0.923202
(0.9,0.9)	3.183966	-0.166481	0.933343	0.947513

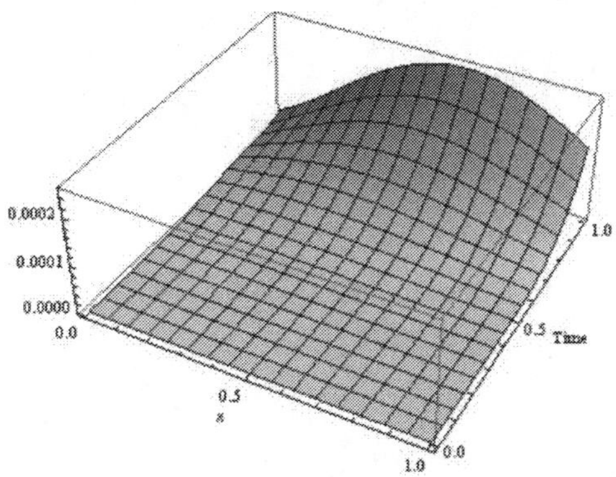

(a) Fourth order PIA-1 solution.

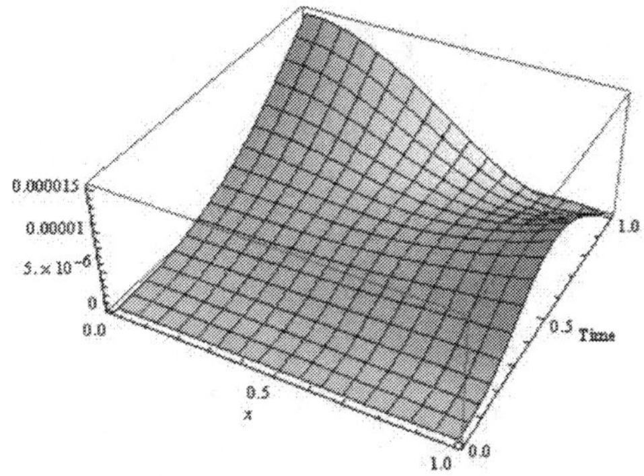

(b) Fourth order PIA-2 solution.

Figure 1: Absolute errors obtained by PIM for Example 2.

subject to the constant initial condition

$$u(x,0) = \left(1 + e^{(3x/2)}\right)^{-1/3}. \tag{64}$$

(63) is considered as

$$F(u_{xx}, u_t, u, \varepsilon) = S + R = u_t + \varepsilon\left(-u_{xx} - u + u^6\right) = 0 \tag{65}$$

where $R = \varepsilon\left(-u_{xx} - u + u^6\right)$,$S = u_t$ and $a = b = 1, \alpha = 6$. Exact solution of this problem is given as [33]

$$u(x,t) = \left(\frac{1}{2} - \frac{1}{2}\tanh\left[\frac{3}{4}\left(x - \frac{5t}{2}\right)\right]\right)^{1/3}. \tag{66}$$

PIA-1 solution:

For this problem, PIA-1 is obtained as:

$$((u_c)_n)_t = (u_n)_{xx} + (u_n) - (u_n)_t - (u_n)^7. \tag{67}$$

With the aid of the equations (13),(64) and (67), we get

$$u_0 = C_0 = \left(1 + e^{(3x/2)}\right)^{-1/3} \tag{68}$$

$$u_1 = \left(1 + e^{(3x/2)}\right)^{-1/3} + \frac{5e^{3x/2}t}{4\left(1 + e^{3x/2}\right)^{4/3}} \tag{69}$$

$$u_2 = u_1 + \frac{25t^2 e^{\frac{3x}{2}}\left(\frac{1}{e^{\frac{3x}{2}}+1}\right)^{28/3}}{131072}\left[\begin{array}{l} -12288 + 16384e^{\frac{21x}{2}} + 4096e^{12x} - 3584e^{3x}(8+5t)^2 - \\ 8192e^{\frac{3x}{2}}(10+7t) - 3584e^{\frac{9x}{2}}\left(96 + 160t + 100t^2 + 25t^3\right) \\ -2240e^{6x}\left(128 + 256t + 240t^2 + 120t^3 + 25t^4\right) + \cdots \end{array}\right] \tag{70}$$

PIA-2 solution:

Proceeding as in the other examples, PIA-2 can be formed as:

$$((u_c)_n)_t + 7\,(u_n)^6\,(u_c)_n = (u_n)_{xx} + (u_n) - (u_n)_t - (u_n)^7 \tag{71}$$

The term u_n in the left hand side of (71) is taken as $\frac{1}{2^{1/3}}$ due to

$$u_0 = C_0 = \left(1 + e^{(3x/2)}\right)^{-1/3} = \frac{1}{2^{1/3}} - \frac{x}{4 \cdot 2^{1/3}} - \frac{x^2}{16 \cdot 2^{1/3}} + O[x]^3. \tag{72}$$

Therefore, (71) turns into simplified form:

$$((u_c)_n)_t + (7/4)\,(u_c)_n = (u_n)_{xx} + (u_n) - (u_n)_t - (u_n)^7\,. \tag{73}$$

After making the relevant calculations, the following results are obtained:

$$u_0 = C_0 = \left(1 + e^{(3x/2)}\right)^{-1/3} \tag{74}$$

$$u_1 = \left(1 + e^{(3x/2)}\right)^{-1/3} - \frac{5}{7}e^{\frac{3x}{2}}\left(-1 + e^{-\frac{7t}{4}}\right)\left(\frac{1}{1 + e^{\frac{3x}{2}}}\right)^{4/3} \tag{75}$$

$$u_2 = u_1 + \frac{5\left(\frac{1}{e^{\frac{3x}{2}}+1}\right)^{28/3} e^{\frac{3x}{2}-\frac{49t}{4}}}{17294403}\left[\begin{array}{l} 2823576e^{\frac{21t}{2}} - 2823576e^{\frac{49t}{4}} - 31250e^{9x} + \\ 21176820e^{\frac{35t}{4}+\frac{3x}{2}} + 15529668e^{\frac{21t}{2}+\frac{3x}{2}} - \\ 36706488e^{\frac{49t}{4}+\frac{3x}{2}} - 12605250e^{7t+3x} + \\ 181515600e^{\frac{35t}{4}+3x} - 8168202e^{\frac{21t}{2}+3x} - \\ 160742148e^{\frac{49t}{4}+3x} + 6002500e^{\frac{21t}{4}+\frac{9x}{2}} \\ -86436000e^{7t+\frac{9x}{2}} + 622339200e^{\frac{35t}{4}+\frac{9x}{2}} + \cdots \end{array}\right] \tag{76}$$

and so on. Table 4 shows a comparison between the four terms approximate solutions obtained by ADM [29], PIA-1 and PIA-2. Figure 2 represents the absolute errors of five terms PIM solutions.

5 Discussion

In this study, perturbation iteration method (PIM) is improved and new iteration algorithms are constructed to solve nonlinear partial differential equations (PDEs). These algorithms are employed specifically to nonlinear generalized Fisher's equations to get more effective and approximate solutions. The obtained solutions are compared with exact solutions as well as Adomian decomposition method (ADM), variational iteration method (VIM) and homotopy perturbation method (HPM). Approximate results have been also displayed in the tables and figures to justify the efficiency and accuracy of the proposed algorithms. Finally, we can say that this paper presents a detailed study of the newly developed perturbation iteration technique for PDEs and it will be encouraging for further studies.

Table 4: The three terms approximate solutions of Fisher equation (63) by ADM and PIM.

(x, t)	ADM	PIA-1	PIA-2	Exact Solution
(0.1,0.1)	0.822572	0.822003	0.822209	0.822296
(0.2,0.2)	0.850714	0.845988	0.846902	0.848388
(0.3,0.3)	0.879949	0.872917	0.872255	0.871810
(0.4,0.4)	0.912240	0.907152	0.892304	0.892509
(0.5,0.5)	0.949442	0.928362	0.919901	0.910540
(0.6,0.6)	0.993089	0.898014	0.915520	0.926041
(0.7,0.7)	1.044232	0.880825	0.908841	0.939211
(0.8,0.8)	1.103356	0.901478	0.926074	0.950285
(0.9,0.9)	1.170372	0.915556	0.928809	0.959515

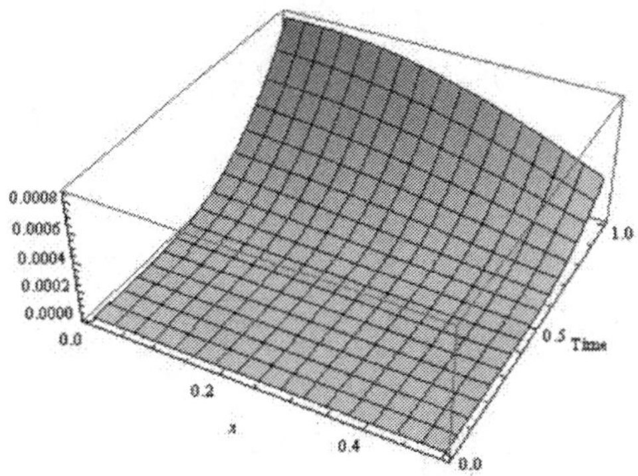

(a) Fifth order PIA-1 solution.

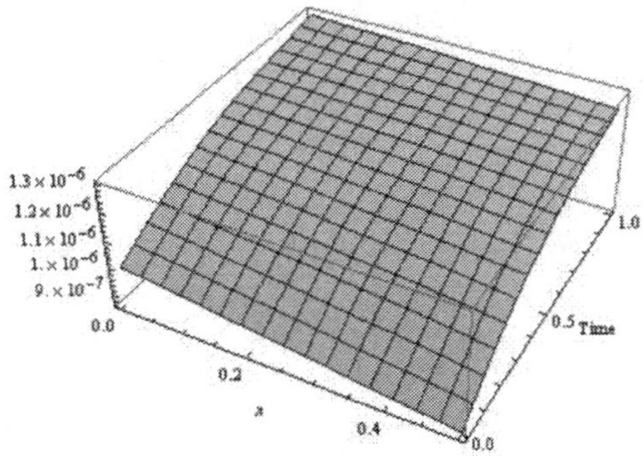

(b) Fifth order PIA-2 solution.

Figure 2: Absolute errors obtained by PIM for Example 3.

Acknowledgement

The authors would like to thank Manisa Celal Bayar University for their financial support.

References

[1] A Abbaoui, K., and Y. Cherruault. "Convergence of Adomian's method applied to nonlinear equations." *Mathematical and Computer Modelling* 20.9 (1994): 69-73.

[2] Evans, David, and Hasan Bulut. "A new approach to the gas dynamics equation: An application of the decomposition method." *International journal of computer mathematics* 79.7 (2002): 817-822.

[3] He, Ji-Huan. "Variational iteration method–a kind of non-linear analytical technique: some examples." *International journal of non-linear mechanics* 34.4 (1999): 699-708.

[4] Deniz, S., Bildik, N.,"Comparison of Adomian decomposition method and Taylor matrix method in solving different kinds of partial differential equations", *International Journal of Modelling and Optimization*, Vol.4(4): pp.292-298, 2014.

[5] Bildik, N., Deniz, S., "Comparison of solutions of systems of delay differential equations using Taylor collocation method, Lambert W function and variational iteration method." *Scientia Iran-*

ica. Transaction D, Computer Science & Engineering and Electrical Engineering, Vol.22(3): pp.1052-1058, (2015).

[6] Deniz, S., "Optimal perturbation iteration method for solving nonlinear heat transfer equations." *Journal of Heat Transfer-ASME*, 139:37,074503-1, (2017).

[7] Bildik, N., Deniz, S., "Modification of perturbation-iteration method to solve different types of nonlinear differential equations", *AIP Conf. Proc.*, 1798, 020027 (2017).

[8] Deniz, S., Bildik, N., "Applications of optimal perturbation iteration method for solving nonlinear differential equations", *AIP Conf. Proc.*, 1798, 020046 (2017).

[9] Abbasbandy, S. "A new application of He's variational iteration method for quadratic Riccati differential equation by using Adomian's polynomials." *Journal of Computational and Applied Mathematics* 207.1 (2007): 59-63.

[10] Öziş, Turgut, and Deniz Ağırseven. "He's homotopy perturbation method for solving heat-like and wave-like equations with variable coefficients." *Physics Letters A* 372.38 (2008): 5944-5950.

[11] Turkyilmazoglu, M. "Convergence of the homotopy perturbation method." *International Journal of Nonlinear Sciences and Numerical Simulation* 12.1-8 (2011): 9-14.

[12] Abbasbandy, Saeid. "Numerical solutions of the integral equations: Homotopy perturbation method and Adomian's decomposition method." *Applied Mathematics and Computation* 173.1 (2006): 493-500.

[13] Liao, Shijun. "On the homotopy analysis method for nonlinear problems." *Applied Mathematics and Computation* 147.2 (2004): 499-513.

[14] Abbasbandy, S. "Homotopy analysis method for the Kawahara equation." *Nonlinear Analysis: Real World Applications* 11.1 (2010): 307-312.

[15] Inc, Mustafa. "Application of homotopy analysis method for fin efficiency of convective straight fins with temperature-dependent thermal conductivity." *Mathematics and Computers in Simulation* 79.2 (2008): 189-200.

[16] Shivanian, Elyas, and Saeid Abbasbandy. "Predictor homotopy analysis method: Two points second order boundary value problems." *Nonlinear Analysis: Real World Applications* 15 (2014): 89-99.

[17] Marinca, Vasile, and Nicolae Herişanu. "Application of optimal homotopy asymptotic method for solving nonlinear equations arising in heat transfer." *International Communications in Heat and Mass Transfer* 35.6 (2008): 710-715.

[18] Marinca, Vasile, et al. "An optimal homotopy asymptotic method applied to the steady flow of a fourth-grade fluid past a porous plate." *Applied Mathematics Letters* 22.2 (2009): 245-251.

[19] Herişanu, N., and Vasile Marinca. "Accurate analytical solutions to oscillators with discontinuities and fractional-power restoring force by means of the optimal homotopy asymptotic method." *Computers & Mathematics with Applications* 60.6 (2010): 1607-1615.

[20] Iqbal, S., and A. Javed. "Application of optimal homotopy asymptotic method for the analytic solution of singular Lane–Emden type equation." *Applied Mathematics and Computation* 217.19 (2011): 7753-7761.

[21] Gupta, A. K., and S. Saha Ray. "Comparison between homotopy perturbation method and optimal homotopy asymptotic method for the soliton solutions of Boussinesq–Burger equations." *Computers & Fluids* 103 (2014): 34-41.

[22] Bildik, Necdet, and Sinan Deniz. "A new efficient method for solving delay differential equations and a comparison with other methods." The European Physical Journal Plus 132.1 (2017): 51.

[23] Tang, S., and R. O. Weber. "Numerical study of Fisher's equation by a Petrov-Galerkin finite element method." *The Journal of the Australian Mathematical Society. Series B. Applied Math-*

ematics 33.01 (1991): 27-38.

[24] Mavoungou, T., and Y. Cherruault. "Numerical study of Fisher's equation by Adomian's method." *Mathematical and computer modelling* 19.1 (1994): 89-95.

[25] Tyson, John J., and Pavel K. Brazhnik. "On traveling wave solutions of Fisher's equation in two spatial dimensions." *SIAM Journal on Applied Mathematics* 60.2 (2000): 371-391.

[26] Aksoy, Yiğit, and Mehmet Pakdemirli. "New perturbation–iteration solutions for Bratu-type equations." *Computers & Mathematics with Applications* 59.8 (2010): 2802-2808.

[27] Aksoy, Yigit, et al. "New perturbation-iteration solutions for nonlinear heat transfer equations." *International Journal of Numerical Methods for Heat & Fluid Flow* 22.7 (2012): 814-828.

[28] Itoh, Shigeru. "Random fixed point theorems with an application to random differential equations in Banach spaces." *Journal of Mathematical Analysis and Applications* 67.2 (1979): 261-273.

[29] Wazwaz, Abdul-Majid, and Alice Gorguis. "An analytic study of Fisher's equation by using Adomian decomposition method." *Applied Mathematics and Computation* 154.3 (2004): 609-620.

[30] Jone, D. S., and B. D. Sleeman. "Differential Equations and Mathematical Biology. 2003." Chapman& Hall/CRC, New York.

[31] Matinfar, M., and M. Ghanbari. "Solving the Fisher's equation by means of variational iteration method." *International Journal of Contemporary Mathematical Sciences* 4.5-8 (2009): 343-348.

[32] Ağırseven, Deniz, and Turgut Öziş. "An analytical study for Fisher type equations by using homotopy perturbation method." *Computers & Mathematics with Applications* 60.3 (2010): 602-609.

[33] Wang, X. Y. "Exact and explicit solitary wave solutions for the generalized Fisher equation." *Physics Letters A* 131.4-5 (1988): 277-279.

PISA QUESTION AND REASONING SKILL

Esen ERSOY[1,*], *Belgin* Bal-Incebacak [2]
[1]Education Faculty, Mathematics Department, 55139 Samsun, Turkey
[2]Education Faculty, Primary School Department, 55139 Samsun, Turkey

Abstract. The objective of the study is to determine the level of the reasoning skills of the secondary school students. This research has been conducted during the academic year of 2015-2016 with the participation of 51 students in total, from a province in the Black Sea region of Turkey by using random sampling method. Case study method has been used in this study, since it explains an existing situation. In this study, content analysis from the qualitative research methods was carried out. In order to ensure the validity of the scope, agreement percentage formula was used and expert opinions were sought.The problem named Holiday from the Chapter 1 of the normal units in Problem Solving Questions from PISA (Program for International Student Assessments) [35] are used as the data collection tool for the study. The problem named Holiday consists of two questions. Applied problems were evaluated according to the mathematical reasoning stages of TIMSS (2003). The findings suggest that the students use proportional reasoning while solving the problems and use the geometric shapes to facilitate the solution of the problem. When they come across problems related to each other, it is observed that they create connections between the problems based on the results of the previous problem. In conclusion, the students perform crosscheck to ensure that their solutions to the problems are accurate.

Keywords: Mathematics, Secondary School, Reasoning.

1. INTRODUCTION

The ability to reason is the most important feature that distinguishes individuals from other living things. Because, with the ability to think, it is possible to make sense out of events and rearrange conditions according to yourself [41]. [40] refers to reasoning as deduction, to try to find a solution, to solve the problem. When the literature for reasoning is examined, [34] interpreted it as a process involving many forms of thinking, whereas [41] described it as a process emerging in the later stages of thinking, and [27] defines it as producing an idea and defending this idea. Generally speaking, in the light of this information, reasoning can be expressed as analyzing a problem well to solve it, presenting an idea and making suggestions to support this idea, testing the suggestions it is available,

* Corresponding author: belginbal33@gmail.com

and deciding the correctness of the result. In this process it can be expressed as using mathematical reasoning skills.

Recent research in mathematics education reveals the need to resolve real life problems associated with everyday life, either from abstract applications or from imaginary questions [16]. This process, referred as mathematical literacy, has become an educational reform in many countries. The Program for International Student Assessments (PISA) administered by the Organization for Economic Co-operation and Development (OECD) and spreading to a wide geographical area is the result of this (Kabael and Barak, 2016). Almost most of OECD countries participate in this test. This exam requires that the levels of students are determined. OECD describes mathematical literacy as formulating, using and interpreting. Students are expected to use these three situations to solve their problems in order to be problem solvers [30]. In other words, students are expected to use skills such as expressing problem situations mathematically and reasoning, etc. to solve existing problems. In this respect, these skills are considered to be important skills as indicators of mathematical literacy. For this purpose, PISA reports the educational conditions of the countries. In the literature, there are studies that reveal the situation of students, teachers, and prospective teachers against the PISA questions [1, 4,5,8, 25,26,37,43]. Opinions were taken in these studies and compared to the existing problems in existing books, the level of practitioners or the problem solving levels of the students were examined.

[37] conducted a study aimed at revealing the difficulties that 140 mathematics teacher candidates had in solving the PISA questions. As a result of the study, it was seen that the prospective teachers were having difficulty to express mathematical arguments, focused on the solution, and struggling to obtain a result without using mathematical reasoning. [43] stated that by asking prospective teachers two PISA questions, it will be important in terms of teaching how they will explain this subject. Teacher candidates cannot fully use mathematical expressions and experiences in the process, but if they are provided with training, they will be a fully equipped, qualified teacher. In situations where prospective teachers are having difficulty approaching problems, determining the approach style of the students to problems will help future teachers to have information about the potential of the students. In this way, teachers will have the opportunity to know what is missing in their students and to have the opportunity to make up their existing deficiencies. Starting from this, it is thought that it will contribute to the field in determining how students use their reasoning skills in PISA questions. It is believed that revealing the reasoning skill levels of students and having an opinion about it will be instructive in the future PISA examinations. From the answers given by the students in the process, it is aimed to examine in detail how they approach the problem and how they reached the solution. It is the purpose of this work to determine how students will make a reasoning for the solution of the problem by leaving them alone with the problem they may encounter in real life. In the process, the mathematical reasoning processes of each student and their approaches to problems are analyzed.

2. METHOD

Case study, which is one of the qualitative research designs, was preferred for the research. Case study is a type of study that examines in depth the solution of the problems that can be encountered in everyday life [11]. The limitation of the existing situation in such studies is to determine the facts in the study, to determine the datasets, to reveal the findings, to interpret and conclude according to the result of this [10].

2.1. Study Group

The study was carried out with a total of 51 students selected by a random sampling method in the Black Sea region of Turkey in the fall semester of the 2015-2016 academic year. [15] and [32] have stated that all units have equal chances for the random sampling method. Here, all the units to be applied are sorted and one of them is randomly selected. All the schools in the Black Sea region of Turkey were randomly sorted and a school was randomly selected for applications to be made with permission from the Directorate of National Education.

2.2. Data Collection Tool

In the study, the problem called "Holiday" in section 1 of the normal units was used from "PISA (Program for International Student Assessments) Sample Problem Solving Questions (2015)". Holiday problem consists of two questions. These questions serve as the data collection tool of the research. The data were collected by interviewing each student individually. In the study, a total of 51 students were interviewed in order to try to find out how the students approached the question and how they used metacognitive considerations based on the data given in the PISA question. Interviews were planned in a quiet environment so that the student will feel comfortable. The students were asked 2 questions, respectively. They are asked to voice what they think when solving the questions and to solve the problem in a loud voice. The interviews were both noted and recorded by the researcher and the process was taken under the record. Each student was interviewed for 15-30 minutes. The data were collected for 2 months. In the process, quotes of students were used to try to show the stages of reasoning and how they approach the problem.

Problemler

 DOI: 10.1051/itmconf/20171301002

TATİL

Bu problem, tatile giderken en iyi ve rahat olan yolun seçimesiyle ilgilidir .

Şekil 1 bölgenin haritasını, Şekil 2 kasabalar arasındaki uzaklıkları göstermektedir.

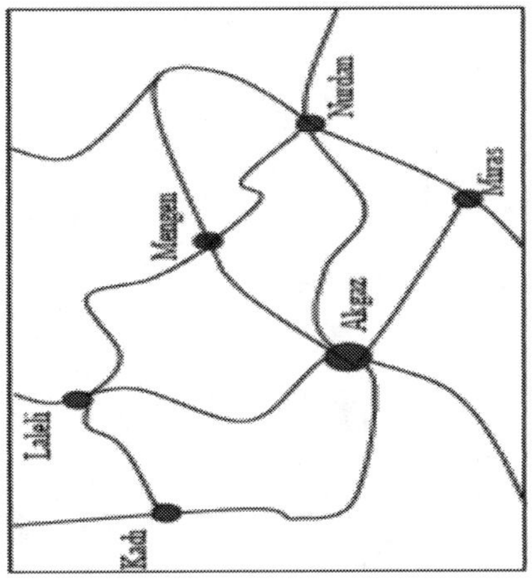

Şekil 1: Kasabalar arasındaki yolların haritası.

	Akgaz	Kadı	Laleli	Mengen	Nurdan	Miras
Akgaz						
Kadı	550					
Laleli	500	300				
Mengen	300	850	550			
Nurdan	500		1000	450		
Miras	300	850	800	600	250	

Şekil 2: Kasabaların kilometre olarak birbirlerine uzaklıkları.

Soru 1: TATİL

Nurdan ve Kadı arasındaki kara yolu ile en kısa uzaklığı hesaplayınız.

Uzaklık: .... Kilometre.

TATİL PUANLAMA 1

Tam Puan

1050 kilometre

Sıfır Puan

Diğer yanıtlar.

- Nurdan – Akgaz – Kadı, uzaklık verilmemiş.

Boş.

Figure 1. Problem one and point scoring system

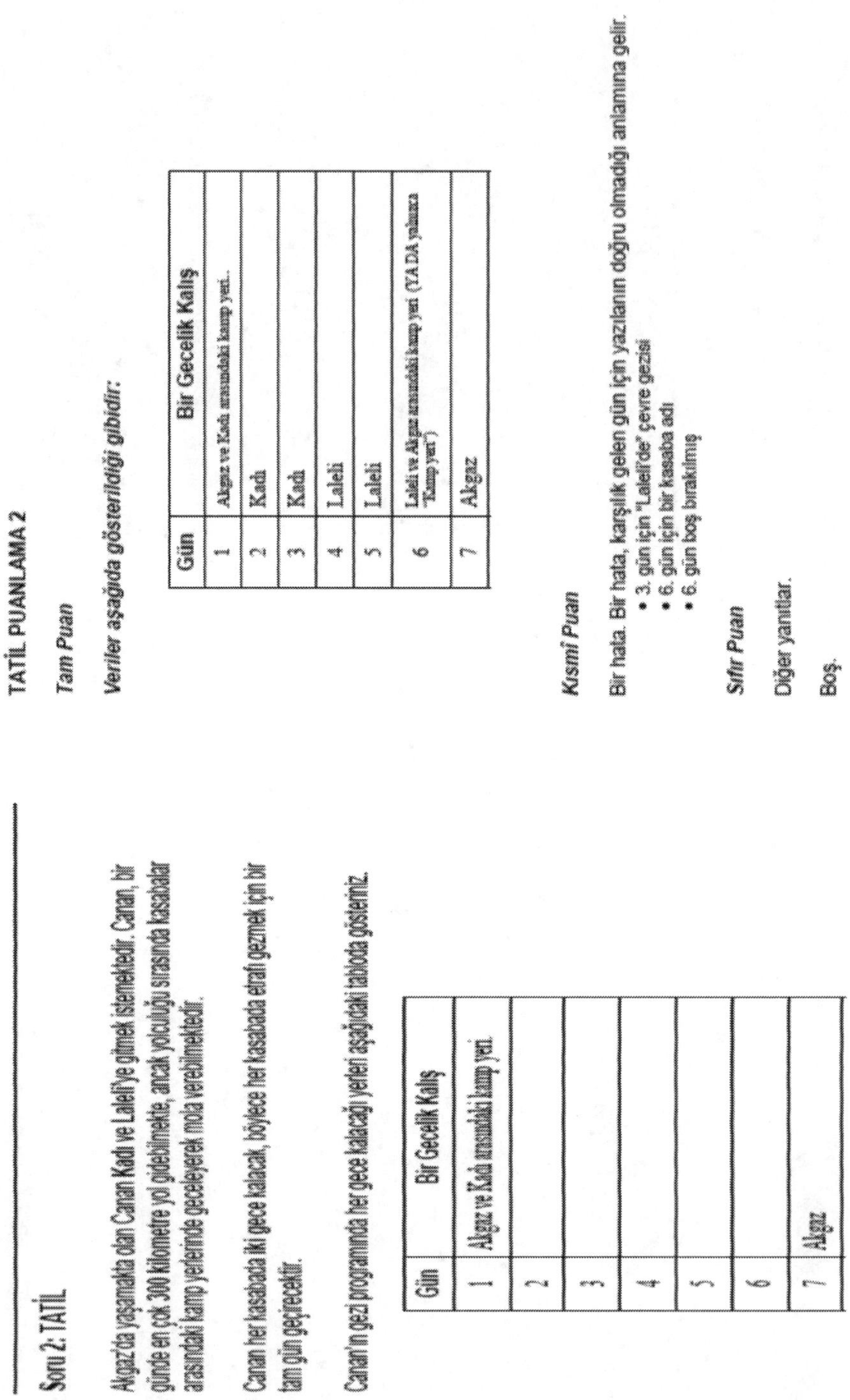

Soru 2: TATİL

Akgaz'da yaşamakta olan Canan **Kadı** ve **Laleli**'ye gitmek istemektedir. Canan, bir günde en çok **300 kilometre** yol gidebilmekte, ancak yolculuğu sırasında kasabalar arasındaki kamp yerlerinde geceleyerek mola verebilmektedir.

Canan her kasabada **iki gece** kalacak, böylece her kasabada etrafı gezmek için bir tam gün geçirecektir.

Canan'ın gezi programında her gece kalacağı yerleri aşağıdaki tabloda gösteriniz.

Gün	Bir Gecelik Kalış
1	Akgaz ve Kadı arasındaki kamp yeri.
2	
3	
4	
5	
6	
7	Akgaz

TATİL PUANLAMA 2

Tam Puan

Veriler aşağıda gösterildiği gibidir:

Gün	Bir Gecelik Kalış
1	Akgaz ve Kadı arasındaki kamp yeri..
2	Kadı
3	Kadı
4	Laleli
5	Laleli
6	Laleli ve Akgaz arasındaki kamp yeri (YA DA yalnızca "Kamp yeri")
7	Akgaz

Kısmî Puan

Bir hata. Bir hata, karşılık gelen gün için yazılanın doğru olmadığı anlamına gelir.
- 3. gün için "Laleli'de" çevre gezisi
- 6. gün için bir kasaba adı
- 6. gün boş bırakılmış

Sıfır Puan

Diğer yanıtlar.

Boş.

Figure 2. Problem two and point scoring system

2.3. Analysis of data

In order to achieve the purpose of the study, students used the problem called "Holiday" from part 1 of the normal units from the Sample Problem Solving Questions (2015) of PISA (Program for International Student Assessments). Holiday problem consists of two questions. The applied problems were evaluated according to the mathematical reasoning stages of [39] and the points to consider are given below.

According to TIMMS (2003) Mathematical reasoning is examined in 5 dimensions. First, students need to analyze, generalize, make connections, make decisions and solve non-routine problems [36]. When these stages are examined;

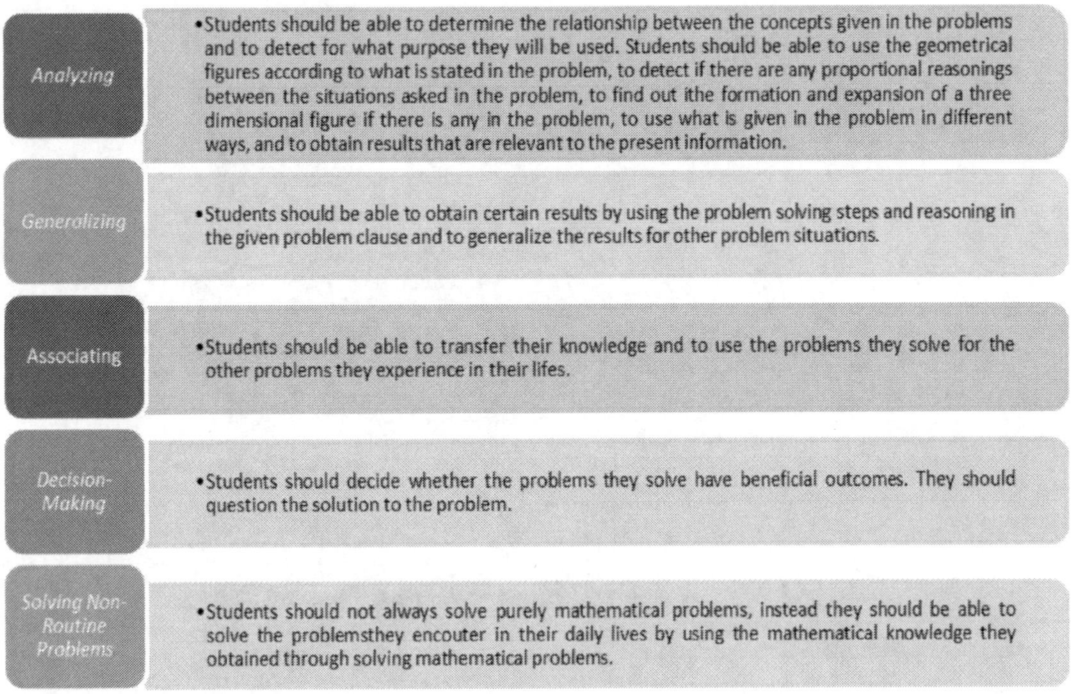

The data obtained were resolved by two researchers according to Creswell's research approach, according to data analysis and presentation technique [11]. This technique primarily starts with data organization. In this process, files are created and edited for the data. Then the text is read and notes are taken on the edges, which at the same time form the pre-codes of the work. Then the structures that constitute its context and the situation are determined. Interpretations are used, with quotations from participants. Natural generalization of what is "learned" is made. Presented using tables and figures that describe the situation in depth. Analyzes were made with careful attention to these steps

3. FINDING

3.1. Results and Comments belonging to the solution of the first problem

The findings of the mathematical reasoning skills of the first problem are given below in order.

3.1.1.Analyzing

The first step in mathematical reasoning is the step of analyzing. The expectations from the student in analyzing; determining the relationship between the concepts available in the problems given to students and identifying the purpose and objective. Able to use geometric shapes according to the given problems, if there is a proportional reasoning between the situations asked in the question, it should identify these and if a three dimensional visual is found in the given problem, it should be able to determine the formation and expansion of this visual, use the available data in the problems in different ways and obtain valid results with the available information.

For the first question, student quotations for the analysis step are listed below.

> *S_1: "first of all, I need to interpret the table. I have been given the distances of the towns to each other in the question, if I write these distances first, I could tell them more easily. Akgaz Miras is 300. Between Akgaz Nurdan is 500... Kadı Miras is 850, if this is already 300, then the other side remains 550... after placing all the numbers this is already visible 500 + 550, so it is 1050. Total roads are given here and we are expected to subtract. "*

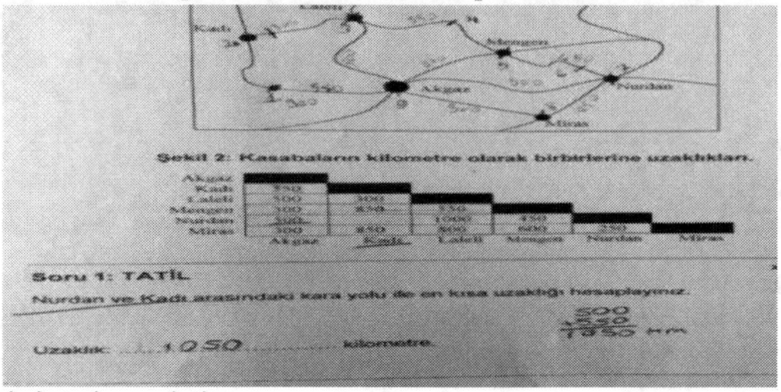

Figure 4: S_1 coded student solutions

> S_{18}:"*If the road between Nurdan and Kadı is required, we need to calculate other roads from the table given to us." The roads to calculate are the roads to go between Nurdan and Kadı. I can both do Nurdan Mengen Laleli Kadı or Nurdan Akgaz Kadı. That is to say, there are other roads, too. According to data given. I calculate these individually, respectively..."*

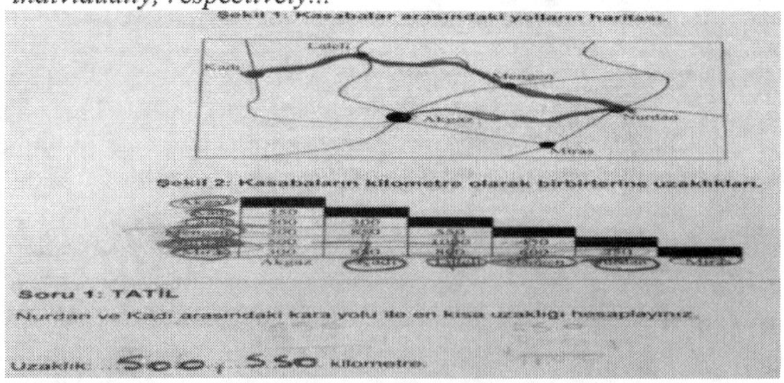

Figure 5: S_{18} coded student solutions

> S_{22}: "*distances between some towns are given. Some are not given. I can find the ones that are not given by placing the given ones. I can find between Kadı and Nurdan from known, that is, given road distances. Mengen Nurdan 450, Mengen Laleli 550, I sum these both and get 1000. Then Akgaz Kadı 550, Akgaz Nurdan 500, if I sum these, 1050 then the answer is 1050. "*

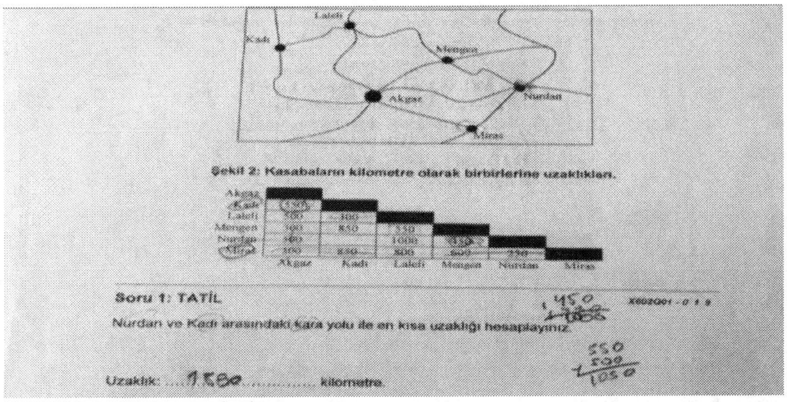

Figure 6: S_{22} coded student solutions

S_{17}: "*table is given and distance between two locations are given. The distances between the towns are written in kilometers. According to the table, between Kadı and Nurdan is 500-300, 200 km. So the answer is 200.* "

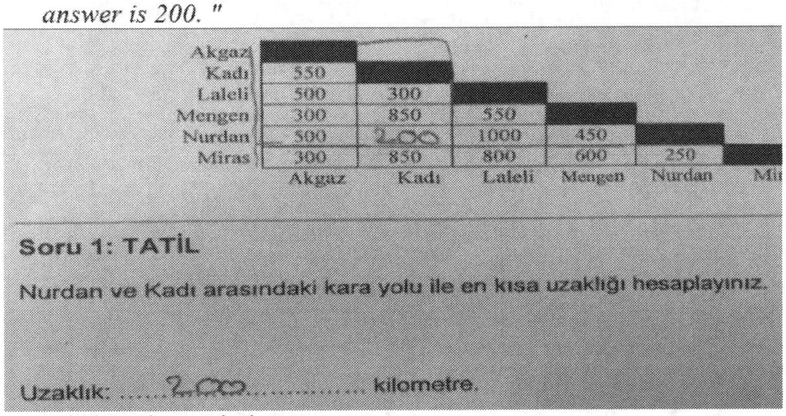

Figure 7: S_{17} coded student solutions

S_{27}: "*...according to the table, between Kadı and Miras is 850, Nurdan Miras 250, that is 1100, so the answer is 1100...*"

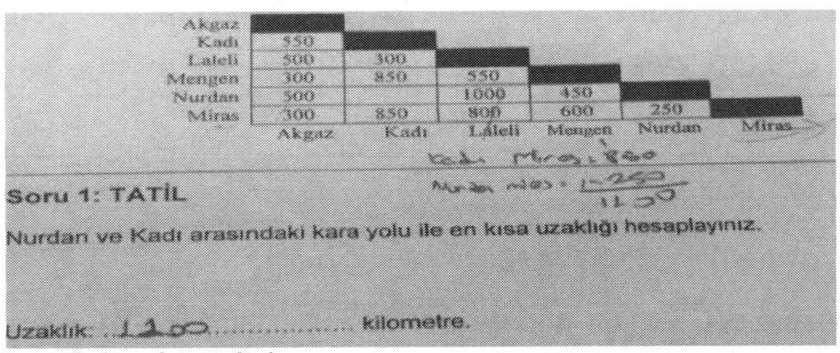

Figure 8: S_{27} coded student solutions

S_9: "*... I read the question. Also looked at the map. I did not understand. I do not want to do.*"

S_{23}: "*...this is the first time I have seen such a table. I did not understand this. I cannot do. Between Kadı and Nurdan is not written, I cannot calculate.* "

S_{32}: "*...I read it. But I do not understand anything. I cannot do it teacher. I want to return it...* "

When the quotations of the students were examined, it was determined that a large number of the students analyzed the data correctly. Too many students tried to solve by randomly adding the numbers on the table. Some of them did not consider the shortest distance but calculated the numbers between two places. From the quotations of the students, it is seen that only 7 students were able to interpret well and place the given data on the visuals. This makes us think that students can analyze the existing information. The rest of the students did not use the information given in the table correctly even if they understood the given. It is understood that 5 students did not answers the question from their papers and quotations. In addition, the students stated that they could not interpret and understand the table because they found it different, and they did not feel the need for an answer because they thought it was a different style problem.

3.1.2. Generalizing

According to mathematical reasoning steps of TIMSS (2003), the second step is the step of generalizing. In the generalization step; Students should be able to achieve certain results by reasoning in the case of problems given to them and by using the steps of problem solving and to generalize their results in other situations. When we look at student quotations, it is seen that Polya have given place to problem solving steps. Student quotations for the analysis step are listed below.

S_{28}: "*...if we look at what is given to us in the problem, there are a lot of things. The distance between the towns is given in the table but some parts are not given I think it wants us to find. There is also a map and there are marked places on the map. I understand what is required when I read the root of the question. It says to calculate the shortest distance. For the shortest route I take a quick look at the shortest route on the map and understand directly. The fact that the map gave us the air distance, it is obvious which route is the shortest. I need to go through Nurdan Kadı to Akgaz because it is the shortest. I need to calculate the distance between them...* "

S_{48}: "*...to solve the question, I need to read it well, what is given, what is required. Distances of the towns and a map are given. Shortest path is requested. Then I make calculations on the map. I check how many km it is by checking from the table. It will be easier to calculate the route from the map in such cases. If there is already an image in the question, that question is solved easier ...* "

It is noteworthy that when student quotations are examined in terms of generalizing, the students have mentioned Polya's problem solving steps. First of all, they realize the step of understanding the problem by determining what is given and what is desired in the problem. Then, it is seen that the students were processing what is given on the figure and went on to application of the problem. Even if they have not indicated how they choose a strategy for solving the problem, it is seen that they ended up using the given data. They mentioned that other problems could be solved like this as well. We see that they use the information they learned in their other lessons in the problem. They reported that air flight distance can be useful in measuring distance. Also, indicating what to do makes us think that there is also the ability to make decisions.

3.1.3. Making Connections

According to mathematical reasoning steps of TIMSS (2003), the third step is making connections. At this stage, the expectation from the students is being able to use the solved problem in other problems in their lives and transfer the knowledge.
The students have made the expected solution for the first question at this stage, but when they solved the other problem they used the information they got for the first question. They did not make any knowledge transfer for the first question. Therefore, student quotes were not included.

3.1.4. Making Decisions

According to mathematical reasoning steps of TIMSS (2003), the fourth step is making decisions. At this stage, students are expected to decide whether the problem they solved has beneficial results. They need to test the result of the problem. Student quotations for this step are given below.

> S_{51}: "*...I calculated all the roads in the problem. I wrote on the map. I have calculated the possible roads to be taken between Kadı and Nurdan. I have determined the simplest three roads. Can go from Kadı to Laleli, Laleli to Mengen and then to Mengen. I already wrote the distances between them. Now I will add them together. I add the numbers 450, 550, 300 and the total is 1300 km distance. Then there is another way. I go by Nurdan Miras, from Miras to Akgaz, Akgaz Kadı. If I calculate it, it's 250, 300, 550 and adding them is 1100. Also, there is Nurdan, Akgaz Kadı route. If I calculate them, the distance between 500 and 550 is 1050 km in total. Of course I can go longer by other ways but now I look at the question is the shortest way so I need not to extend the path. Then the answer is 1050. I can confirm my answer on the figure. Addition operations also show that I am doing it right. I think it's the right answer.* "

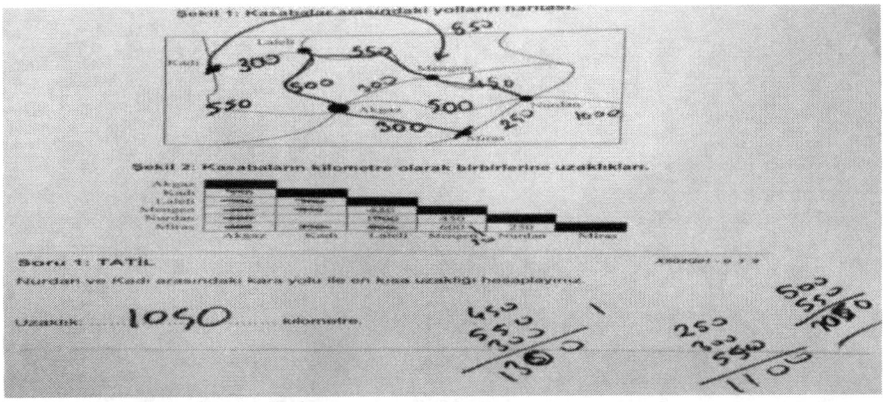

Figure 9: S_{51} coded student solutions

> S_{11}: "*....I calculate the distances between the two distances on the table are already written on the table I think between the two is 1000 km. Because between Akgaz and Kadı is 500, and my result is correct as understood from the table.*

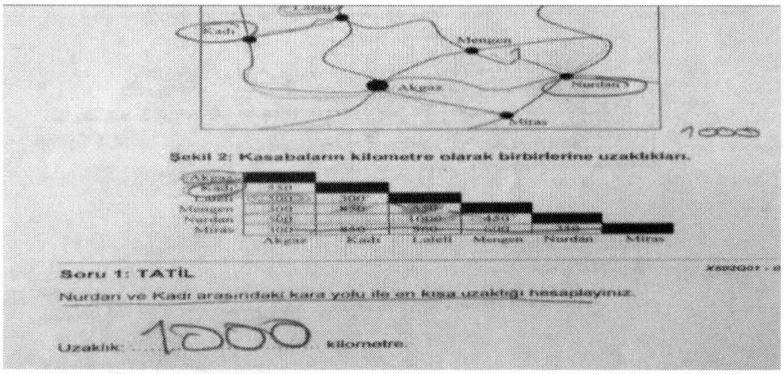

Figure 10: S_{11} coded student solutions

Examples of incorrect student papers are listed below.

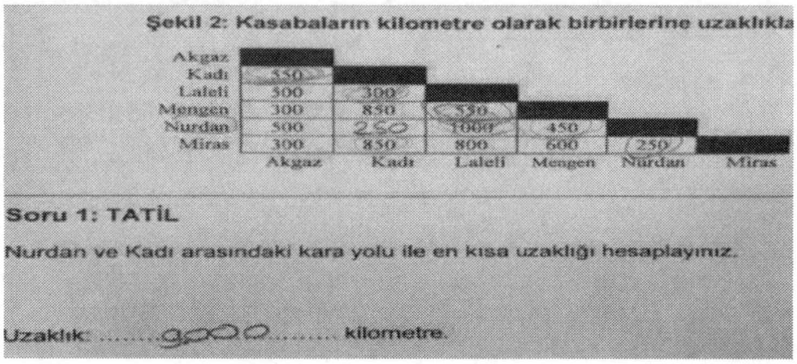

Figure 11: S_{19} coded student solutions

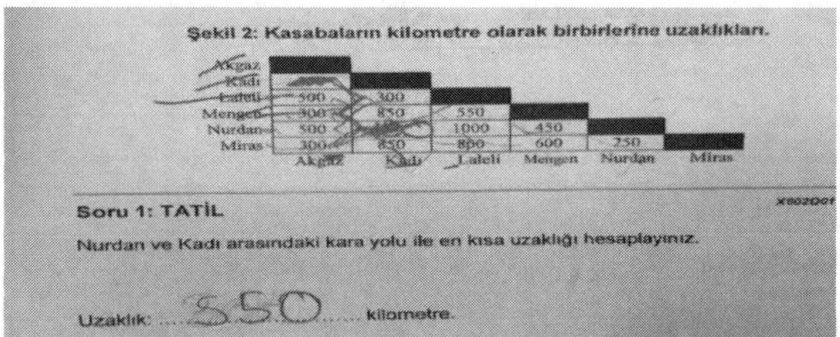

Figure 12: $Ö_{13}$ coded student solutions

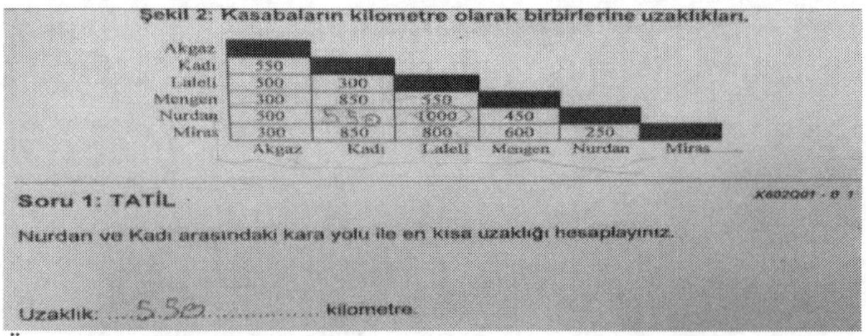

Figure 13: $Ö_{21}$ coded student solutions

When the student quotes were examined and when the expressions of the other students were analyzed, many of the students backtracked their processes and tested their answers. While some students seemed to be able to control over the table, some students only made checks on the map. While some of them reached the right conclusion and made the right decisions, some students believe that their solution is correct even if they solved it incorrectly. They tried to mathematically prove their mistakes.

3.1.5. Solving non-routine problems

The expectations from the students in the solving non-routine problems step, which is the final evaluation step of the problem; students should be able to solve the problems encountered in daily life instead of solving pure mathematical problems at all times by using available mathematical knowledge. Here, students are analyzed on the basis of solving the problem given to them by reaching the right result.

> S_{10}: *"... The shortest way to find between Nurdan Kadı is to go through Akgaz. So we need to calculate the location between the two. Nurdan Akgaz is 500, Akgaz Kadı is 550 and Kadı Nurdan is 550 and 500 on the other hand. It's 1050 between the two. "*

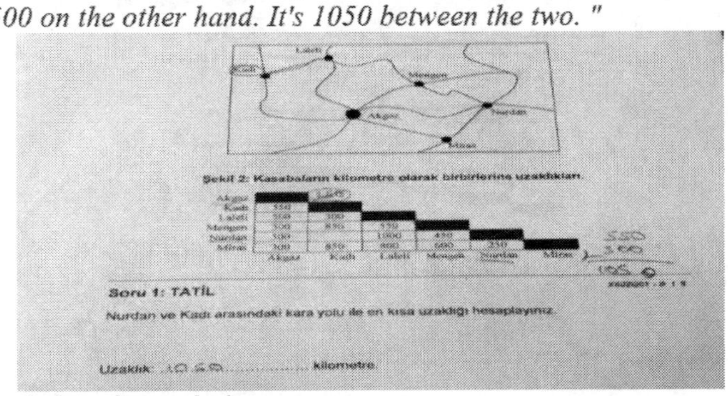

Figure 14: S_{10} coded student solutions

> S_{38}: *"... Akgaz Kadı is 550, Akgaz Laleli is 500, Laleli Mengen is 550, Nurdan is 250, I will go between the Nurdan Kadı. 550 + 450 + 300 = 1300 km. How did I do I found the distances between them then I added them. Because it says to calculate the distance to go. "*

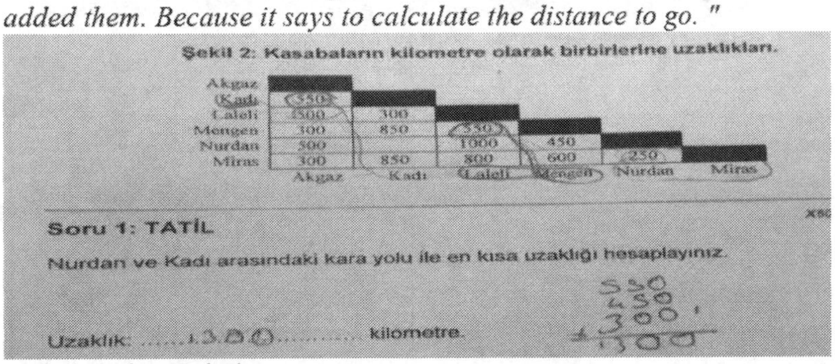

Figure 15: S_{38} coded student solutions

> S_{50}: *"... Laleli Nurdan is 1000, Laleli Mengen is 550, Kadı Miras is 550, Kadı Mengen is 800, if we sum them up, I can calculate all the roads. I think this was wrong. I went the wrong way. Laleli Nurdan was given 1000 in the table, there is that. Then Nurdan Akgaz is 500, Nurdan Mengen is 450 and if we sum them up, it is 1950. "*

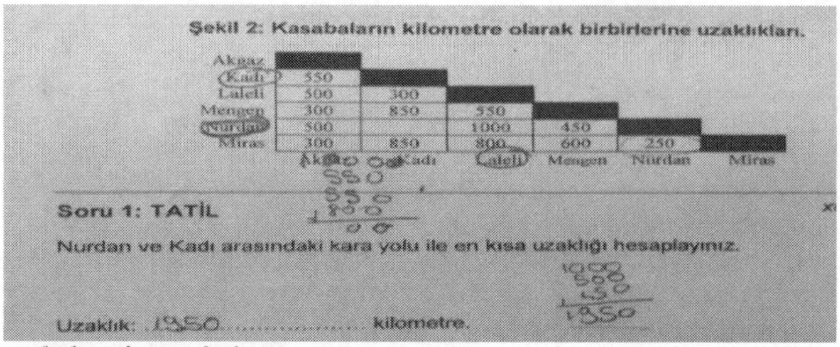

Figure 16: S_{50} coded student solutions

Examples of incorrect student papers are listed below.

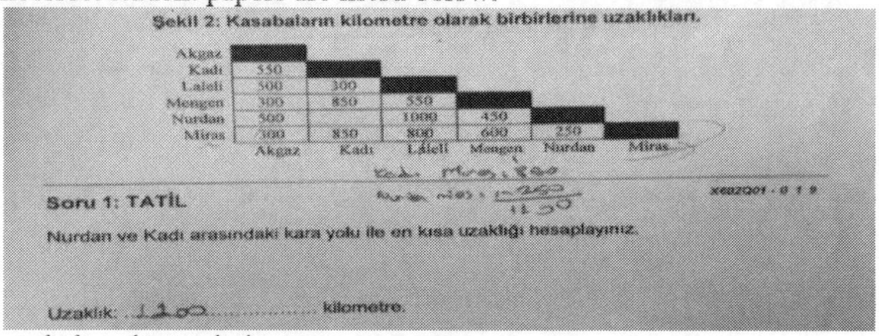

Figure 17: S_{29} coded student solutions

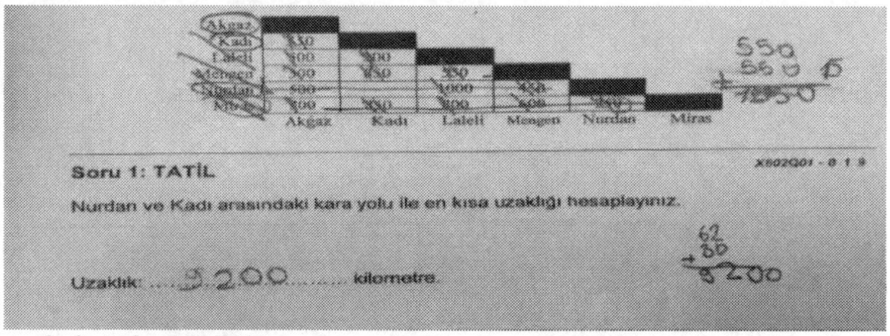

Figure 18: S_{49} coded student solutions

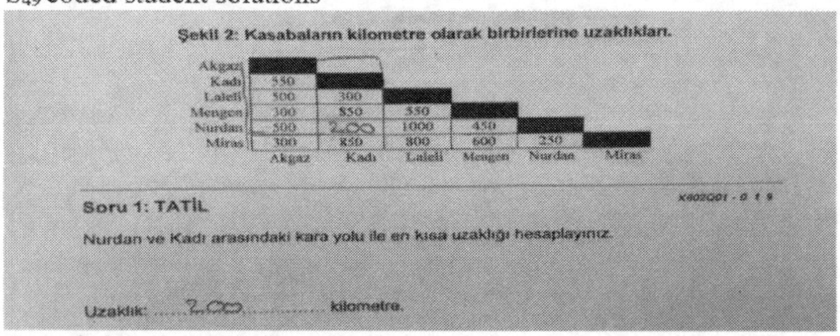

Figure 19: S_{41}coded student solutions

When the student papers are examined, students coded S_1, S_{13}, S_{18}, S_{22}, S_{28}, S_{48}, S_{51} have solved the first question of the problem by using the correct logic. It is seen that they have identified what is given and required in the problem, reading and interpreting the table, have the ability to do operations from the table. Some students who answered correctly were seen to correlate the problem situation with situations that exist in everyday life. It is seen that students coded S_7, S_9, S_{23}, S_{32}, S_{41} cannot understand what they are given in case of problems, cannot analyze required and given data, and cannot read and interpret the table correctly. Therefore, these students did not make any judgments about the solution of the problem. The remaining students did not reach the correct result because of reasons such as not being able to understand what is given in the problem, unable to read and interpret the table, not being able to use the data properly, and not being able to choose the path between the distance to be traveled correctly and appropriately. It is understood that very few students have solved the problem correctly.

3.2. Results and Comments belonging to the solution of the second problem

The analysis steps applied in the first problem are also valid for the second question. According to these steps, the findings of student papers and quotations are given below.

3.2.1. Analysis

The analysis step was briefly analyzed with careful attention to detail, such as distinguishing the given and required items in the problem, making sense of the problem, and giving clues as to which reasoning to use. Student quotations are listed below.

> S_{24}, S_{27}, S_{38}: *"...Canan, who lives in Akgaz, wants to go to Kadı and Lale. She walks maximum 300 km in one day. She's taking a break when she is tired. She needs to spend two days in every town... "*
>
> S_{19}: *"...first of all, the camp locations need to be identified. Since she travels 300 km, she needs to move 300 km and go around the village... "*

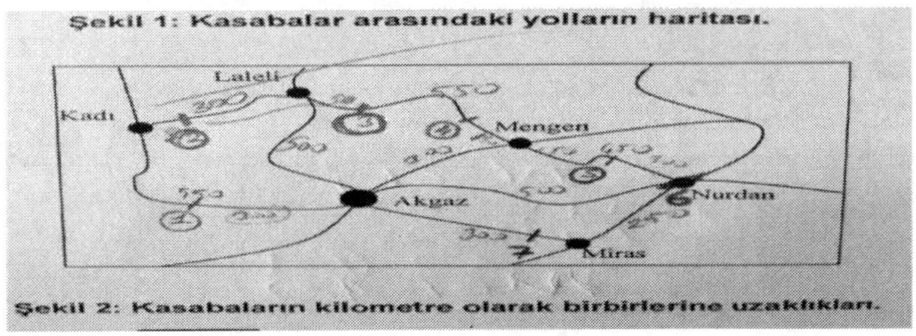

Figure 20: S_{19} coded student solutions

> S_2, S_{34}, S_{42},: "*...camp locations need to be identified. Kadı and Laleli will be visited. But she travels 300 km. She also had to stay for two nights. I need to make the chart of Canan's traveling schedule... "*
>
> S_{32}, S_{39}, S_{41}, S_{47}, "*Canan, who lives in Akgaz, wants to go to Kadı and Laleli. Canan can go maximum of 300 km. She can camp. She will stay two nights in every town. We are required to show the places where Canan will stay during her traveling schedule on the following table. The requirement is to fill the table..."*
>
> S_7, S_9: *"... there is filling on the table. She will visit everywhere... I did not understand. That's why I do not want to solve it. "*

When the student quotes are examined, it is seen that most of the students clearly read the problem, and almost most of them expressed the given and required. The students are aware that Kadı and Laleli needs to be visited. They stated that a maximum of 300 km should be traveled per day. It is also seen that express the rule for staying for 2 nights in every town. Two students stated that they did not want to solve the problem by expressing that they did not understand the question, just like in the first question. Since our study was based on voluntariness, the students were not forced and their blank papers were taken back. It was stated that if they want to solve later, they will be interviewed again. But both of the students did not want to solve the question. The other 3 students who left blank in the first question tried to solve the problem. But as a solution, they read the question and then wrote the names of the places on the table in order by using the statement "if villages are to be visited, first location is selected and all village is visited in turn". It is seen here that students cannot analyze the problem situation in general.

3.2. 2. Generalizing

According to mathematical reasoning steps of TIMSS (2003), the second step is the step of generalizing. The things to pay attention in this step for the second question; Students being able to analyze by reasoning what they are doing in problem situation, to obtain certain results using the problem-solving steps and to analyze the existing information for other situations. Examples of student quotations are given below.

> S_{11}: *"...we will calculate in km and we need to calculate for the place to go. We looked at the given. What is required of us is to determine where we will camp. We go by choosing from the map. We have to start from Akgaz and return to Akgaz. And then we have to go everywhere twice so I'll take it one forward and bring it back and do what is required... "*

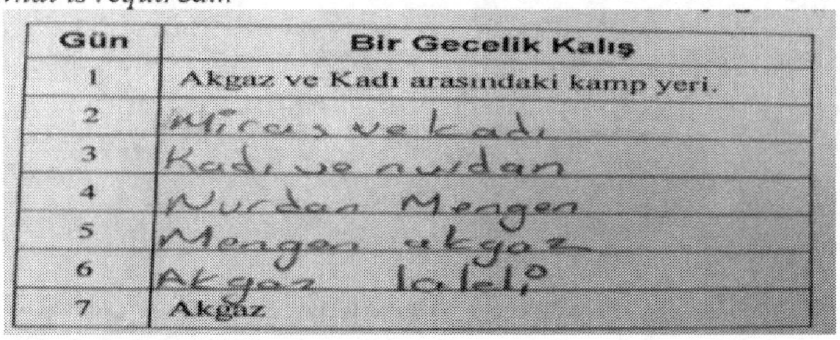

Gün	Bir Gecelik Kalış
1	Akgaz ve Kadı arasındaki kamp yeri.
2	Miras ve kadı
3	Kadı ve nurdan
4	Nurdan Mengen
5	Mengen akgaz
6	Akgaz laleli
7	Akgaz

Figure 21: S_{11} coded student solutions

> S_{13}: *"...we can solve the problem by using the data from first problem. The things that are given to us and required from us are obvious we are going to go from Akgaz to camp sites and visit the entire village. I think we have to go starting from Akgaz as already done in the problem. If we do Akgaz Laleli, Akgaz Mengen, Akgaz Nurdan, Akgaz Miras, Akgaz Kadı, we can return to where we started. In this way we will be visiting the whole village. In the end, this question can be solved by reasoning... "*

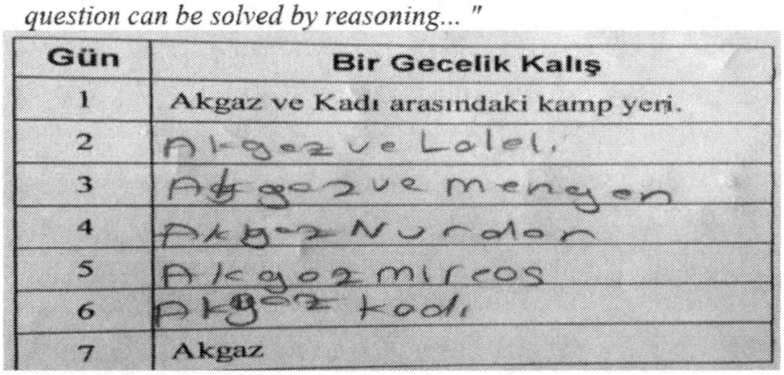

Gün	Bir Gecelik Kalış
1	Akgaz ve Kadı arasındaki kamp yeri.
2	Akgaz ve Laleli
3	Akgaz ve mengen
4	Akgaz Nurdan
5	Akgaz mircas
6	Akgaz kadı
7	Akgaz

Figure 22: S_{13} coded student solutions

When we look at the statement of the students, it is obvious that they do not understand clearly and fail to distinguish according to the problem solving steps. This is because even if the students have indicated what are given and required in the problems, they could not use the information given in the implementation of the solution properly and could not focus on the result with a wrong strategy because they do not understand the desired information clearly. When it is identified that there are two students who understood the question clearly, it is determined that the students could not analyze the process well. Because of that, they have done wrong applications with wrong strategy.
hould be set to single line spacing.

3.2.3. Making Connections

 DOI: 10.1051/itmconf/20171301002

According to mathematical reasoning steps of TIMSS (2003), the third step is making connections. At this stage, the expectation from the students is being able to use the solved problem in other problems in their lives and transfer the knowledge. Since there are two problem situations here and the problems are related to each other, the process should be analyzed as follows. Student quotations for the analysis step are listed below.

> S_{35}: *"...The question is asking us to determine the places to visit. She can go 300 km. In the first question, we have already calculated the distance between the two villages and the other villages. The distance between some places were already written in the table. I will answer by looking at the map accordingly. first of all, the camp locations need to be identified. Since she travels 300 km, she needs to move 300 km and go around the village. On the first day, the camp site between Akgaz and Kadı, then the camp site between Kadı and Laleli, the camp site between Laleli and Mengen on the third day, the camp site between Mengen Nurdan on the fourth day, the camp site between Nurdan Miras on the fifth day, and she arrives at Akgaz on seventh day. In this way she can visit all village..."*

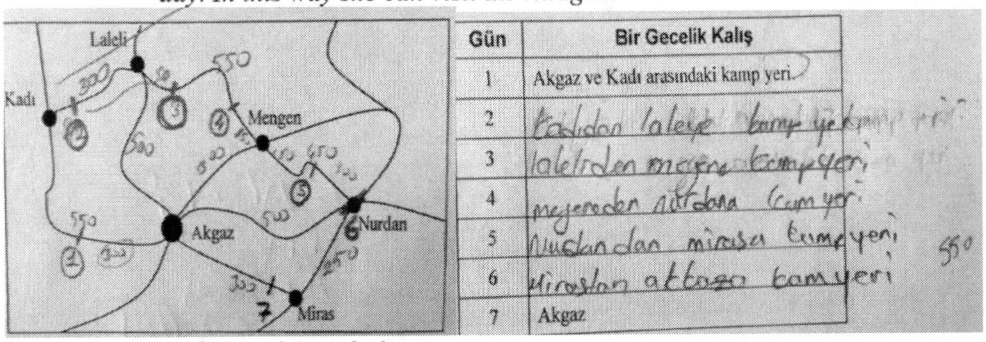

Gün	Bir Gecelik Kalış
1	Akgaz ve Kadı arasındaki kamp yeri.
2	Kadıdan laleye kamp yeri
3	lalelıden mengene kamp yeri
4	mengenden nurdana camp yeri
5	nurdandan mirasa kamp yeri
6	mirastan akgaza kamp yeri
7	Akgaz

Figure 23: S_{35} coded student solutions

> S_{29}: *"...In the first question, we have calculated distances in km. Now we can write these on days using these numbers. First, Laleli Kadı 300 km, then Mengen Laleli 550 km, 250 km between Nurdan and Miras, then Mengen Laleli 550 km, Kadın Mengen 550, Akgaz Kadı 550 km ..."*

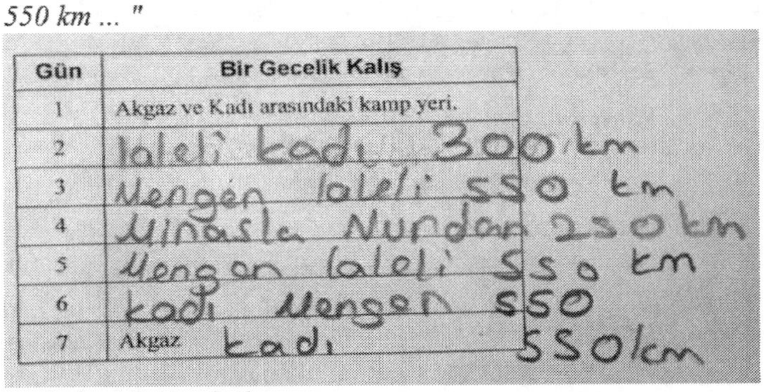

Gün	Bir Gecelik Kalış
1	Akgaz ve Kadı arasındaki kamp yeri.
2	laleli kadı 300 km
3	Mengen laleli 550 km
4	Mirasla Nurdan 250 km
5	Mengen laleli 550 km
6	kadı Mengen 550
7	Akgaz kadı 550 km

Figure 24: S_{29} coded student solutions

> S_{33}: *"... in km, we can write the distances by looking at the table in the first question. The camp site between Akgaz and Kadı is 550 km, Kadı and Laleli 450 km, Laleli and Nurdan 250 km, Akgaz and Mengen 600 km, Akgaz and Laleli 550 km, Laleli and Miras 550 km, Akgaz and Miras 800 km ..."*

DOI: 10.1051/itmconf/20171301002

Gün	Bir Gecelik Kalış
1	Akgaz ve Kadı arasındaki kamp yeri. 550
2	Kadı ve Laleli 650 km
3	Laleli ve Nurdan 250 km
4	Akgaz ve menge 600 km
5	Akgaz ve Laleli 550 km
6	Laleli ve Miras 550 km
7	Akgaz ve miras 850 km

Figure 25: S_{33} coded student solutions

When student papers are examined, the students are aware that in the process, they should use the information in the first question. They did not recalculate the distances when solving the problem and tried to solve the problem through the table or the previous ones. It is seen here that students made connections with the other problem. But, because they do not understand correctly what is required in the problem, they have realized the solution by choosing the wrong way to solve. It has been found that the students were careless, switched to solution without understanding the question and tend to complete as soon as possible.

3.2.4. Making Decisions

According to mathematical reasoning steps of TIMSS (2003), the fourth step is making decisions. At this stage, students are expected to decide whether the problem they solved has beneficial results. The students need to test the result of the problem. Student quotations for this step are given below.

> S_{44}: *"...Akgaz Nurdan must be a camp site. When we check on the map, she visits everywhere. Only between the Mengen Nurdan remains. That is 250 km, not 300 km. So she needs another day. I think the problem is correct because she will have visited the village like this, but if she wants to visit everywhere, I may have done something wrong because she needs one day ... "*

Gün	Bir Gecelik Kalış
1	Akgaz ve Kadı arasındaki kamp yeri.
2	Laleliyle Kadı kamp yeri
3	mengenle lale kamp yeri
4	laliyle miras kamp yeri
5	laliyle mengen kamp yeri
6	mirasla nurdağı var.
7	Akgaz nurdan ve damp yeri

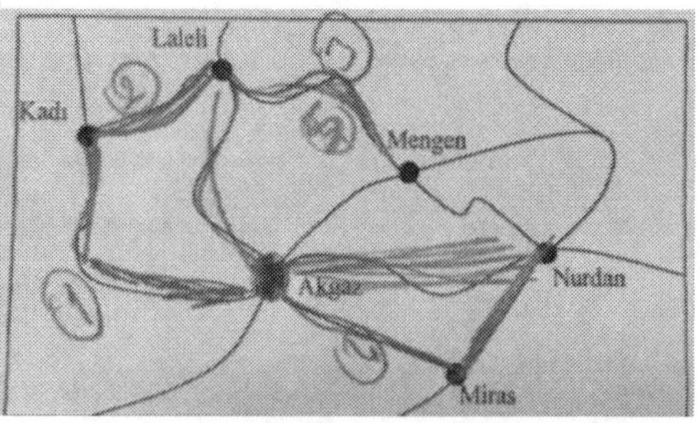

Figure 26: S_{44} coded student solutions

> S_{26}: *"...since she started camp between Akgaz Kadı, she has made 300 km. From there she goes to Kadı Laleli. Because 300 kilometers of 550 kilometers were traveled. She camps between Kadı Laleli. So she spends two days in Laleli. Traveling is done from Laleli to Mengen, from Mengen to Nurdan, from Nurdan to Laleli, and from Miras to Akgaz. Everywhere will be stayed at for two nights. When we look at the table in order, we see she stays everywhere for two nights. It means we have solved the question correctly ... "*

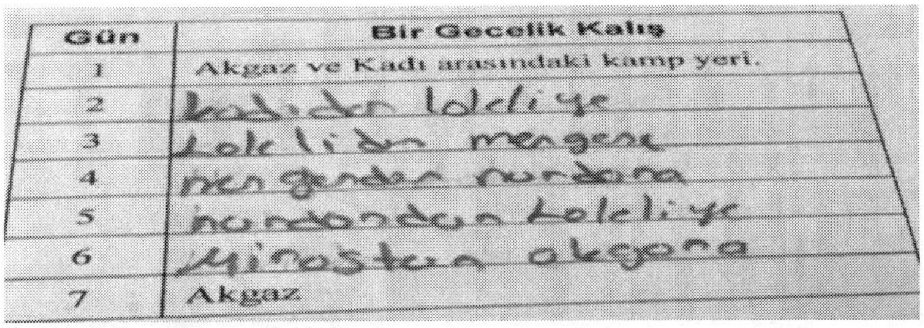

Figure 27: S_{26} coded student solutions

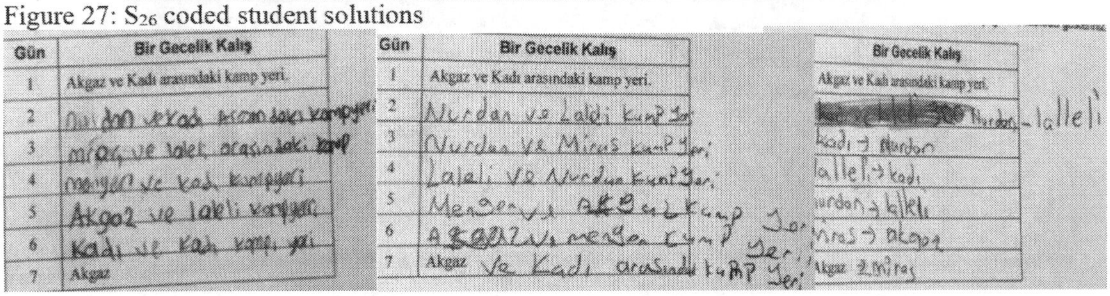

Figure 28: Incorrect student solutions (respectively, S_{16}, S_3, S_{48})

When student answers are examined, it is determined that the students tested the results of their problems. Even if they do not answer the questions correctly, they have used this step in their problems, confident that their answers are correct. Some students did not show any test on the correctness or incorrectness of their problem. It has been determined that they did not use this stage. The reason for this is that they approach the question with a tendency to solve the problem as soon as possible. Students are in an effort to achieve a result rather than the correctness of the problem. Therefore, it is thought that they skipped the testing of the problem stage. It seems that they think that they have solved the problem by writing the places to go in order.

3.2.5. Solving non-routine problems

The expectations from the students in the solving non-routine problems step, which is the final evaluation step of the problem; students should be able to solve the problems encountered in daily life instead of solving pure mathematical problems at all times by using available mathematical knowledge. Here the students are analyzed on the basis of the solution situations that will lead to the right result.

> S_{24}: *"...Canan, who lives in Akgaz, wants to go to Kadı and Laleli. She walks maximum 300 km in one day. She's taking a break when she is tired. She needs to spend two days in every town. If camped between Akgaz and Kadı, first day. The second day goes 300 km from here. That is, she continues traveling between Akgaz and Kadı. She camps between Kadı Laleli when she arrives at Kadı. She will spend another night in Kadı and make another camp. Then goes to between Laleli Akgaz and camps there. Then she camps between Laleli Akgaz again because she will stay for two days. Then goes to Akgaz and the journey. In total she may have made a total of 2100 km with 300 km a day for 7 days ... "*

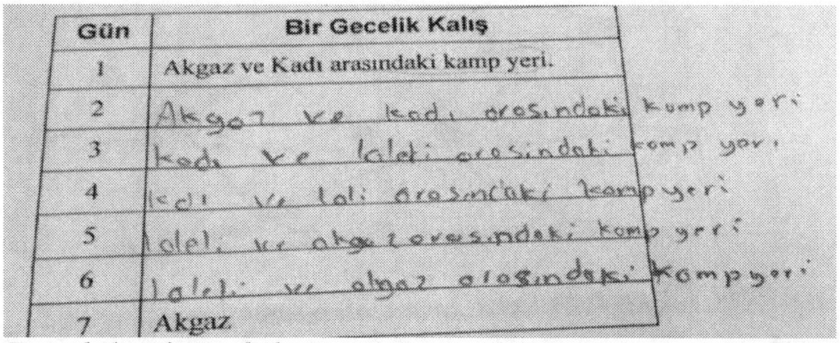

Gün	Bir Gecelik Kalış
1	Akgaz ve Kadı arasındaki kamp yeri.
2	Akgaz ve kadı arasındaki kamp yeri
3	kadı ve laleli arasındaki kamp yeri
4	kadı ve lali arasındaki kamp yeri
5	laleli ve akgaz arasındaki kamp yeri
6	laleli ve akgaz arasındaki kamp yeri
7	Akgaz

Figure 29: S_{24} coded student solutions

S_{37}: *".. if she starts camping between Akgaz and Kadı, the first day will be spent there. Then she has to leave Kadı and to Laleli and has to camp there. After Laleli, she must go to Mengen. After Mengen, she needs to set up camp at Nurdan. Then she has to go to Miras from Nurdan. She goes to Akgaz from Miras to complete her camp. In other words, she will stay in Kadı for two nights, then Laleli for two nights, then in Nurdan, then in Miras... "*

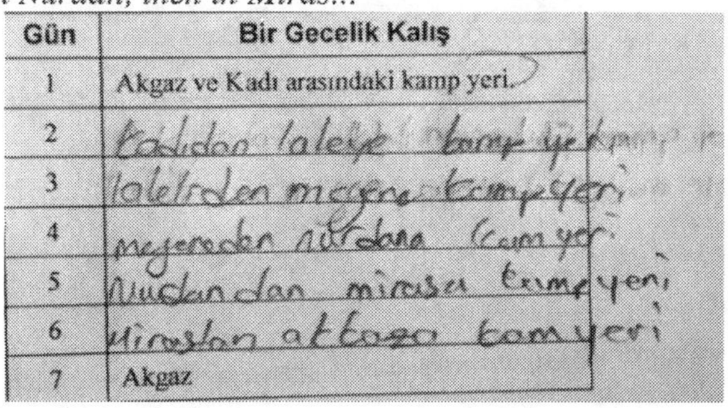

Gün	Bir Gecelik Kalış
1	Akgaz ve Kadı arasındaki kamp yeri.
2	kadıdan laleye kamp yeri
3	laleliden megene kamp yeri
4	megeneden nurdana kamp yeri
5	nurdandan mirasa kamp yeri
6	mirastan akgaza kamp yeri
7	Akgaz

Figure 30: S_{37} coded student solutions

S_{51}: *"...if we check the map on the first page, we understand how to act on camp site. If the camp site was first made between Akgaz and Kadı, it would mean she went 300 km. And she will have gone one day.. Then she goes to Kadı and stays there for two days because she would stay in each village for two days. Then she departs for Laleli and takes a break in Laleli. She stays there. She stays for two nights.. Because my teacher she would stay for two nights. Then she must come to Akgaz before extending the way. So goes between Laleli Akgaz and 7 days will be completed ... "*

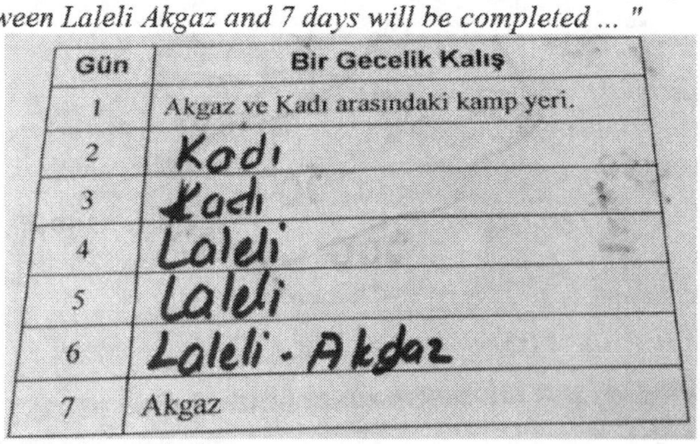

Gün	Bir Gecelik Kalış
1	Akgaz ve Kadı arasındaki kamp yeri.
2	Kadı
3	Kadı
4	Laleli
5	Laleli
6	Laleli - Akgaz
7	Akgaz

Figure 31: S_{51} coded student solutions

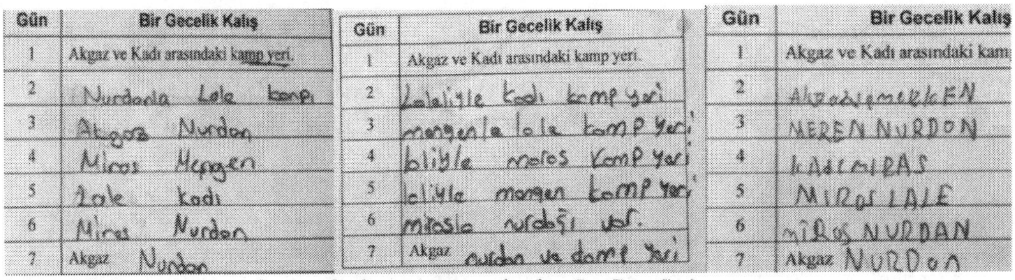

Figure 32: Incorrect student solutions (respectively, S_5, S_{42}, S_{47})

When student quotes and papers are examined, it is seen that students solve problems by using information they have available in the course of solution of non-routine problems. It has been determined that some students analyze well the situations given to them and try to use all the available information on the problem. Some students missed details while using available information. Not paying attention to these details have caused them to solve the problem incorrectly. Almost all of the students in the process made the mistake of thinking the entire village needs to be visited. For this reason, some have forgotten to look at the days, and some at the kilometer information. Solutions were always wrong in the process due to not being analyzed well.

3. DISCUSSION AND CONCLUSIONS

According to the findings, since students cannot use the problem solving steps clearly, it is generally determined that they perform the arithmetic operations in order as they do them in the class. [19,20] investigated the problem-solving steps of teachers in the classroom and as a result they reached the conclusion that students solved the steps of each problem differently and did not include all the steps required in the problem-solving process. This suggests that the students have not sufficiently understood the problem-solving steps and have experienced difficulties in implementing these steps. If the teacher is always following the same problem-solving steps, it is very likely that the students will solve the problems using those steps. The study made contains similar results. Students tend to try solving a problem by following the same steps as with the arithmetic order they solve the problems in their classrooms.

[31] interpreted in his study researching the failures in PISA questions as not reading the problem correctly. For success, he stated that students should be left face to face with different problems which they will use problem solving strategies. In this study, it was determined that students solved the problem incompletely due to not completely reading the problem or not paying attention to all the information given. [14] stated that they focus on a certain numbers and words in the problem-solving process, do not think much about the results rather than process, and tend to obtain a concrete result as a result of arithmetic operations. It has been seen that the students who are in the effort to achieve the result have solved the questions without paying attention to the given and required.

The students read the question from beginning to end and done this by emphasizing given and required were obtained as a result of interviews conducted. However, it has been determined that students have difficulty in implementing the remainder of problem solving process. The students tried to solve the problem without paying attention to most of the steps in the process. [29] and [12] point out that it is important for the students to find out the answer to the problem by short way rather than the problem-solving process when the Turkish Education System has an evaluation process consisting of multiple-choice items. This suggests that students have a tendency to get results in the shortest way rather than analyzing and thinking the problem. It turned out that for both problem situations, students tend to add the numbers or do what is required last in the question root rather than do a good analysis. [21] points out that students focus on the results since the grades students will be getting from the exams are more important than learning the problem-solving steps. This indicates that the difficulties students encounter in solving problems may be partly due to the steps teachers are taking to solve the problem. In other words, when compared with the result here, it is possible that the students cannot perform the required steps clearly because of the teachers, not themselves.

[37] and [3] have concluded in their studies that participants have failed to read and interpret the graphics and tables in the PISA questions. Similarly, it is understood that students cannot read and interpret the table correctly. Especially for the second question, many students tried to solve the question without paying attention to the information given to them in the table while reading the first question correctly and filling the second question according to the points that should be noted in the table.

While [2] and [26] report that they correctly read and express the concepts and mathematical expressions in problem situations, in the study of [37], it was identified that they were not able to use mathematical skills and reasoning skills clearly in their work. Similar results were obtained for all three studies. The reason for this is that it is understood from the student quotations that they express the concepts given in case of problems but do not use mathematical skills clearly in using them in practice and do not use reasoning skills. In a study by [37], participants noted that they were trying to solve the problem with daily life information instead of mathematical reasoning. Similarly, it has been determined that the students approach the questions by giving examples from their own lives and try to solve them.

[24] reports that conducting activities oriented to test technique in Turkish Education System causes not to think about the reasoning of the solution. They also stated that students generally prefer the approach of "reading the roots of the question first, then read the problem". This process in turn causes the students to fail in the solving phase of the problem. The study conducted also supports this result. Because many of the students have tried to solve the problem without paying attention to what was given earlier and paying attention to the root of the last question.

[29] stated that teachers usually do not use the step of verifying and checking the problem because of the time constraint. They have stated that the students who were trained this way tend not to use this step. It has been observed in this study that students generally use this stage and tend to confirm their results on the map or on the table. It is thought that the use of this stage may have been due to the fact that they were in an effort to confirm that they were doing it correctly during the interview. Because students stated that they felt the need to get a confirmation after every operation in the process of solving the problem. This was felt in all of the students even though no student was confirmed in the process.

[9] stated that with the teacher's anxiety to make the deadline with the curriculum, they prefer teaching problem solutions in only one way. When the answers of the students were examined, it was determined that almost all of them tended to solve the problem in the same way.

[22] and [28] have obtained the result that students were indecisive in being confident in proving solutions to non-routine problems. Once students solved the problem, they were not very successful with both problems in the course of checking the validity. In the study by [42], some students stated that they used excellent strategies to solve the problems but they could not reach the right result by making simple mistakes. [13] stated that instead of focusing on the result, they should focus on the process [trns. 42] Similarly, even if students have identified the right strategy for solving the problem, the simple mistakes they make and the percentage of achieving the right result remain at a very low level.

[17] have concluded, according to the results of the study they made for determination of table and graphic reading levels of teacher candidates, that the higher the performance of interpretation of tables and graphics, the higher the ability of reasoning to judge. However,

in our study, the result was that the students did not have good reasoning skills even though they read the table correctly. Especially for the second question, most students cannot reason about the problem because they do not understand the question clearly. [18] Mathematical reasoning is used in the decision-making process to determine what meaning to obtain from reading. He stated that it is a high-level thinking ability. At this stage, the result can be obtained that the group in this study does not clearly understand what they are reading. In general, Güler (2013) stated that the students' ability to reason in the face of PISA questions is weak. This result is supported by the work done. [7] stated in their study that the PISA questions were inadequate in terms of improving the ability to think effectively, communicate, perceive and problem solving in the Turkish education system, and thus the achievement is lower compared to other countries. Altun, et al., (2012) suggested that teaching programs should be supported by teaching situations that address vital problems and that students should be able to develop and defend proposals for specific purposes in order to be successful in solving PISA questions. The work done may have a similar suggestion because it is a fact that students have a low success in real life problems. Altun, et al. (2012) stated in the study they made students should be supported with teaching conditions addressing vital skills in teaching programs and support that students to develop suggestions in the face of such problems and defend their suggestions. They stated that this way failure will be avoided.

[38] stated that PISA questions are questions that require high-level thinking skills, and even students who succeeded in the LGS exams may fail in the process. The reason for this is that LGS questions were abstract and based on memorization. Students were also not successful due to not structuring the problem well. Even the most successful students in the class could not correctly answered both problems. Only one student from 51 students answered both problems correctly.

[23] investigated the inadequacies in solving real life problems. He stated that it is necessary to provide environments that will allow students to become problem solvers. When the literature is examined, it is revealed that the education given to the students to be good problem solvers contributes to the development of the students [6,14,33,44,45]. [43] stated participants should be left alone with PISA questions which they will make solutions by using mathematical reasoning. He said that when faced with such problems, there is an opportunity to use the reasoning skill. As a result, it is seen that the students have low level of reasoning ability and they have failed in the process due to the difficulties they have experienced in interpreting and reading the tables.

References

1. Akyüz, G., Pala, N. M. The effect of student and class characteristics on mathematics literacy and problem solving in PISA 2003. *Elementary Education Online,* **9**(2), 668–678 (2010).
2. Altun, M., Akkaya, R. Matematik öğretmenlerinin PISA matematik soruları ve ülkemiz öğrencilerinin düşük başarı düzeyleri üzerine yorumları. *Hacettepe Üniversitesi Eğitim Fakültesi Dergisi,* **29**(1), 19-34 (2014).
3. Altun, M., Aydın, N., Akkaya, R., Uzel, D. *PISA perspektifinden ilköğretim 8. sınıf öğrencilerinin matematik başarı düzeyinin tahlili.* https://doktora2012.files.wordpress.com/2012/10/zpisa-kuyeb.doc 10 Nisan 2017 (2012).
4. Anagün, Ş. S. The impact of teaching-learning process variables to the students' scientific literacy levels based on PISA 2006 results. *Education and Science.* **36**(162), 84-102 (2011).

5. Anıl,D.Uluslararası Öğrenci Başarılarını Değerlendirme Programı (PISA)'nda Türkiye'deki Öğrencilerin Fen Bilimleri Başarılarını Etkileyen Faktörler, *Eğitim ve Bilim (Education and Science), 34*(152), 87-100 (2009)
6. Arslan, Ç., Altun, M. Rutin olmayan matematiksel sözel problemlerin çözümünü öğrenme. *İlköğretim Online E-Dergi, 6*(1), 50-61, (2007).
7. Aydın, A., Sarıer, Y., Uysal, Ş. Sosyoekonomik ve Sosyokültürel Değişkenler Açısından PISA Matematik Sonuçlarının Karşılaştırılması. *Eğitim ve Bilim*. **37**(164), 20-30 (2012).
8. Aydoğdu İskenderoğlu, T., Baki, A. İlköğretim 8. sınıf matematik ders kitabındaki soruların PISA Matematik yeterlilik düzeylerine göre sınıflandırılması. *Eğitim ve Bilim, 36* (161), 287-301. (2011).
9. Bağcı, N., Gülçiçek, Ç., Moğol, S. Fizik konularının öğretiminde alternatif çözümlerin öğrenci başarısına etkisi. *Fırat Üniversitesi Fen ve Mühendislik Bilimleri Dergisi,* **16**(1), 49-59, (2004).
10. Bassey, M. *Case study research in educational settings.* USA: Open University (1999).
11. Creswell, J. W. Nitel Araştırma Yöntemleri, (Çev. Etd: Mesut Bütün- Selçuk Beşir Demir). *Beş Nitel Araştırma Yaklaşımı* (s. 69-110), Ankara: Siyasal Kitapevi (2013).
12. Çakan, M. Öğretmenlerin ölçme değerlendirme uygulamaları ve yeterlik düzeyleri: İlk ve ortaöğretim. *Ankara Üniversitesi Eğitim Bilimleri Fakültesi Dergisi,* **37**(2), 99-114 (2004).
13. Çakmak, M. *Matematik Derslerinde Problem Çözme Yaklaşımının Değerlendirilmesi.* Matematikçiler Derneği Bilim Köşesi.www.matder.org.tr. (2003).
14. Çelik, D., Güler, M. İlköğretim 6. sınıf öğrencilerinin gerçek yaşam problemlerini çözme becerilerinin incelenmesi. *Dicle Üniversitesi Ziya Gökalp Eğitim Fakültesi Dergisi,* **20**, 180-195,(2013).
15. Dawson B., Trapp, R.G. *Probability&related topics for making inferences about data. Basic&Clinical Biostatistics.* 3rd Edition, Lange medical Books/McGraw-Hill Medical Publishing Division, 69-72, (2001).
16. De Corte, E. Mainstreams and perspectives in research on learning mathematics from instruction. *Applied Psychology, 53*, 279–310, (2004).
17. Dündar, S., Yaman, H. Sınıf öğretmeni adaylarının matematiksel muhakeme becerilerine göre tablo ve grafikleri yorumlama başarılarının incelenmesi. *K. Ü. Kastamonu Eğitim Dergisi* **23**(4), 1695-1710, (2014).
18. Erdem, E. Relationship between mathematical reasoning and reading comprehension: The case of the 8th grade. *Necatibey Faculty of Education Electronic Journal of Science and Mathematics Education,* **10**(1), 393-414 (2016).
19. Eryılmaz Toksoy, S., Akdeniz, A. R. *Hacettepe Üniversitesi Eğitim Fakültesi Dergisi (H. U. Journal of Education)* **32**(1), 185-208 doi: 10.16986/HUJE.2016016668 (2017)
20. Eryılmaz, S., Akdeniz, A.R. 10. Sınıfta yer alan "Kuvvet ve Hareket" ünitesiyle ilgili problemleri çözerken öğretmenlerin sergiledikleri adımlar. *Hacettepe Üniversitesi Eğitim Fakültesi Dergisi,* **28**(3), 166-181 (2013).
21. Gök, T. Real time assesment of problem solving of physics students using computer based technology. *Hacettepe Eğitim Fakültesi Dergisi,* **43**, 210-221, (2012).
22. Gökkurt, B., Soylu, Y. Üniversite öğrencilerinin matematiksel ispat yapmaya yönelik görüşleri. *Eğitim ve Öğretim Araştırmaları Dergisi,* **1**(4), 56-64 (2012).
23. Greer, B. Modeling reality in mathematics classrooms: The case of word problems. *Learning and Instruction*, **7**(4), 293-307 (1997).

24. Güler, H. K. Türk öğrencilerin PISA'da karşılaştıkları güçlüklerin analizi *Uludağ Üniversitesi Eğitim Fakültesi Dergisi* **26** (2), 501-522 (2013).
25. Gürsakal, S. PISA 2009 öğrenci başarı düzeylerini etkileyen faktörlerin değerlendirilmesi. *Süleyman Demirel Üniversitesi İktisadi ve İdari Bilimler Fakültesi Dergisi,* **17** (1), 441-452 (2012).
26. Kabael, T., Barak, B. Ortaokul matematik öğretmeni adaylarının matematik okuryazarlık becerilerinin PISA soruları üzerinden incelenmesi. *Turkish Journal of Computer and Mathematics Education,* **7** (2), 321-349 (2016).
27. Lithner, J. A research framework for creative and imitative reasoning. *Educational Studies in Mathematics*, **67**(3), 255-276 (2008).
28. Moralı, S., Uğurel, I., Türnüklü, E., Yeşildere, S. Matematik öğretmen adaylarının ispat yapmaya yönelik görüşleri. *Kastamonu Eğitim Dergisi*, **14**(1), 147-160 (2006).
29. Nakiboğlu, C. V., Kalın, Ş. Ortaöğretim öğrencilerinin kimya derslerinde problem çözme güçlükleri I: (deneyimli kimya öğretmenlerine göre). *Kastamonu Eğitim Dergisi,* **11**(3), 305-317 (2003).
30. OECD. (2010). *Draft PISA 2012 mathematics framework*. http://www.oecd.org
31. Okur, S. *Students' Strategies, Episodes and Metacognitions in the Context of PISA 2003 Mathematical Literacy Items*. Yayımlanmamış Yüksek Lisans Tezi. ODTÜ Ortadoğu Teknik Üniversitesi. Ankara (2008).
32. Özdamar K. *SPSS ile Biyoistatistik*, Ankara: Nisan Kitapevi, (2001).
33. Palm, T. Impact of authenticity on sense making in word problem solving. *Educational Studies in Mathematics,* **67**(1), 37-58 (2008).
34. Peresini, D., Webb, N. *Analyzing Mathematical Reasoning in Students' Responses Across Multiple Performance Assessment Tasks. Developing Mathematical Reasoning in Grades K-12/ Lee V. Stiff,* Editör, National Council of Teachers of Mathematics, Reston, Virginia (1999).
35. PISA. Uluslararası Öğrenci Başarılarını Değerlendirme Programı - Örnek Problem Çözme Soruları. Ankara Milli Eğitim Bakanlığı: Ölçme Değerlendirme ve Sınav Hizmetleri Genel Müdürlüğü. 1-77. (2015).
36. Pilten, P. *Üstbiliş Stratejileri Öğretimin İlköğretim Besinci Sınıf Öğrencilerinin Matematiksel Muhakeme Becerilerine Etkisi*. (Yayımlanmış doktora tezi), Gazi Üniversitesi/Eğitim Bilimleri Enstitüsü, İlköğretim Anabilim Dalı: Ankara. (2008)
37. Saenz, C. The role of contextual, conceptual and procedural knowledge in activating mathematical competencies (PISA). *Educational Studies in Mathematics,* **71,** 123–143. (2009).
38. Savran, N. Z. PISA projesinin Türk eğitim sistemi açısından değerlendirilmesi. *Türk Eğitim Bilimleri Dergisi.* **2**(4), 397-414, (2004).
39. TIMSS. IEA's TIMSS 2003 International Report on Achievement in the Mathematics Cognitive Domains: Findings from a Developmental Project International Association for the Evaluation of Educational Achievement. TIMSS & PIRLS International Study Lynch School of Education, Boston College, (2003).
40. Türk Dil Kurumu [TDK] http://www.tdk.gov.tr/index.php?option=com_gts&arama=gts&guid=TDK.GTS.58eb353432cac0.73153177 (2017).
41. Umay, A. Matematiksel muhakeme yeteneği, *Hacettepe Üniversitesi Eğitim Fakültesi Dergisi*, **24**, 234–243. (2003).

42. Uysal, O. İlköğretim II. Kademe Öğrencilerinin Matematik Dersine yönelik Problem Çözme Becerileri, Kaygıları ve Tutumları Arasındaki İlişkilerin Değerlendirilmesi. Dokuz Eylül Üniversitesi, Eğitim Bilimleri Enstitüsü, İzmir (2007).
43. Widjaja, W. Towards mathematical literacy in the 21st century: perspectives from Indonesia. *Southeast Asian mathematics education journal*, **1**(1),75-84. (2011)
44. Xin, Z., Lin, C., Zhang, L., Yan, R. The Performance of Chinese Primary School Students on Realistic Arithmetic Word Problems. *Educational Psychology in Practice* **23**, 145–159, (2007).
45. Yazgan, Y., Bintaş, J. İlköğretim dördüncü ve beşinci sınıf öğrencilerinin problem çözme stratejilerini kullanabilme düzeyleri: bir öğretim deneyi. *Hacettepe Üniversitesi Eğitim Fakültesi Dergisi,* **28**, 210-218. (2005).

DEVELOPING AN ACHIEVEMENT TEST FOR FRACTION TEACHING: VALIDITY AND RELIABILITY ANALYSIS

Belgin Bal-Incebacak[1,*], *Esen* ERSOY[2]

[1]Education Faculty, Primary School Department, 55139 Samsun, Turkey
[1]Education Faculty, Mathematics Department, 55139 Samsun, Turkey

Abstract. The aim of the study is to develop an achievement test that can be used to measure the achievement status of elementary school 4th grader students in terms of their fractions learning in mathematics courses. Examining the literature, it is visible that there are 8 development stages of the achievement test.. ITEMAN 3.5 package software was employed to analyze the validity and reliability of the test. Statistical analysis of the test reveals that KR-20 Internal Consistency and KR-21 Internal Consistency are 0.90. This value indicates that the consistency of the test is high. The average discrimination of the test is 0.80, which indicates that the test is highly distinctive among students. The average strength of the test has been calculated as 0.67. Therefore, it has been determined that the test should be both easy and distinctive when it comes to measuring the effectiveness of the teaching method. The result of the analysis suggest that when consistency and distinctiveness values are considered, the developed test is not only easy to use but also highly distinctive. Finally, it has been discovered that the academic achievement test for fraction teaching is valid and reliable.

1. Introduction

Education institutions are institutions that aim to provide students with goals and behaviors that the state considers appropriate. The purpose of education is to give an individual terminal behavior in schools, which is a well-known definition [25]. For this reason, teachers are trying to provide students to obtain information by planning learning experiences and activities. Teachers need to perform assessments to determine the level of knowledge they bring. Assessment is the most important step in the curriculum [12]. Questions such as "How much students could achieve the objectives?" and "How much knowledge, skills and attitudes are brought to students?" are most important questions of the assessment [24]. To perform these assessments, many methods are referred to. Measurement tools such as written examinations, open-ended questions, multiple choice tests, oral examinations, scale types, observation and interviewing are used by researchers and teachers to meet the mentioned demand [57] Achievement is an indication of how

* Corresponding author: belginbal33@hgmail.com

much the participants benefit from a course or academic program. As for school success, it can be regarded as the average grade or score of the student from the academic courses in the school [46]. Common measurement tools used in contemporary education systems are multiple choice (standard achievement tests) tests [60]. By using assessment and evaluation methods, achievement is tried to be measured as reliable and valid as possible. The most common tool used for this measurement is multiple-choice tests from achievement tests. While developing an achievement test, there are various stages for a valid and reliable achievement test. The achievement test measures the competence of an individual including knowledge and skills in a field or subject [27]. By another definition, tools developed to measure how much an individual learns in an educational, instructional and learning environment are called achievement tests [57] With these tests, it is tried to determine the level of knowledge of students about a teaching subject.

When we think about mathematics teaching, it can be said that fractions are one of the most challenging topics. [4] stated that the teaching of mathematics is very important because of this difficulty of fractions. When [50] and [62] studies are examined, it is stated that students have low level of comprehension of fractions and mostly lacking knowledge. Fraction teaching is an important issue for mathematics. Among the reasons for being important is that it is related to many subjects in mathematics courses. It is involved in many math related subjects such as measurement, data, ratio and proportion, natural numbers, etc. Fractions are handled as part-whole, measurement, division, processing and ratio. Fractions are given the most part in part-whole relation and there is not much emphasis on its other concepts.

Many studies conducted show that students are unsuccessful in the subject of fractions. [13,53] stated that the students are lacking in operations with fractions. [32,43,51] stated that students have misconceptions regarding the fractions. [31] stated that there are problems with dividing fragments into equal parts and writing the defined fractions, [3] stated that there are problems in understanding basic concepts in every topic of fractions, and [8] stated that there are problems regarding problem solving about fractions. Even if concrete materials were used, they failed in fraction teaching. In short, it has been pointed out by many researchers that students have been unsuccessful in teaching the concept of fractions and many concepts related to fractions, and these subjects have been among the most difficult mathematics subjects to learn [23,33,38,40,49,52,54,58,61]. Şiap and Duru [56] indicate that the reason for this difficulty in fractions arises from the fact that students see the dividend and denominator of the fractions as two separate numbers. The difficulties in the concept of fractions are difficult for students because each process is unique and abstract. Fractions teaching is also very important to assess because it is intertwined with many subjects that make up a great part of mathematics. Correct assessment and evaluation is the most important step in the teaching process, since it will determine whether the desired subject is understood.

The aim of the assessment and evaluation in schools is to determine the achievement status of the students, how successful they are if successful, how unsuccessful they are if unsuccessful, which students have leveled up to next step or which students need to repeat the program [37] It is important for the procedures to be done in a reliable manner because they have an important place in the lives of the students. The data to be collected for this measurement work need an error-free, adequate, valid and reliable measuring tool. Since there are many difficulties about fractions, it is thought that an accurate measurement tool will be needed. For this reason, in the elementary school 4th grade Mathematics course, it was needed to develop an achievement test on fractions, and with this study, it is tried to shed light on new researches and researchers.

2. Method

The research was conducted by survey method. Survey research are studies to determine specific characteristics of a group [17]. According to Karasar [35] survey models are an approach aimed at describing the present situation in its both present and past form.

2.1. Study Group

In the selection of the participants in the study, purposeful sampling was used which allowed the researcher to select the people to answer the research questions [21]. The sample of the study consisted of 431 students who are studying in Samsun city center in 2015-2016.

Although the test developed in the research is directed to the fourth grade fractions unit, the inclusion of the fifth grade students in the sample of the research is to consider that the students in the 4th grade should not have studied this unit and that they will tend to leave the questions unattended when the test is implemented. In this research, the sample was formed of primary school 5th grade students for the purpose of minimizing this problem and ensuring all questions are answered.

The distribution of the sample by schools and gender is shown in Table 1.

Table 1: The distribution of the sample by schools and gender

School Type	Class	Gender and Numbers of Students			
		Famale	Male	Total	Total
Normal State School (Secondary School)	A	11	12	23	170
	B	12	12	24	
	C	10	10	20	
	D	15	15	30	
	E	17	21	38	
	F	20	15	35	
YIBO (Boarding Elementary School Regional Middle School)	A	12	13	25	125
	B	14	16	30	
	C	20	20	40	
	D	16	14	30	
Imam-Hatip Secondary School	A	20	20	40	136
	B	22	24	46	
	C	9	10	19	
	D	16	15	31	
Total		214	217	431	431

When the table was examined, a total of 431 students were studied. Of these students, 214 are female and 217 are male. According to the school types, studies were carried out with 3 different schools: normal public school, regional boarding middle-school and vocational religious middle school. Our study group consists of 170 students from normal public school, 125 from regional boarding middle school and 136 from vocational religious middle school.

2.2. Data Analysis

All the steps of the Achievement Test for Fraction Teaching, which is planned to be developed in the scope of the study, are described in detail below. The achievement test was developed by following the process steps specified by [6].

2.2.1. The Area to be used for Test Scores

The purpose of the test to be applied in the test development process should be clearly stated. Every work to be done after the determination of the purpose is to realize this purpose in effective and efficient manner [44] For this purpose, we aim to develop the achievement test to determine the knowledge of students on the subject of fractions and to determine whether all achievements have been achieved at the end of the training.

2.2.2. Determining the behaviors representing the area or the statement

At this stage of the test development, learning areas, sub-learning areas and achievements related to these learning areas are determined within the scope of the test. In this research, the achievements of learning and sub-learning areas taken from the Curriculum of Elementary School Mathematics Lesson (1, 2, 3 and 4th Grades) dated 30/06/2005 and No. 2575 (2015) and presented in Table 2 were included within the scope of achievement test. This table of specifications was arranged in accordance with the "Cognitive Domain Taxonomy" prepared by Bloom et al [14]

Table 2. Table of specifications

			Time	Cognitive Domain Steps			
Learning Area	Sub-Learning Area	Achievements		Recall	Understanding	Application	Total Number of Questions Written
Numbers	Fractions	M4.1.27. Recognizes simple, compound, and integer fractions and presents with models.	12	x	x	x	8
		M4.1.28. Compares and orders unit fractions.			x	x	5
		M4.1.29. Determines a specified simple fraction of a quantity.			x		1
		M4.1.30. Compares up to three fractions with same or multiple denominators.			x	x	3
	Operations with Fractions	M4.1.31. Performs addition and subtraction operations with equal denominator fractions.	5			x	4
		M4.1.32. Solves problems requiring addition and subtraction operations with fractions.		x	x	x	5
	Decimal Notation	M4.1.33. Specifies when a whole is divided into 10 and 100 equal parts that the resulting units of fraction can be expressed in decimal notation.	12	x	x		3
		M4.1.34. Writes a fraction with a denominator of 10 and 100 by using decimal notation.		x	x	x	5
		M4.1.35. It specifies the whole part, fraction part and step names of the decimal notation.			x	x	3
		M4.1.36. Compares two numbers which the decimal notations are given and presents the relation between them by larger, smaller or equal symbol.			x	x	6
		Total	29				43

The programs showed way to Bloom's Cognitive Domain Taxonomy, which was developed in 1956, when they were being developed [16, 28]. The six-stage original bloom taxonomy consists of categories of knowledge, comprehension, application, analysis, synthesis and evaluation, from simple to complex, in accordance with the cognitive domain [14, 15]. According to the table of specifications, Numbers were discussed in 10 achievements according to old and 13 achievements according to new program, consisting of 3 sub-learning fields as fractions, operations in fractions and decimal notation (Names fractions, of which the dividend and denominator are maximum two digit natural numbers, obtaining from the units of the fraction. Shows the fractions, of which the dividend and denominator are maximum two-digit number on the numerical axis. Compares fractions. Orders a maximum of four fractions with equal denominator, from largest to smallest or smallest to largest. Orders a maximum of four fractions with equal dividends and different denominators, from largest to smallest or smallest to largest. Determines a specified simple fraction of a quantity. Specifies when a whole is divided into 10 and 100 equal parts that the resulting units of fraction are fractions. Writes decimal fractions using commas. It specifies the whole part, fraction part and step names of the decimal fractions. Compares two decimal fractions and shows the relation between them with a bigger, smaller and equal symbol. Performs addition operation with equal denominator fractions. Performs subtraction operations with equal denominator fractions. Solves and poses problems requiring addition and subtraction operations with fractions.) A total of 43 questions were written in accordance with these achievements.

2.2.3. Writing Test İtems

At this stage, as the achievement test was developed for the 4th grades in primary school, questions suitable for the dimensions of remembering, understanding and application. Analysis, synthesis and evaluation are considered as metacognitive levels while knowledge, understanding, and practice are considered sub-levels[5]. When these questions are being created, care has been taken to ensure that they are appropriate for achievement and grade level. Attention has been paid to the rules of item writing and the types of items while the questions are being created. If the appropriate item type is not selected, the item will have low validity, since the behavior to be measured with the item cannot be measured sufficiently or not at all [6]. When writing the test items, attention was paid to the average difficulty of the test and the difficulty distribution of the test items. The positions of the questions have been changed in accordance with expert opinions by the difficulty and easiness and by achievement.

2.2.4. Reviewing The Test İtems

The test items created need to be developed and reviewed in accordance with certain criteria. It is emphasized in the literature that it should be reviewed in terms of compliance with the following qualifications:

- For validity, determining whether the behavior to be measured is worth measuring,
- Determination of the accuracy in terms of science,
- Determination of whether it is comprehensible in terms of language and there are grammar and spelling errors,
- Determination of whether there are technical defects,
- Determination of whether it is suitable for development characteristics of students [11,39,59] should be done.

Table 3. Expert opinion table

Question No: 1			
Evaluation Criteria	Yes	No	Suggestion (if any)
Question text clear, understandable			
Question is appropriate for 4. grade			
Distractors are well prepared			
Question is appropriate for achievement			
Which cognitive process dimension should the question be placed according to Bloom's taxonomy.	Recall (Knowledge)	Understanding (Comprehension)	Application

Compliance with the criteria was determined by the questions in Table 3 prepared. This table was prepared for each question and opinions were taken from the experts in Table 4. Achievement test, consisting of 43 test items, were presented to a total of 17 experts, 12 academicians, 3 elementary mathematics teachers and 2 class teachers, in order to determine the compliance to the above mentioned qualifications.

Achievement test, consisting of 43 test items, were presented to a total of 17 experts, 12 academicians, 3 elementary mathematics teachers and 2 class teachers, in order to determine the compliance to the above mentioned qualifications.

Table 4. Expert opinion table

Kod	Sex	University/Shocol	Degree	Education
1	Male	Sinop University	Associate Professor	Mathematics Education
2	Male	Ondokuz Mayıs University	Associate Professor	Science Education
3	Male	İstanbul University	PhD, Assistant Professor	Mathematics Education
4	Female	Dokuz Eylül University	PhD, Assistant Professor	Mathematics Education
5	Female	İstanbul University	PhD, Assistant Professor	Mathematics Education
6	Female	Ondokuz Mayıs University	PhD, Assistant Professor	Mathematics Education
7	Female	Ondokuz Mayıs University	PhD, Assistant Professor	Mathematics Education
8	Female	Ondokuz Mayıs University	PhD, Assistant Professor	Mathematics Education
9	Female	İstanbul University	Research Assistant PhD	Mathematics Education
10	Female	Ondokuz Mayıs University	Research Assistant PhD	Science Education
11	Male	Mersin University	Research Assistant PhD	Assesment and evaluation
12	Female	Ondokuz Mayıs University	Research Assistant	Mathematics Education
13	Male	Ministry of Education	Master Degree Teacher	Primary School
14	Female	Ministry of Education	Teacher	Primary School
15	Female	Ministry of Education	Master Degree Teacher	Mathematics Education
16	Male	Ministry of Education	Master Degree Teacher	Mathematics Education
17	Male	Ministry of Education	Master Degree Teacher	Mathematics Education

It has been examined in terms of scope validity with the expert group formed by teachers and academic members. Such procedures increase the validity and reliability of the test [1,47].

2.2. 5.Preparing The Test Form

After the expert opinions are obtained, necessary arrangements are made and questions should be sorted in a certain order. Below are the steps to be considered when preparing the test form. These steps consist of three stages;

1. Distribution of the items in the test form
2. Writing of test instructions
3. Writing of items [6]

According to the opinions taken from the experts, the test was shaped by correcting the errors. Formally unsuitable questions, unreadable, incomplete and hard-to-comprehend questions are corrected. In addition, while the questions were being edited, [6] emphasized that the items should be grouped according to the subjects and the questions should be ordered from the easy to difficult, without disturbing the objective unity. Font size in the literature for the fourth grade students was determined as 12. According to this information the question texts are adjusted to be size 12 fonts. After these adjustments are made, information such as "Switch to the next page" or "Switch to the other page" should be placed. At the end of the test, there should be information such as "Test is over" and "Check your answers" should be placed [6].

2.2. 6.Putting The Test on A Trial Implementation

Before the moving on to application, it is necessary to determine the duration for the test. In this context, a random class was selected from a school with a low level of socio-economic level at the 5th grade level and a pilot application of the test was conducted on a total of 31 students without a time limit. As a result of the pilot application, it was determined that the best student in the class finished in 35 minutes, and a student who was lower in level finished within 50 minutes. Students at the average level finished in 45 minutes. In this context, the duration of the test was determined as 45 minutes.

The final form was given to the test, the application duration was determined as 45 minutes and trial application was performed. Accordingly, the test was applied to the 5th grade students so that the final test can be formed. As the number of students in the test group increased, the closeness of statistics obtained would get nearer to the actual statistics, so the test was applied to 400 students.

2.2.7.Selecting Materials by Analyzing Them According to The Trial Implementation

After the test application, item analysis was carried out so that the desired test can be formed, and suitable items for the final test were selected. IBM-SPSS, FINESSE, ITEM and hand calculations can be done to analyze the validity and reliability of a test. It is preferred to use ITEMAN 3.5 package program in order to obtain the most accurate and correct analyzes. The reason for preferring this program is to give an idea about the statistics of each question as well as their options. The distinctiveness and difficulty indices of the items are examined in the item selection [39] and also the distractor characteristics of the questions to be selected are reviewed. The ITEMAN program calculates p (item difficulty) and r (distinctiveness power) values for each question. Table 5 shows the item difficulty and distinctiveness index of each item in the item analysis for the developed achievement test.

Table 5. Item difficulty and distinctiveness index of each item in the item analysis

Seg No.	Number of Replies	Prop. Correct	Disc. Index (Point-to-Double Series Correlation)	item reliability	Item Varyans	Item Standard deviation	item correlation (Pearson)	item difficulty index evaluated	Item distinctiveness index evaluated
	Ij	pj	rjx	rj	sj2	sj	rjy		
1	27	0,87	0,74	0,25	0,11	0,34	0,67	Easier *	Good item
2	30	0,97	0,80	0,14	0,03	0,18	0,57	Easier *	Good item
3	20	0,65	0,61	0,29	0,23	0,48	0,59	Easier *	Good item
4	25	0,81	0,55	0,22	0,16	0,40	0,51	Easier *	Good item
5	30	0,97	0,62	0,11	0,03	0,18	0,45	Easier *	Good item
6	12	0,39	0,54	0,26	0,24	0,49	0,53	Difficult ***	Good item
7	27	0,87	0,74	0,25	0,11	0,34	0,67	Easier *	Good item
8	19	0,61	0,66	0,32	0,24	0,49	0,65	Easier *	Good item
9	19	0,61	0,59	0,29	0,24	0,49	0,57	Easier *	Good item
10	20	0,65	0,62	0,30	0,23	0,48	0,60	Easier *	Good item
11	13	0,42	0,70	0,34	0,24	0,49	0,69	Medium difficulty **	Good item
12	16	0,52	0,29	0,15	0,25	0,50	0,29	Medium difficulty **	Analyze-correct
13	14	0,45	0,39	0,19	0,25	0,50	0,38	Medium difficulty **	Not distinctive
14	22	0,71	0,75	0,34	0,21	0,45	0,72	Easier *	Good item
15	5	0,16	-0,35	-0,13	0,14	0,37	-0,35	Difficult ***	Not distinctive
16	13	0,42	0,60	0,30	0,24	0,49	0,59	Medium difficulty **	Good item
17	5	0,16	-0,32	-0,12	0,14	0,37	-0,32	Difficult ***	Not distinctive
18	3	0,10	0,14	0,04	0,09	0,30	0,14	Difficult ***	Not distinctive

19	17	0,55	0,72	0,36	0,25	0,50	0,71	Medium difficulty **	Good item
20	6	0,19	-0,04	-0,02	0,16	0,40	-0,04	Difficult ***	Not distinctive
21	9	0,29	0,02	0,01	0,21	0,45	0,02	Difficult ***	Not distinctive
22	7	0,23	-0,40	-0,17	0,17	0,42	-0,40	Difficult ***	Not distinctive
23	4	0,13	-0,09	-0,03	0,11	0,34	-0,09	Difficult ***	Not distinctive
24	3	0,10	0,20	0,06	0,09	0,30	0,20	Difficult ***	Analyze-correct
25	5	0,16	0,35	0,13	0,14	0,37	0,35	Difficult ***	Distinctive
26	17	0,55	0,31	0,16	0,25	0,50	0,31	Medium difficulty **	Distinctive
27	22	0,71	0,44	0,20	0,21	0,45	0,42	Easier *	Good item
28	17	0,55	0,76	0,38	0,25	0,50	0,75	Medium difficulty **	Good item
29	18	0,58	0,46	0,23	0,24	0,49	0,45	Medium difficulty **	Good item
30	16	0,52	0,74	0,37	0,25	0,50	0,73	Medium difficulty **	Good item
31	17	0,55	0,76	0,38	0,25	0,50	0,75	Medium difficulty **	Good item
32	17	0,55	0,31	0,16	0,25	0,50	0,31	Medium difficulty **	Distinctive
33	22	0,71	0,44	0,20	0,21	0,45	0,42	Easier *	Good item
34	17	0,55	0,76	0,38	0,25	0,50	0,75	Medium difficulty **	Good item
35	18	0,58	0,46	0,23	0,24	0,49	0,45	Medium difficulty **	Good item
36	16	0,52	0,74	0,37	0,25	0,50	0,73	Medium difficulty **	Good item
37	17	0,55	0,76	0,38	0,25	0,50	0,75	Medium difficulty **	Good item
38	17	0,55	0,31	0,16	0,25	0,50	0,31	Medium difficulty **	Distinctive
39	22	0,71	0,44	0,20	0,21	0,45	0,42	Easier *	Good item
40	17	0,55	0,76	0,38	0,25	0,50	0,75	Medium difficulty **	Good item
41	18	0,58	0,46	0,23	0,24	0,49	0,45	Medium difficulty **	Good item
42	17	0,55	0,31	0,16	0,25	0,50	0,31	Medium difficulty **	Distinctive
43	22	0,71	0,44	0,20	0,21	0,45	0,42	Easier *	Good item

The distinctiveness index value can be found through the internal consistency Cronbach Alpha item-total table calculated by SPSS software. The item difficulty index takes values between 0 and 1. The difficulty index approaching to 0 indicates that the item is getting difficult, approaching to 1 indicates that the item is getting easier, being 0.50 indicates that the item has medium difficulty [18,19,21,45]. Item distinctiveness index is an important decision criterion when items are included in the test. It allows to distinguish between the informed and the uninformed student. The item difficulty index takes values between -1 and +1 [48]. It can be said that the item difficulty is related to the reliability, one of the qualifications required in measurement tool" [10,57]. The fact that the difficulty of any item in an achievement test is moderate, is a sign that that item is a good item.

The criteria to be used by Ebel (1995) in determining which of the test items should be tested is shown in Table 6 [57]

Table 6. Item selection criteria based on item distinctiveness indices

Item difficulty index (p)	Item distinctiveness index (r)	
0.90 >	-	- Preferred if there is an effective teaching
0.60-0.90	r>0.20	- A typical good item

0.60-0.90	r<0.20	- An item needs to be worked on
p<0.60	r>0.20	- A difficult but distinctive item (This question is good if you have high standards)
p<0.60	r<0.20	A difficult but not distinctive item (This item cannot be used)

The questions were evaluated according to the above criteria. In this respect, the distribution of 43 questions according to the item difficulty index and item distinctiveness index given in table 6 is shown in Table 7:

Table 7. Item ranking according to item difficulty and distinctiveness index

Item difficulty *index* (p)	*Item distinctiveness index* (r)	Comment
0.90 >	-	--
0.60-0.90	r>0.20	1,2,3,4,5,7,8,9,10,14,27,33,39,43
0.60-0.90	r<0.20	--
p<0.60	r>0.20	6,11,12,13,16,19,22,24,25,26,28,29,30,31,32,34,35,363,37,38,40,41,42
p<0.60	r<0.20	15,17,18,20,21,23

When the Table 7 was examined, it was seen that the test was a test with questions of high standards. [44] points out that the range of difficulty indices of test items for achievement tests should generally range from 0.20 to 0.80. The test is in this range, but there are a lot of questions that are difficult and distinctive. Accordingly, the items need to be made easier.

In the item score analysis, item difficulty and distinctiveness were examined for analysis of optional items. After analyzing the results of the item analysis which was coded in the ITEMAN program of the test consisting of 43 items, it was determined that the item correlations of 15 items were low and removed from the test. ITEMAN program has provided us with the following information for each question. In light of this information, final test was created with 28 questions in total

Table 8. Item analysis results and evaluation for test item

Item Statistics					Alternative Statiscs					
Seq. No.	Scale-item	Prop. Cerrect	Disc. index	Point Biser.	Alt.	Prop. Total	Endorsing low	Endorsing High	Point Biser.	Key
1	1-1	.81	.56	.80	A	.03	.09	.00	-.28	
					B	.04	.15	.00	-.32	
					C	.03	.03	.03	-.06	
					D	.81	.41	.97	.80	*

p: item difficulty index (p) D: Item distinctiveness index (r)

Table 8 shows the results of the item analysis of the 1st test item conducted by ITEMAN program. It is a question that upper group was successful and some of the subgroup were successful. The question in the table, as a result of the analysis, is an easy answer (p = 0.81) with a good distinctiveness power (D = 0.56). Since 1st item is a typical good item, it can be included in the test as it is.

2.2.8. Prognosis of the selected items that generate the statistics of the final test

When the literature is examined, the test-retest reliability expressing consistency and continuity of the scores obtained from a test emphasizes that, it can be determined by the relationship coefficients between the before and after measurements [7,18,19,20,35,45]. The reliability coefficient is between 0 and 1. As the correlation coefficient approaches 1, it can be said that the scores obtained from the applications are close to each other and closer to 0, respectively. The calculation of the coefficient values as 0.70 and over is interpreted as the measurement tool making stable measurement [9,10]

A test's reliability is found when a test is run twice on the same group at a certain time interval. The time elapsed is a factor that affects the consistency of the test. For this reason, the time must not be too long or too short [20] Reported that ranges of 2-6 weeks are appropriate for this reliability.

The main test statistics such as average, standard deviation, distinctiveness and difficulty indices and reliability coefficient of the achievement test formed after the item analysis and selection of the items to be included in the test are given in Table 9:

Table 9. Final test statistics.

not be indented; subsequent paragraphs should be indented by 5 mm.

The use of sections to divide the text of the paper is optional and left as a decision for the author. Where the author wishes to divide the paper into sections the formatting shown in Table 2 should be used.

NUMBER OF ITEMS IN THE TEST	28	
NUMBER OF ANSWERERS TAKING THE TEST	115	
TEST AVERAGE	20,06	
TEST VARIANCE	44,42	
STANDARD DEVIATION OF TEST	6,66	
HIGHEST SCORE	30,00	
LOWEST SCORE	2,00	
MEDIAN	21,00	
RANGE	28,00	
SKEWNESS	-1,09	Skewed to Left
KURTOSIS	1,26	Acute
RELATIVE VARIABILITY COEFFICIENT	33,22	Kurtic and heterogeneous
KR-20 INTERNAL CONSISTENCY	0,90	High
KR-21 INTERNAL CONSISTENCY	0,90	High
SPLIT HALF CONSISTENCY	0,80	High
STND. ERROR OF AVERAGE	2,08	
AVERAGE DISTINCTIVENESS OF TEST	0.48	Very good
AVERAGE DIFFICULTY OF TEST	0,67	Easy
CRONBACH'S ALPHA VALUE	,901	

When the statistical results of the test are looked at, it is seen that KR-20 Internal Consistency and KR-21 Internal Consistency are 0.90. This shows that the consistency of the test is high. The average distinctiveness of the test is seen as 0.80. This also suggests that the test is distinctive between students at a very good level. Average difficulty of the test is seen as 0,67. This states that when it is desired to measure the effectiveness of the effective teaching method of the test, it should be easy but highly distinctive. When the values of the test are examined, it is seen that the test is easy, highly distinctive.

3. DISCUSSION AND CONCLUSIONS

Many achievement tests have been developed in the literature to measure the achievement of students. In the literature, there are achievement tests especially for the science [2,22,29,30,55] and geography lessons [60]. It was seen that there are very few achievement tests accessed as articles. Not many achievement test articles concerning mathematics were accessed. It is thought that the developed achievement test is a valid and reliable achievement test for mathematics teaching in elemental school.

Achievement tests generally consist of multiple choice questions, and [36] noted that multiple-choice tests are often used to measure student achievement. [42] stated that many questions were asked at the same time and that tests were used as a means of measuring the whole of the desired courses in a short period of time. For this purpose, the achievement test is a reliable measurement tool for making comparisons between classes and students. Developing an achievement test is actually obtaining a valid and reliable test [41]. The developed achievement test is a measurement tool that measures all achievements with sufficient questions, useful for elementary schools with small age groups. Fonthal (2004) reported that schools with more consistent curricula in reading and mathematics areas were more successful in evaluating students. In other words, he emphasized the importance of preparing a question to cover the subjects taught. [34] stated that developing a test that measures all achievements will be more effective in achievement measurement. For this purpose, questions including all the achievements have been prepared.

In short, 431 students who are studying in Samsun city center in 2015-2016 were worked with in the study. Questions including all the achievements given in the table of specifications have been prepared in accordance with the purpose. Expert opinions were obtained for the items prepared for testing and adjustments were made. Test form was prepared and a pilot study was conducted with 31 students to determine how many minutes the test would take. Here, the average of the first and last completing students is taken and an average duration is calculated. After that, 400 students were asked these questions and item analysis were made. Test questions consisting of 43 items were reduced to 28 questions after the item analysis. Then, test-retest application was made for the final results and reliability of the test. The test was applied to 115 students first and then two weeks later, applied again to the same 115 students and the consistency between the two tests was examined. When the statistical results of the test are examined, it is seen that KR-20 Internal Consistency and KR-21 Internal Consistency are 0.90. This result indicates that the test is highly reliable. The average of the test was calculated as 20,06, the variance as 44,42, the standard deviation as 6,66, the median as 21, and the range as 28. Moreover, average difficulty of the test was found as better than the values in the literature with 0,67, and an achievement test with high distinctiveness as 0,48 was obtained.

This study is a unique study showing that KÖYBT has valid and reliable results. Moreover, since the KÖYBT is prepared by following the test development steps, the method can serve as an example for other researches. This work is original in terms of inventory collection and construction and is parallel to the results of other scientific studies.

References

1. Abraham, M.R., Williamson, V.M., Westbrook, S.L. A., *Cross-Age Study of the Understanding Five Concepts,* Journal of Research in Science Teaching, **31**, 2, 147-165 (1994).
2. Akbulut, H. İ., Çepni, S. Bir Üniteye Yönelik Başarı Testi Nasıl Geliştirilir?: İlköğretim 7. Sınıf Kuvvet ve Hareket Ünitesi. *Amasya Üniversitesi Eğitim Fakültesi Dergisi,* **2**(1), 18-44 (2013).
3. Aksu, M., Student performance in dealing with fractions. *The Journal of Educational Research,* **90**(6), 375-380 (1997).
4. Alacaci, C. Öğrencilerin kesirler konusundaki kavram yanılgıları. E. Bingölbali ve M. F. Özmantar (Ed.), *İlköğretimde karşılaşılan matematiksel zorluklar ve çözüm önerileri* (1. Baskı). Ankara: Pegem Akademi Yayınları (2009).

5. Arı, A. Bilişsel Alan Sınıflamasında Yenilenmiş Bloom, SOLO, Fink, Dettmer Taksonomileri ve Uluslararası Alanda Tanınma Durumları. *Uşak Üniversitesi Sosyal Bilimler Dergisi,* **6** (2), 259-290 (2013).
6. Atılgan, H. Eğitimde Ölçme ve Değerlendirme. Hakan Atılgan (Ed.), *Test Geliştirme* içinde (s.315- 348). Ankara: Anı Yayıncılık (2013).
7. Balcı, A. Sosyal Bilimlerde Araştırma Yöntem, Teknik ve İlkeler. Ankara: Pegem A Yayıncılık (2015).
8. Başgün, M., Ersoy, Y. Sayılar ve Aritmatik I: Kesir ve Ondalık Sayıları Öğretilmesinde Bazı Güçlükler ve Yanılgılar, IV. *Fen Bilimleri Eğitimi Kongresi Bildiri Kitabı*: 604-608, MEB Yay., Ankara (2000).
9. Başol, G. *Eğitimde Ölçme ve Değerlendirme.* Ankara: Pegem A Yayıncılık (2013).
10. Baştürk, S. *Eğitimde Ölçme ve Değerlendirme*. (Edt. Savaş Baştürk). Ölçme Araçlarının Taşıması Gereken Nitelikler (s. 23-53). Ankara: Nobel Yayıncılık (2014).
11. Baykul, Y. *Eğitimde ve Psikolojide Ölçme: Klasik Test Teorisi ve Uygulanması*. Ankara: ÖSYM (2000).
12. Bayrak, B., Erden, A. M. Fen bilgisi öğretim programının değerlendirilmesi. *Kastamonu Eğitim Dergisi,* **15**(1), 137-154, (2007).
13. Behr, M. J., Lesh, R., Post, T., Silver, E. A. Rational number concepts. In R. Lesh, M. Landau (Eds.), *Acquisitions of mathematics concepts and processes* (pp. 92–126). New York: Academic Press (1983).
14. Bloom, B.S. *Taxonomy of Educational Objectives: The Classification of Educational Goals*, Handbook I: The Cognitive Domain. New York: David Mckay Company Inc (1956).
15. Bloom, B.S., Hastings, J.T., Madaus, G.F. *Handbook on Formative and Summative Evaluation of Student Learning*. New York: McGraw- Hill (1971).
16. Bümen, N.T. Program Geliştirmede Bir Dönüm Noktası: Yenilenmiş Bloom Taksonomisi. *Eğitim ve Bilim,* **31**(142), 3-14 (2006).
17. Büyüköztürk, Ş. *Bilimsel Araştırma Yöntemleri* Ankara: Pegem Akademi (2009).
18. Büyüköztürk, Ş. *Veri analizi El Kitabı.* Ankara: Pegam A Yayıncılık (2011).
19. Can, A. *SPSS ile Bilimsel Araştırma Sürecinde Nicel Veri Analizi.* Ankara: Pegem Yayıncılık (2014).
20. Christensen, L. B., Johnson, R. B., Turner, L. A. Research Methots Desing and Analysis (Çev Edt: Ahmet Alpay). Değişkenlerin Ölçümü ve Örneklem (s.149-180). Ankara: Anı Yayıncılık (2015).
21. Cohen, L., Manion, L., Morrison, K. *Research methods in education*. (6th edition). London: Routledge Falmer (2007).
22. Çalık, M., Ayas, A. Çözeltilerde kavram başarı testi hazırlama ve uygulama. *Pamukkale Üniversitesi Eğitim Fakültesi Dergisi*, **2** (14), 1-16 (2003).
23. De Castro, B. Cognitive models: the missing link to learning fraction multiplication and division. *Asia Pacific Education Review*, **9**(2), 101-112 (2008).

24. Erden, M. *Eğitimde Program Değerlendirme*, Ankara: Pegem Yayıncılık (1993).
25. Ertürk, S. *Eğitimde Program Geliştirme*. Ankara: Meteksan (1997).
26. Fonthal, G. *Alignment of State Assessments and Higher Education Expectations: Definition and Utilization of an Alignment Index*. University of California, Irvine (2004).
27. Gay, L. R., Geoffrey E. Mills, Airasian, P. *Educational Research: Competencies for Analysis and Applications*. Boston: Pearson (2009).
28. Gezer, M., Şahin, İ. F., Öner Sünkür, M., Meral, E. 8. Sınıf Türkiye Cumhuriyeti İnkılâp Tarihi ve Atatürkçülük Dersi Öğretim Programı Kazanımlarının Revize Edilmiş Bloom Taksonomisine Göre Değerlendirilmesi. *Bartın Üniversitesi Eğitim Fakültesi Dergisi*, **3** (1), 433 – 455 (2014).
29. Gönen, S., Kocakaya, S., Kocakaya, F. Dinamik konusunda geçerliliği ve güvenilirliği sağlanmış bir başarı testi geliştirme çalışması. *Yüzüncü Yıl Üniversitesi, Eğitim Fakültesi Dergisi*, **8** (1), 40-57 (2011).
30. Güven, E. Çevre sorunları başarı testinin geliştirilmesi ve öğretmen adaylarının bilgi düzeylerinin belirlenmesi. *Trakya Üniversitesi Eğitim Fakültesi Dergisi*,**3**(2), 114-127 (2013).
31. Haser, Ç., Ubuz, B., İlköğretim 5.sınıf Öğrencilerinin Kesirler Konusunda Kavramsal Anlama ve İşlem Yapma Performansı. *IV. Fen Bilimleri Eğitimi Kongresi*, s: 609-612 MEB Yay., Ankara (2001).
32. Işık, C., Kar, T. 7. sınıf öğrencilerinin kesirlerde toplama işlemine kurdukları problemlerin analizi. *İlköğretim Online*, **11**(4), 1021-1035. (2012).
33. İşeri, A., *Diagnosis on Students' Misconceptions on Decimal Numbers* (unpublished master teshiss), ODTÜ, Ankara (1997).
34. Kara, Y., Çepni, S. Investigation the Alignment between School Learning and Entrance Examinations through Item Analysis. *Journal of Baltic Science Education*, **10**(2),73-86 (2011).
35. Karasar, N. *Bilimsel araştırma yöntemleri*. Ankara: Nobel Yayıncılık (2009).
36. Kempa, R. *Assessment in Science. Cambridge University Press*, Cambridge, London (1986).
37. Küçükahmet, L. *Öğretimde Planlama Değerlendirme*. Ankara: Nobel Yayın Dağıtım (2004).
38. Malcolm, P. S. Understanding Rational Numbers. *Mathematics Teacher*. **80**, 518-521 (1987).
39. Metin, M. Kuramdan uygulamaya eğitimde bilimsel araştırma yöntemleri. (Edt). *Nicel Veri Toplama Araçları* (s. 161-2014). Ankara: Pegem A Yayıncılık. (2015).
40. Moss, J., Case, R. Developing children's understanding of the rational numbers: a new model and experimental curriculum. *Journal for Research in Mathematics Education*, **30**(2), 122 -147. (1999).
41. Narlı, S., Başer, N. Küme, Bağıntı, Fonksiyon" Konularında Bir Başarı Testi Geliştirme ve Bu başarı testi İle Üniversite Matematik Bölümü 1. Sınıf Öğrencilerinin Bu Konulardaki Hazırbulunuşluklarını Betimleme Üzerine Nicel Bir Araştırma. *Dokuz Eylül Üniversitesi Buca Eğitim Fakültesi Dergisi*, **24**, 147-158 (2008).
42. Ogan Bekiroğlu, F. *Ne kadar Başarılı?, Klasik ve Alternatif Ölçme- Değerlendirme Yöntemleri ve Fizikte Uygulamalar* (1. baskı). Ankara: Nobel Yayın Dağıtım. (2004).
43. Olkun, S., Toluk-Uçar, Z. *İlköğretimde etkinlik temelli matematik öğretimi*. Ankara: Anı Yayıncılık (2012).

44. Özçelik, D. A. *Eğitimde Ölçme ve Değerlendirme*, Ankara: ÖSYM Yayınları (1992)
45. Özdamar, K. *Paket Programları İle İstatiksel Veri Analizi –I (MINITAB 16- IBM SPSS 21)*. Eskişehir: Nisan Kitapevi (2013).
46. Özgüven, İ. E. *Bireyi Tanıma Teknikleri,* Ankara: Pdrem Yayınları (1998).
47. Peterson, R.F., Treagust, D.F. Grade-12 Students' Misconceptions of Covalent Bonding and Structure, *Journal of Chemical Education,* **66** (6), 459-460. (1989)
48. Popham, W. J. *Classroom assessment: What teachers need to know*. Boston: Allyn Bacon (1995).
49. Post, T., Fractions and Other National Numbers. *Aritmetic Teacher*. **37**, 3, 28 (1989).
50. Sowder, J.T., Wearne, D. What do we know about eighth-grade student achievement? *Mathematics Teaching in Middle School,* **11**(6), 285-293. (2006).
51. Soylu, Y., Soylu, C. İlköğretim beşinci sınıf öğrencilerinin kesirler konusundaki öğrenme güçlükleri: Kesirlerde sıralama, toplama, çıkarma, çarpma ve kesirlerle ilgili problemler. *Erzincan Eğitim Fakültesi Dergisi*, **7**(2), 101-117 (2005).
52. Soylu, Y. Öğrencilerin kesirler konusundaki hata ve yanlış anlamaları ve sınıf öğretmen adaylarının tahmin edebilme becerileri. *Çağdaş Eğitim Dergisi,* **33**(356), 31-39 (2008).
53. Stafylidou, S., Vosniadou, S. The development of students' understanding of the numerical value of fractions. *Learning and Instruction*, **14,** 503-518. (2004).
54. Sweetland, R. Understanding Multiplication of Fractions, *Arithmetic Teacher*. **32,** 48-52. (1984).
55. Şen, H. C., Eryılmaz, A. Bir başarı testi geliştirme çalışması: Basit elektrik devreleri başarı testi geçerlik ve güvenirlik araştırması. *Yüzüncü Yıl Üniversitesi, Eğitim Fakültesi Dergisi,* **8** (1), 1-39. (2011).
56. Şiap, İ., Duru, A. Kesirlerde geometriksel modelleri kullanabilme becerisi. *Kastamonu Eğitim Dergisi, 12* (1), 89-96 (2004).
57. Tekin, H. *Eğitimde Ölçme ve Değerlendirme*. Ankara: Yargı Yayınevi (2009).
58. Tirosh, D. Enhancing prospective teachers' knowledge of children's conceptions: the case of division of fractions. *Journal for Research in Mathematics Education*, **31**(1), 5-25 (2000).
59. Turgut, M. F., Baykul, Y. Eğitimde Ölçme ve Değerlendirme. Ankara: Pegem Akademi (2010).
60. Uzunöz, A., Buldan, İ. ortaöğretim coğrafya dersi doğal sistemler konu alanı atmosfer ve iklim ünitesi başarı testi geliştirme çalışması. *Kastamonu Eğitim Dergisi, 20* (1), 291-312 (2012).
61. Ünlü, M., Ertekin, E. Why do pre-service teachers pose multiplication problems instead of division problems in fractions? *Procedia - Social and Behavioral Sciences, 46*, 490-494 (2012).
62. Wearne, D., Kouba, V.L. Rational numbers. In E.A. Silver and P.A. Kenney (Eds.), *Results from the seventh mathematics assessment of the National Assessment of Education Progress* (pp. 163-191). Reston, VA: NCTM (2000).

ITM Web of Conferences **13**, 01004 (2017) DOI: 10.1051/itmconf/20171301004
CMES2017

Asymptotically Statistical Equivalent of Order α in Amenable Semigroups

Yavuz ALTIN[1,⋆], *Mikail* ET[1,⋆⋆], and *Hifsi* ALTINOK[1,⋆⋆⋆]

[1] *Department of Mathematics, Fırat University, 23119 Elazığ, Turkey*

Abstract. In this study we introduce the concepts of asymptotically statistical equivalent functions of order α and strong asymptotically equivalent functions of order α defined on discrete countable amenable semigroups.

1 Introduction

The idea of statistical convergence was given by Zygmund [30] in the first edition of his monograph published in Warsaw in 1935. The concept of statistical convergence was introduced by Steinhaus [29] and Fast [13] and later reintroduced by Schoenberg [27] independently. Later on it was further investigated from the sequence space point of view and linked with summability theory by Altin et al. [2], Connor [1], Çinar et. al. [3], Çolak [4], Et et. al. ([8],[9],[10],[11],[12]), Fridy [14], Işık [15], Küçükaslan and Yılmaztürk [16], Mursaleen [20], Nuray and Rhoades ([22],[23],[24]), Salat [26], Şengül and Et [28] and many others.

The idea of statistical convergence depends upon the density of subsets of the set $\mathbb{N}$ of natural numbers. The density of a subset $\mathbb{E}$ of $\mathbb{N}$ is defined by

$$\delta(\mathbb{E}) = \lim_{n\to\infty} \frac{1}{n} \sum_{k=1}^{n} \chi_{\mathbb{E}}(k), \quad \text{provided that the limit texists.}$$

A sequence $x = (x_k)$ is said to be statistically convergent to L if for every $\varepsilon > 0$, $\delta(\{k \in \mathbb{N} : |x_k - L| \geq \varepsilon\}) = 0$.

Marouf [19] introduced definitions for asymptotically equivalent sequences and asymptotic regular matrices. Patterson [25] extend these concepts by presenting an asymptotically statistically equivalent analog of these definitions and natural regularity conditions for nonnegative summability matrices.

Suppose that G be a discrete countable amenable semigroup with identity in which both right and left cancelation laws hold, and $w(G)$ and $m(G)$ denote the spaces of all real valued functions and all bounded real functions on G, respectively. $m(G)$ is a Banach space with the norm

$$\| f \|_{\infty} = sup\{|f(g)| : g \in G\}.$$

⋆e-mail: yaltin23@yahoo.com
⋆⋆e-mail: mikailet68@gmail.com
⋆⋆⋆e-mail: hifsialtinok@yahoo.com

Nomika [21] showed that, if G is a countable amenable group, there exists a sequence $\{S_n\}$ of finite subsets of G such that

i) $G = \cup_{n=1}^{\infty} S_n$,

ii) $S_n \subset S_{n+1}, \quad n = 1, 2, 3, ...$

iii) $\lim\limits_{n\to\infty} \dfrac{|S_n g \cap S_n|}{|S_n|} = 1, \quad \lim\limits_{n\to\infty} \dfrac{|gS_n \cap S_n|}{|S_n|} = 1, \quad$ for all $g \in G$.

Here $|A|$ denotes the number of elements the set A.

Any sequence of finite subsets of G satisfying (i), (ii) and (iii) is called a Folner sequence for G. For a detailed account of amenable semigroups one may refer to ([5],[6],[7],[17],[18])

2 Main Results

Definition 1 *Let α be any real number such that $0 < \alpha \leq 1$. We define the upper and lower Folner α−density of the subset S of G by*

$$\overline{\delta}_\alpha(S) = \limsup_{n\to\infty} \frac{1}{|S_n|^\alpha} |\{g \in S_n : g \in S\}|$$

and

$$\underline{\delta}_\alpha(S) = \liminf_{n\to\infty} \frac{1}{|S_n|^\alpha} |\{g \in S_n : g \in S\}|$$

respectively. If $\overline{\delta}_\alpha(S) = \underline{\delta}_\alpha(S)$, then

$$\delta_\alpha(S) = \lim_{n\to\infty} \frac{1}{|S_n|^\alpha} |\{g \in S_n : g \in S\}|$$

is called Folner α−density of S. For $\alpha = 1$, we have Folner density of $S \subset G$ which were defined an studied by Nuray and Rhoades ([22],[23],[24]).

Definition 2 *Let G be a discrete countable amenable semigroup with identity in which both right and left cancelation laws hold and let α be any real number such that $0 < \alpha \leq 1$. The function $f \in w(G)$ is said to be statistically convergent to s of order α for any Folner sequence $\{S_n\}$ for G if*

$$\lim_{n\to\infty} \frac{1}{|S_n|^\alpha} |\{g \in S_n : |f(g) - s| \geq \varepsilon\}| = 0.$$

The set of all statistically convergent functions of order α will be denoted by $S^\alpha(G)$.

Definition 3 *Let G be a discrete countable amenable semigroup with identity in which both right and left cancelation laws hold, α be any real number such that $0 < \alpha \leq 1$. The function $f \in w(G)$ is said to be strongly summable of order α to s for any Folner sequence $\{S_n\}$ for G if*

$$\lim_{n\to\infty} \frac{1}{|S_n|^\alpha} \sum_{g\in S_n} |f(g) - s| = 0.$$

The set of all strongly summable functions of order α will be denoted by $w^\alpha(G)$.

Definition 4 *Let G be a discrete countable amenable semigroup with identity in which both right and left cancelation laws hold and let α be any real number such that $0 < \alpha \leq 1$. Two nonnegative functions $f, h \in w(G)$ are said to be asymptotically statistical equivalent of order α, for any Folner sequence $\{S_n\}$ for G if, for every $\varepsilon > 0$,*

$$\lim_{n\to\infty} \frac{1}{|S_n|^\alpha} \left| \left\{ g \in S_n : \left| \frac{f(g)}{h(g)} - 1 \right| \geq \varepsilon \right\} \right| = 0.$$

In this case we write $f \overset{S^\alpha}{\sim} g$.

Definition 5 *Let G be a discrete countable amenable semigroup with identity in which both right and left cancelation laws hold and let α be any real number such that $0 < \alpha \leq 1$. Two nonnegative functions $f, h \in w(G)$ are said to be strong asymptotically equivalent of order α, for any Folner sequence $\{S_n\}$ for G if, for every $\varepsilon > 0$,*

$$\lim_{n\to\infty} \frac{1}{|S_n|^\alpha} \sum_{g\in S_n} \left| \frac{f(g)}{h(g)} - 1 \right| = 0.$$

In this case we write $f \overset{w^\alpha}{\sim} g$.

Theorem 6 *Let G be a discrete countable amenable semigroup with identity in which both right and left cancelation laws hold and suppose also that the parameters α and β are fixed real numbers such that $0 < \alpha \leq \beta \leq 1$, then $f \overset{S^\alpha}{\sim} g$ implies $f \overset{S^\beta}{\sim} g$.*

Proof. Proof follows from the inequality

$$\frac{1}{|S_n|^\beta} \left| \left\{ g \in S_n : \left| \frac{f(g)}{h(g)} - 1 \right| \geq \varepsilon \right\} \right| \leq \frac{1}{|S_n|^\alpha} \left| \left\{ g \in S_n : \left| \frac{f(g)}{h(g)} - 1 \right| \geq \varepsilon \right\} \right|.$$

From Theorem 6 we have the following. ■

Corollary 7 *Let G be a discrete countable amenable semigroup with identity in which both right and left cancelation laws hold and let α be any real number such that $0 < \alpha \leq 1$, then $f \overset{S^\alpha}{\sim} g$ implies $f \overset{S}{\sim} g$.*

Theorem 8 *Let G be a discrete countable amenable semigroup with identity in which both right and left cancelation laws hold and suppose also that the parameters α and β are fixed real numbers such that $0 < \alpha \leq \beta \leq 1$, then $f \overset{w^\alpha}{\sim} g$ implies $f \overset{w^\beta}{\sim} g$.*

Proof. Omitted. ■

From Theorem 8 we have the following.

Corollary 9 *Let G be a discrete countable amenable semigroup with identity in which both right and left cancelation laws hold and let α be any real number such that $0 < \alpha \leq 1$, then $f \overset{w^\alpha}{\sim} g$ implies $f \overset{w}{\sim} g$.*

Theorem 10 *Let G be a discrete countable amenable semigroup with identity in which both right and left cancelation laws hold and suppose also that the parameters α and β are fixed real numbers such that $0 < \alpha \leq \beta \leq 1$, then $f \overset{S^\alpha}{\sim} g$ implies $f \overset{w^\beta}{\sim} g$.*

Proof. Omitted. ■

From Theorem 10 we have the following results.

Corollary 11 *Let G be a discrete countable amenable semigroup with identity in which both right and left cancelation laws hold and let* α *be any real number such that* $0 < \alpha \leq 1$, *then* $f \overset{S^\alpha}{\sim} g$ *implies* $f \overset{w^\alpha}{\sim} g$.

Corollary 12 *Let G be a discrete countable amenable semigroup with identity in which both right and left cancelation laws hold and let* α *be any real number such that* $0 < \alpha \leq 1$, *then* $f \overset{S^\alpha}{\sim} g$ *implies* $f \overset{w}{\sim} g$.

Corollary 13 *Let G be a discrete countable amenable semigroup with identity in which both right and left cancelation laws hold, then* $f \overset{S}{\sim} g$ *implies* $f \overset{w}{\sim} g$.

References

[1] Connor, J. S. The statistical and strong $p-$Cesàro convergence of sequences. Analysis **8(1-2)** (1988), 47–63.

[2] Altin, Y. ; Et, M. and Çolak, R. Lacunary statistical and lacunary strongly convergence of generalized difference sequences of fuzzy numbers. Comput. Math. Appl. **52(6-7)** (2006), 1011–1020.

[3] Çinar, M. ; Karakaş, M. and Et, M. On pointwise and uniform statistical convergence of order α for sequences of functions, Fixed Point Theory Appl. , **2013:33** (2013), 11 pp.

[4] Çolak, R. Statistical convergence of order α, Modern Methods in Analysis and Its Applications, New Delhi, India: Anamaya Pub, 2010: 121–129.

[5] Day, M. Amenable semigroups, Illinois J. Math. **1** (1957), 509-544.

[6] Douglass, S. A. On a concept of Summability in Amenable Semigroups, Math. Scand. **28** (1968),96-102.

[7] Douglass, S. A. Summing Sequences for Amenable Semigroups, Michigan Math. J. **20** (1973), 169-179.

[8] Et, M. Generalized Cesàro difference sequence spaces of non-absolute type involving lacunary sequences, Appl. Math. Comput. **219(17)** (2013), 9372–9376.

[9] Et, M. ; Çınar, M. and Karakaş, M. On $\lambda-$statistical convergence of order α of sequences of function, J. Inequal. Appl., **2013:204** (2013), 8 pp.

[10] Et, M. and Şengül, H. Some Cesaro-type summability spaces of order α and lacunary statistical convergence of order α, Filomat **28(8)** (2014), 1593–1602.

[11] Et, M. ; Altinok, H. and Altin, Y. On some generalized sequence spaces, Appl. Math. Comput. **154(1)** (2004), 167–173.

[12] Et, M. ; Mursaleen, M. and Işık, M. On a class of fuzzy sets defined by Orlicz functions. Filomat **27(5)** (2013), 789–796.

[13] Fast, H. Sur la convergence statistique, Colloq. Math. **2** (1951) 241-244

[14] Fridy, J. On statistical convergence, Analysis **5** (1985) 301-313.

[15] Işik, M. Strongly almost $(w, \lambda, q)-$summable sequences, Math. Slovaca **61(5)** (2011), 779–788.

[16] Küçükaslan M. and Yılmaztürk M. On deferred statistical convergence of sequences, Kyungpook Math. J. **56** (2016), 357-366.

[17] Mah, P. F. Matrix summability in amenable semigroups, Proc. Amer. Math. Soc. **36** (1972), 414-420.

[18] Mah, P. F. Summability in amenable semigroups, Trans.Amer. Math. Soc.**156** (1971), 391-403.

[19] Marouf, M. S. Asymptotic equivalence and summability, Internat. J. Math. Math. Sci. **16(4)** (1993), 755–762.

[20] Mursaleen, M. $\lambda-$ statistical convergence, Math. Slovaca, **50(1)** (2000), 111 -115.

[21] Nomika, I. Folner's conditions for amenable semigroups, Math. Scand. **15** (1964), 18-28.

[22] Nuray, F. and Rhoades, B. E. Some kinds of convergence defined by Folner sequences, Analysis (Munich) **31(4)** (2011), 381–390.

[23] Nuray, F. and Rhoades, B. E. Asymptotically and statistically equivalent functions defined on amenable semigroups, Thai J. Math. **11(2)** (2013), 303–311

[24] Nuray, F. and Rhoades, B. E. Almost statistical convergence in amenable semigroups, Math. Scand. **111(1)** (2012), 127–134.

[25] Patterson, R. F. On asymptotically statistical equivalent sequences, Demonstratio Math. **36(1)** (2003), 149–153.

[26] Šalát, T. On statistically convergent sequences of real numbers, Math. Slovaca **30** (1980), 139-150.

[27] Schoenberg, I. J. The integrability of certain functions and related summability methods, Amer. Math. Monthly **66** (1959), 361-375.

[28] Şengül, H. and Et, M. On lacunary statistical convergence of order σ, Acta Math. Sci. Ser. B Engl. Ed. **34(2)** (2014), 473–482.

[29] Steinhaus, H. Sur la convergence ordinaire et la convergence asymptotique, Colloquium Mathematicum **2** (1951), 73-74.

[30] Zygmund, A. Trigonometric Series, Cambridge University Press, Cambridge, UK, 1979.

ITM Web of Conferences **13**, 01005 (2017) DOI: 10.1051/itmconf/20171301005
CMES2017

A Note on Sturm-Liouville Problem for Difference Equations

Erdal Bas[1,*], *Ramazan* Ozarslan[2]

[1,2]Firat University, Department of Mathematics, Elazig, Turkey

Abstract. In this study, Sturm-Liouville delay difference equation is considered with initial conditions as follows

$$-\Delta^2 x(n-1)+q(n)x(n-k)=\lambda x(n),$$

$$x(1)=x(2)=1,$$

where $k\in\mathbb{Z}^+$, n is a finite integer. Sum representation is found and from here, behaviors of eigenfunctions are investigated for different k values by illustrating with graphs and tables.

1 Introduction

Difference equations is discrete analogue of differential equations. So, some special type differential equations' discrete counterparts are analyzed paralelly. Sturm-Liouville problems is one of these examples. Sturm-Liouville problem has an important role because of a lot of applications in mathematical physics. Sutrm-Liouville differential and also difference equations were studied by variable authors [1-6]. Oscillation criteria and asymptotic behaviours of delay difference equations were studied by [7-14].

While generally, Sturm-Liouville difference equations are handle with a constant potential function, we handle it as a variable in our studies [3,4]. For the first time, we consider Sturm-Liouville delay difference equation.

In this study, our aim is to investigate behaviors of eigenfunctions for Sturm-Liouville delay difference initial value problem, defined as follows

$$-\Delta^2 x(n-1)+q(n)x(n-k)=\lambda x(n), \tag{1}$$

$$x(1)=x(2)=1, \tag{2}$$

where $k\in\mathbb{Z}^+$, n is a finite integer, Δ is the forward difference operator, λ is the positive spectral parameter, $q(n)$ is a real valued potential function.

* Corresponding author: erdalmat@yahoo.com

2 Preliminaries

Definition 1. [15] Casoratian is defined by

$$w(n)=\begin{bmatrix} x_1(n) & x_2(n) & \dots & x_r(n) \\ x_1(n+1) & x_2(n+1) & \dots & x_r(n+1) \\ \vdots & \vdots & \ddots & \vdots \\ x_1(n+r-1) & x_2(n+r-1) & \dots & x_r(n+r-1) \end{bmatrix},$$

where $\sum_{i=a}^{b-1} x(i)\Delta y(i)=\left[x(i)y(i)\right]_a^b-\sum_{i=a}^{b-1}\Delta x(i)y(i+1)$. are solution functions.

Theorem 1 [17] (*Wronskian-Type Identity*) Let x and y be a solutions of (1). Then,

$$W[x,y](n)=\left[x(n)\Delta y(n-1)-y(n)\Delta x(n-1)\right]$$

is a constant.

Theorem 2 [15] **(Summation by parts)** If $a<b$, then

$$\sum_{i=a}^{b-1} x(i)\Delta y(i)=\left[x(i)y(i)\right]_a^b-\sum_{i=a}^{b-1}\Delta x(i)y(i+1).$$

Theorem 3 [15] *If* $y(n)$ is an indefinite sum of $x(n)$, then

$$\sum_{i=a}^{b-1} y(i)=x(b)-x(a).$$

3 Main Results

In this study, our aim is to investigate behaviors of eigenfunctions for Sturm-Liouville initial value delay difference problem, defined as follows

$$-\Delta^2 x(n-1)+q(n)x(n-k)=\lambda x(n), \tag{1}$$

$$x(1)=x(2)=1, \tag{2}$$

where $k\in\mathbb{Z}^+$, n is a finite integer Δ is the forward difference operator, λ is the positive spectral parameter, $q(n)$ is a real valued potential function.

Theorem 4 The representation of solution of Sturm-Liouville initial value delay difference problem (1)-(2) is found as follows;

$$x(n)=\frac{\sin 2\theta-\sin\theta\left(1+q(1)x(1-k)\right)}{\sin\theta}\cos n\theta+\frac{-\cos 2\theta+\cos\theta\left(1+q(1)x(1-k)\right)}{\sin\theta}\sin n\theta$$
$$-\frac{1}{\sin\theta}\sum_{i=1}^{n-1} q(i)x(i-k)\sin(n-i)\theta,$$

where $\lambda=2-2\cos\theta, \theta\neq k\pi, \sum_{i=a}^{b}.=0,$ if $a>b$.

Proof It is straightforward to obtain the representation of solution by variation of parameters method [15,16]. It is seen that the representation of solution holds for the problem (1)-(2) by writing into place.

3.1 Applications

Application 1 Let $q(n)=\frac{1}{n^{3/2}}, \lambda=1,$ and corresponding eigenfunctions for *k=0* and *k=1* are $x_1(n)$ and $x_2(n)$ respectively, then we can obtain datas following

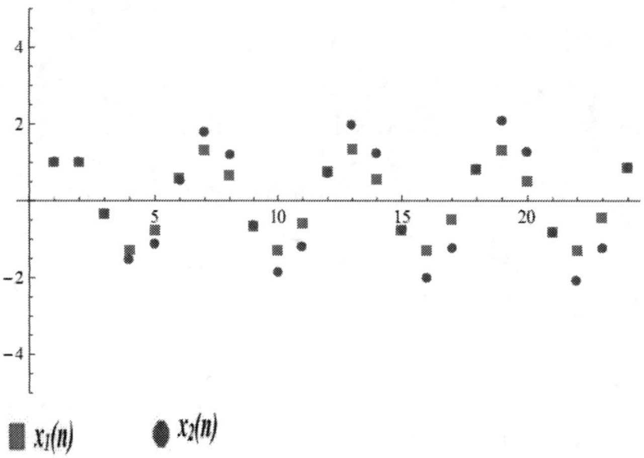

Fig. 1. Comparison of datas in Table 1

Table 1

n	$x_1(n)$	$x_2(n)$
1	1	1
2	1	1
3	−0.35	−0.35
4	−1.28	−1.54
5	−0.77	−1.14
10	−1.31	−1.88
11	−0.59	−1.22
22	1.3	−2.1
23	−0.46	−1.25
24	0.84	0.86

Application 2 Let $q(n)=\frac{1}{n^{3/2}}, \lambda=1,$ and corresponding eigenfunctions are $x_1(n)$ and $x_2(n)$ for *k=0* and *k=2* respectively, then we can obtain datas following

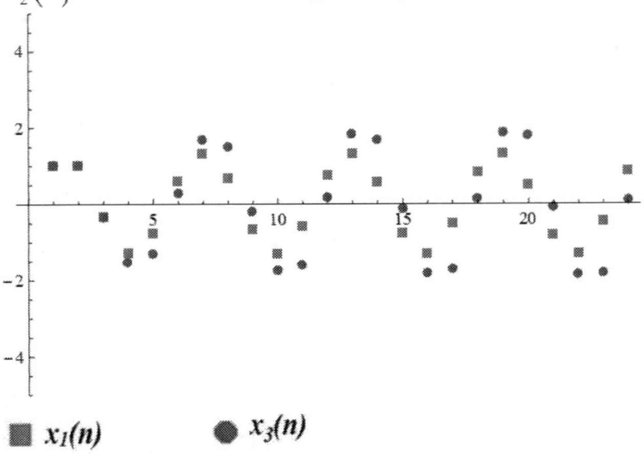

Fig. 2. Comparison of datas in Table 2

Table 2

n	$x_1(n)$	$x_3(n)$
1	1	1
2	1	1
3	−0.35	−0.35
4	−1.28	−1.54
5	−0.77	−1.31
10	−1.31	−1.75
11	−0.59	−1.6
22	1.3	−1.88
23	−0.46	−1.81
24	0.84	0.06

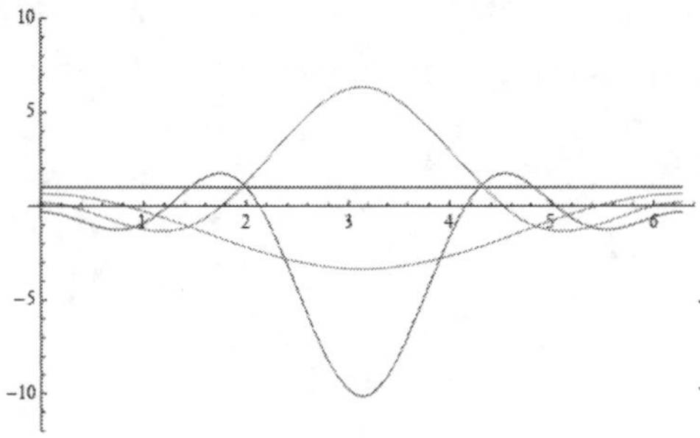

Fig. 3. $x_1(n)$, n=1,2,3,4

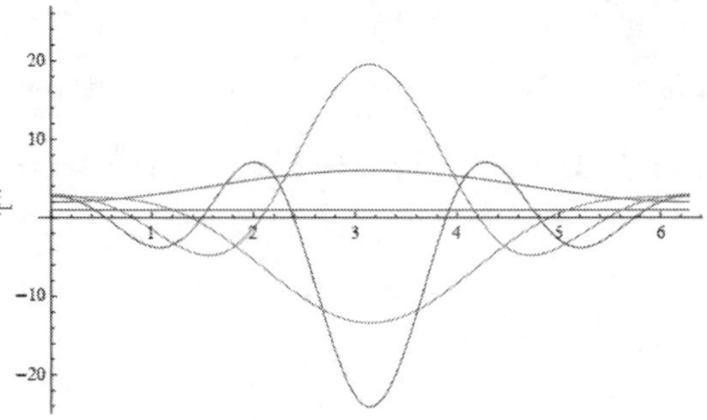

Fig.4. $x_2(n)$, n=1,2,…,5

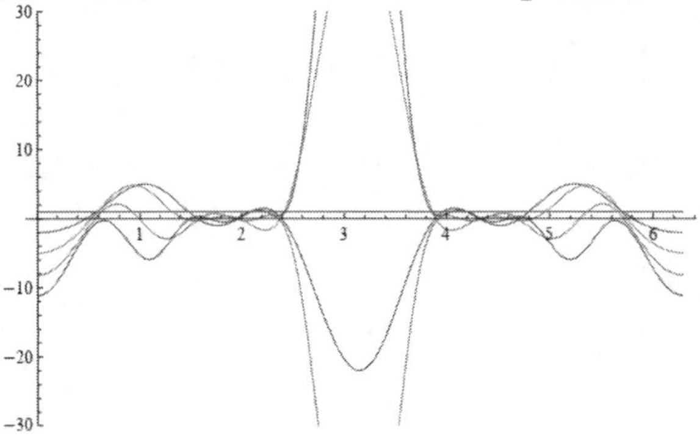

Fig. 5. $x_3(n)$, n=1,2,…,5

Conclusion

In this study, we investigated the behaviors of eigenfunctions for different k values and obtained some numerical results. We showed the results by tables and figures. Obtained results show that the eigenfunctions are shifted for different k values.

References

1. E. Bas, and E. Panakhov. "A New Approximation For Singular Inverse Sturm-Liouville Problem." Thai J. Math., **10**.3 (2012): 685-692.
2. E. Bas, and F. Metin. "Fractional singular Sturm-Liouville operator for Coulomb potential." Adv. Difference Equ. ,2013.1 (2013): 300.
3. E. Bas, and R. Ozarslan. "Sturm-Liouville Problem via Coulomb Type in Difference Equations." *Filomat* **31**.4 (2017): 989-998.
4. E. Bas, R. Ozarslan, Spectral results of Sturm-Liouville difference equation with Dirichlet boundary conditions: ICANAS 2016. AIP Publishing; 2016. p. 020065.
5. B. M. Levitan, I. S. Sargsjan, Introduction to Spectral Theory: Selfadjoint Ordinary Differential Operators, American Mathematical Society, Providence Rhode Island, p.5–8, 1975.
6. R. S. Hilscher, Spectral and oscillation theory for general second order Sturm-Liouville di_erence equations, Adv. Difference Equ. **82** (2012).

7. I. Gyori, and M. Pituk. "Asymptotic Formulae for the Solutions of a Linear Delay Difference Equation−." J. Math. Anal. Appl. **195**.2 (1995): 376-392.
8. Ch G. Philosand, I. K. Purnaras. "On linear Volterra difference equations with infinite delay.", Adv. Difference Equ. 2006.1 (2006): 1-28.
9. H. Péics, and A. Roznjik. "Asymptotic Behavior of Solutions of a Scalar Delay Difference Equations with Continuous Time." *Novi Sad J. Math* **38**.3 (2008): 47-54.
10. R. D. Driver, G. Ladas, and P. N. Vlahos. "Asymptotic behavior of a linear delay difference equation." *Proceedings of the American Mathematical Society* (1992): 105-112.
11. B. Karpuz, "Some oscillation and nonoscillation criteria for neutral delay difference equations with positive and negative coefficients." Comput. Math. Appl. **57**.4 (2009): 633-642.
12. M. F. Aktaş, A. Tiryaki, and A. Zafer. "Oscillation of third-order nonlinear delay difference equations." Turkish J. Math. **36**.3 (2012): 422-436.
13. W. Peiguang, and W. Wang. "Anti-periodic boundary value problem for first order impulsive delay difference equations.", Adv. Difference Equ. 2015.1 (2015): 93.
14. R. P. Agarwal, M. Bohner, S. R. Grace, D.O'Regan, (2005). Discrete Oscillation Theory, Hindawi Publ. *Corporation, NewYork.*
15. W.G. Kelley, A.C. Peterson, Difference Equations: An Introduction with Applications, Academic Press, San Diego (2001).
16. C.M. Bender, S.A. Orszag, Advanced Mathematical Methods for Scientists and Engineers: Asymptotic Methods and Perturbation Theory, Springer-Verlag, Newyork (1999).
17. A. Jirari, Second Order Sturm-Liouville Difference Equations and Orthogonal Polynomials. Memoirs of the American Mathematical Society, Vol. 113, Number 542, Providence Rhode Island; 1995.

ITM Web of Conferences 13, 01006 (2017) DOI: 10.1051/itmconf/20171301006
CMES2017

Trigonometric Function Solutions of Fractional Drinfeld's Sokolov -Wilson System

Zeynep Fidan Koçak[1] and *Gülnur* Yel[2,*]

[1]Department of Mathematics, University of Mugla Sıtkı Koçman, Mugla, Turkey
[2]Department of Mathematics Education,Final International University, Kyrenia, Cyprus

Abstract: In this paper, we construct exact trigonometric solutions of the space-time fractional classical Drinfeld's Sokolov-Wilson system by Modified Trial Equation Method (MTEM). These solutions may explain some physical phenomena and lead to researchers in physics and engineering.

1. Introduction

Studies in recent years on fractional analysis have been increasingly proceed. Fractional analysis describe many scientific phenomena such as physics, chemistry, biology, engineering and so on. Several efficient analytical, semi-analytical and numerical methods are applied for solving fractional equations and equation systems. Some of them are Modified Trial Equation method, Sumudu Transform method, Riccati-Bernoulli sub-ODE method, İmproved Bernoulli sub-equation method, modified Kudryashov method, Functional Variable method, Homotopy Analysis method and so on have been used to find new solutions of fractional equations [1-11]

We apply Modified Trial Equation method (MTEM) for obtain new travelling wave solution to fractional Drinfeld's Sokolov-Wilson system [1,12].

2. The Modified Trial Equation Method

In this subsection we describe structure of the MTEM [2-5].

We consider partial differential equation in two variables and a dependent variable u :

$$P(u,u_t,u_x,u_{xx},u_{xxx},...)=0, \tag{2.1}$$

and take the wave transformation,

$$u(x,t)=u(\xi),\xi=kx-ct, \tag{2.2}$$

where k and c are constants can be determined later. By substituting Eq. (2.2) into Eq. (2.1), a nonlinear ordinary differential equation (NODE) is converted as following:

$$N(U,U',U'',U''',...)=0. \tag{2.3}$$

We assume that the solution can be expressed in the form

$$U'=\frac{F(u)}{G(u)}=\frac{\sum_{i=0}^{n}a_iu^i}{\sum_{j=0}^{l}b_ju^j}=\frac{a_0+a_1u+a_2u^2+\cdots+a_nu^n}{b_0+b_1u+b_2u^2+\cdots+b_lu^l}, \tag{2.4}$$

$$U''=\frac{F(u)\left(F'(u)G(u)-F(u)G'(u)\right)}{G^3(u)}, \tag{2.5}$$

where $F(u)$ and $G(u)$ are polynomials. Substituting above relations into Eq.(2.3) yields an equation of polynomial $\Omega(u)$ of u :

$$\Omega(u)=\rho_su^s+\cdots+\rho_1u+\rho_0=0. \tag{2.6}$$

According to the balance principle, we can get a relationship between n and l. We can compute some values of n and l.

Let the coefficients of $\Omega(u)$ all be zero will yield an algebraic equations system:

$$\rho_i=0,i=0,1,2,\cdots,s. \tag{2.7}$$

By solving this system, we will thus determine the values of $a_0,a_1,\dots,a_n$ and $b_0,b_1,\dots,b_l$.

Reduce Eq.(2.4) to the elementary integral form,

$$\pm(\mu-\mu_0)=\int\frac{G(u)}{F(u)}du. \tag{2.8}$$

* Corresponding author: gulnuryel33@gmail.com

Using a complete discrimination system for polynomial of $F(u)$, we solve Eq.(2.8) with the help of Mathematica 9 and classify the exact solutions to Eq.(2.3). For better explication of results obtained in this way, we can plot two and three dimensional surfaces of the solutions obtained by using suitable parameters.

3. Application

Let consider the the space-time fractional classical Drinfeld's Sokolov-Wilson system as follows [1],

$$\frac{\partial^\alpha u}{\partial t^\alpha}+pv\left(\frac{\partial^\beta v}{\partial x^\beta}\right)=0,$$

$$\frac{\partial^\alpha v}{\partial t^\alpha}+q\left(\frac{\partial^{3\beta} v}{\partial x^{3\beta}}\right)+ru\left(\frac{\partial^\beta v}{\partial x^\beta}\right)+sv\left(\frac{\partial^\beta u}{\partial x^\beta}\right)=0, \tag{3.1}$$

where u and v are the functions of (x,t), $0<\alpha,\beta\le 1, x>0$. We apply following transformations then the Eq. (3.1) can be reduced to ordinary differential equation.

$$u(x,t)=U(\eta), v(x,t)=V(\eta),$$
$$\eta=\frac{wx^\beta}{\Gamma(1+\beta)}-\frac{\lambda t^\alpha}{\Gamma(1+\alpha)}, \tag{3.2}$$

where w and λ are arbitrary constants and $a\neq 1, b\neq 1$ are fractal constants. We get differentiate the function $U(\eta), V(\eta)$ respect to η, then Eq.(3.2) is rearranged, it yields us,

$$-\lambda bU'+pawVV'=0, \tag{3.3}$$

$$-\lambda bV'+qa^3w^3V'''+rawUV'+sawVU'=0. \tag{3.4}$$

We obtain following equalities from Eq.(3.3),

$$U'=\left(\frac{paw}{b\lambda}\right)VV', \tag{3.5}$$

$$U=\left(\frac{paw}{b\lambda}\right)V^2+c_1. \tag{3.6}$$

where c_1 is constant of integration. Embedding Eqs.(3.5,3.6) into Eq.(3.4), we have the nonlinear ordinary differential equation:

$$qa^3w^3V'''+pa^2w^2\left(\frac{r+2s}{2b\lambda}\right)V^2V'+ (rawc_1-b\lambda)V'=0. \tag{3.7}$$

Integrating Eq.(3.7) once,

$$qa^3w^3V''+pa^2w^2\left(\frac{r+2s}{6b\lambda}\right)V^3+ (rawc_1-b\lambda)V-c_2=0. \tag{3.8}$$

where c_1 is constant of integration. When we reconsider the Eq.(3.8) for homogenous balance principle between V'' and V^3, we obtain the following relationship for n and l,

$$2n-2l-1=3\Rightarrow n=l+2, \tag{3.9}$$

Case 1: For the values of $l=0$, $n=2$, we get ;

$$V'=\frac{F(v)}{G(v)}=\frac{\sum_{i=0}^{n}a_iv^i}{\sum_{j=0}^{l}b_jv^j}=\frac{a_0+a_1v+a_2v^2}{b_0}, \tag{3.10}$$

$$V''=\frac{F(v)(F'(v)\,G(v)-F(v)\,G'(v))}{G^3(v)} =\frac{(a_1+2a_2v)(a_0+a_1v+a_2v^2)}{b_0^2}, \tag{3.11}$$

where $a_2\neq 0$ and $b_0\neq 0$. Using the Wolfram Mathematica 9, the algebraic equation system is solved,

$$a_1=0, c_2=0, r=\frac{\lambda bb_0^2-2qa^3w^3a_0a_2}{awb_0^2c_1},$$
$$p=-\frac{12a^2bqw^2\lambda a_2^2c_1}{2qa^3w^3a_0a_2-b_0^2(b\lambda+2aswc_1)}, \tag{3.12}$$

we have coefficients above. By substituting these coefficients in Eq.(3.8), we obtain the solution of $v(x,t)$ and $u(x,t)$

$$v(x,t)=\frac{\sqrt{a_0}Tan\left(\frac{\sqrt{a_0a_2}}{b_0}\left(\frac{wx^\beta}{\Gamma(1+\beta)}-\frac{\lambda t^\alpha}{\Gamma(1+\alpha)}\right)\right)}{\sqrt{a_2}}, \tag{3.13}$$

$$u(x,t)=\frac{apwa_0Tan\left[\frac{\sqrt{a_0a_2}}{b_0}\left(\frac{wx^\beta}{\Gamma(1+\beta)}-\frac{\lambda t^\alpha}{\Gamma(1+\alpha)}\right)\right]^2}{2b\lambda a_2} +c_1 \tag{3.14}$$

* Corresponding author: gulnuryel33@gmail.com

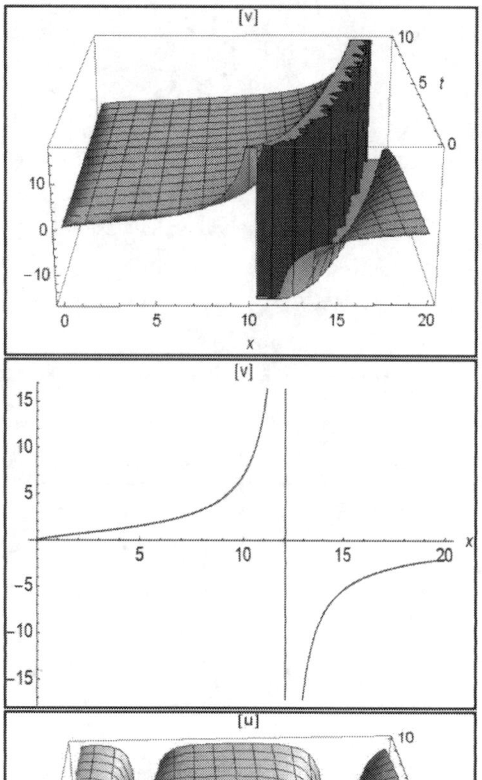

Fig. 1. The 2D and 3D surfaces for the the space-time fractional classical Drinfeld's Sokolov-Wilson system with $b_0 = 0.2, \lambda = 0.3, a = b = 0.1, \alpha = \beta = 0.5,$

$c_1 = 1, w = 0.9, s = 0.04, \quad 0 < x < 20, 0 < t < 10,$

$q = 0.04, a_0 = 0.1, a_2 = 0.9,$ and $t = 0.4$, for 2D surfaces.

4. Conclusions

In this manuscript, we have efficiently and easily practice the Modified Trial Equation Method that is give analytical solution. We can say that this method is applied to nonlinear differential equations and systems. These trigonometric function solutions have been introduced to the literature for the first time. We think that these new solutions lead the way other scientific area.

References

1. A. R. Shehata, E. M. Kamal, and H. A. Kareem, Solutions of the space-Time fractional of some nonlinear systems of partial differential equations using modified Kudryashov method, IJPAM, **101(4),** 477-487. (2015)

2. Z. F. Koçak, H. Bulut, and G. Yel, The Solution of Fractional Wave Equation by using Modified Trial Equation Method and Homotopy Analysis Method, AIP Conference Proceedings, 1637, 504. (2014)

3. H. Bulut, H. M. Baskonus, and Y. Pandir, The modified trial equation method for fractional wave equation and time fractional generalized Burgers equation, Abstract Appl. Anal., Article ID 636802, pp. 1-8. (2013)

4. H. Bulut, G. Yel, H. M. Baskonus, Novel Structure to the Coupled Nonlinear Maccari's System by Using Modified Trial Equation, Advanced Math. Models & Applications, Vol.**2**, No.1, pp.14-19. (2017)

5. G. Yel, On The Analytical and Numerical Solutions of Nonlinear Fractional Differential Equations, Ph.D Thesis, Mugla Sıtkı Koçman University, Mugla, 95 p. (2016)

6. H. Bulut, G. Yel, H. M. Baskonus, An Application of Improved Bernoulli SubEquation Function Method to The Nonlinear Time-Fractional Burgers Equation, Turk. J. Math. Comput. Sci. Vol.**5**,1–7. (2016)

7. H. Bulut, H. M. Baskonus and F.B.M. Belgacem, The Analytical Solutions of Some Fractional Ordinary Differential Equations By Sumudu Transform Method, Abstract Appl. Anal, Volume **2013**, Article ID 203875, 6 pages. (2013)

9. S. T. Demiray, H. Bulut, and F. B. M. Belgacem, Sumudu Transform Method for Analytical Solutions of Fractional Type Ordinary Differential Equations, Mathematical Problems in Engineering, Article ID 131690. (2015)

10. A. Atangana and A. Kılıcman, The use of Sumudu transform for solving certain nonlinear fractional heat-like equations, Abstract Appl. Anal., Vol. **2013**, Article ID 737481, 12 pages. (2013)

8. Z. Hammouch, T. Mekkaoui, Traveling-wave solutions of the Generalized Zakharov Equation with time-space fractional derivatives, MESA, **5(4),** 489-499. (2014)

* Corresponding author: gulnuryel33@gmail.com

11. X. F. Yang, Z. C. Deng and Y. Wei, A Riccati Bernoulli sub-ODE method for nonlinear partial differential equations and its application, Adv. Differ. Equ., **117**, 1-17. (2015)

12. V.G. Drinfel'd and V.V. Sokolov, Equations of Kortweg-de Vries type and simple lie algebras, Sov. Math. Dokl. **23**, 457-462. (1981)

* Corresponding author: gulnuryel33@gmail.com

ITM Web of Conferences **13**, 01007 (2017) DOI: 10.1051/itmconf/20171301007
CMES2017

Generalized Cesàro Summable Difference Sequence Spaces

Mikail ET[1,*]

[1] *Department of Mathematics, Fırat University, 23119 Elazığ, Turkey*

Abstract. Et (*M. Et, On some generalized Cesàro difference sequence spaces, İstanbul Üniv. Fen Fak. Mat. Derg. 55/56 (1996/97), 221–229*) introduced Cesàro difference sequence spaces $C_p(\Delta^m)\,(1 \leq p < \infty)\,, C_\infty(\Delta^m)$ and studied some topological properties. In this paper we continue to examine others relations with related the sequence spaces $C_1(\Delta^m)$ and $C_\infty(\Delta^m)$.

1 Introduction

Let w be the set of all sequences of real or complex numbers and ℓ_∞, c and c_0 be respectively the Banach spaces of bounded, convergent and null sequences $x = (x_k)$ with the usual norm $\|x\|_\infty = \sup |x_k|$, where $k \in \mathbb{N} = \{1, 2, \ldots\}$, the set of positive integers. Also by bs, cs, ℓ_1 and ℓ_p; we denote the spaces of all bounded, convergent, absolutely summable and $p-$absolutely summable series, respectively.

A sequence space X with a linear topology is called a $K-$space provided each of the maps $p_i : X \to \mathbb{C}$ defined by $p_i(x) = x_i$ is continuous for each $i \in \mathbb{N}$, where $\mathbb{C}$ denotes the complex field. A $K-$space X is called an $FK-$space provided X is a complete linear metric space. An $FK-$space whose topology is normable is called a $BK-$space. We say that an $FK-$space X has AK (or has the AK property), if (e_k) (the sequence of unit vectors) is a Schauder bases for X.

Study of difference sequence spaces is a recent development in the summability theory. The approach of using differences of terms does not help for all the problems involving sequences. However, for many sequences, the use of differences can be very profitable and, moreover, the differences themselves prove extremely valuable in many other contexts. Sometimes a situation may arise that we have a sequence at hand and we are interested in sequences formed by its successive differences and in the structure of these new sequences.

The notion of difference sequence spaces was introduced by Kızmaz [16] and the notion was generalized by Et and Çolak [11]. Later on Et and Nuray [12] generalized these sequence spaces to the following sequence spaces. Let X be any sequence space and let m be a non-negative integer. Then,

$$\Delta^m(X) = \{x = (x_k) : (\Delta^m x_k) \in X\}$$

$\Delta^0 x = (x_k)\,, \Delta^m x = \left(\Delta^{m-1} x_k - \Delta^{m-1} x_{k+1}\right)$ and so $\Delta^m x_k = \sum_{i=0}^{m} (-1)^i \binom{m}{i} x_{k+i}$. The sequence spaces $\Delta^m(X)$ are Banach spaces normed by

*e-mail: mikailet68@gmail.com

$$\|x\|_\Delta = \sum_{i=1}^{m} |x_i| + \|\Delta^m x_k\|_\infty .$$

Let X be any sequence space, if $x \in X(\Delta^m)$ then there exists one and only one $y = (y_k) \in X$ such that

$$x_k = \sum_{i=1}^{k-m} (-1)^m \binom{k-i-1}{m-1} y_i = \sum_{i=1}^{k} (-1)^m \binom{k+m-i-1}{m-1} y_{i-m},$$
$$y_{1-m} = y_{2-m} = \cdots = y_0 = 0 \qquad (1)$$

for sufficiently large k, for instance $k > 2m$. Recently, a large amount of work has been carried out by many mathematicians regarding various generalizations of difference sequence spaces. For a detailed account of difference sequence spaces one may refer to ([1-12],[14-15]).

The Cesàro sequence spaces Ces_p and Ces_∞ have been introduced by Shiue [19]. Jagers [13] has determined the Köthe duals of the sequence space Ces_p $(1 < p < \infty)$. It can be shown that the inclusion $\ell_p \subset Ces_p$ is strict for $1 < p < \infty$. Later on the Cesàro sequence spaces X_p and X_∞ of non-absolute type are defined by Ng and Lee ([17],[18]).

2 Topological Properties of $C_1(\Delta^m)$ and $C_\infty(\Delta^m)$

In this section we prove some results involving the sequence spaces $C_1(\Delta^m)$ and $C_\infty(\Delta^m)$.

Definition 1 *Let m be a non-negative integer. We define the following sequence spaces:*

$$C_1(\Delta^m) = \left\{ x = (x_k) : \lim_n \frac{1}{n} \sum_{k=1}^{n} (\Delta^m x_k - L) = 0 \right\},$$
$$C_\infty(\Delta^m) = \left\{ x = (x_k) : \sup_n \left(\frac{1}{n} \sum_{k=1}^{n} \Delta^m x_k \right) < \infty \right\}.$$

The above sequence spaces contain some unbounded sequences for $m \geq 1$, for example let $x = (k^m)$, then $x \in C_\infty(\Delta^m)$, but $x \notin \ell_\infty$.

Theorem 2 *The sequence spaces $C_1(\Delta^m)$ and $C_\infty(\Delta^m)$ are Banach spaces normed by*

$$\|x\|_\Delta = \sum_{i=1}^{m} |x_i| + \sup_n \frac{1}{n} \left| \sum_{k=1}^{n} \Delta^m x_k \right| .$$

Proof. Prof follows from Theorem 2.2 of Et and Nuray [12]. ■

Theorem 3 *$\ell_\infty(\Delta^{m-1}) \subset C_1(\Delta^m)$ and the inclusion is strict.*

Proof. Let $x \in \ell_\infty(\Delta^{m-1})$, then there exists $M > 0$ such that $\left|\Delta^{m-1}x_k\right| \le M$, for all $k \in \mathbb{N}$. On the other hand, we can write

$$\begin{aligned}\left|\frac{1}{n}\sum_{k=1}^{n}\Delta^m x_k\right| &= \frac{1}{n}\left|(\Delta^m x_1 + \Delta^m x_2 + \dots + \Delta^m x_n)\right| \\ &= \frac{1}{n}\left|\left\{\left(\Delta^{m-1}x_1 - \Delta^{m-1}x_2\right) + \left(\Delta^{m-1}x_2 - \Delta^{m-1}x_3\right) + \dots + \left(\Delta^{m-1}x_n - \Delta^{m-1}x_{n+1}\right)\right\}\right| \\ &= \frac{1}{n}\left|\left\{\Delta^{m-1}x_1 - \Delta^{m-1}x_{n+1}\right\}\right| \\ &\le \frac{1}{n}\left|\Delta^{m-1}x_1\right| + \frac{1}{n}\left|\Delta^{m-1}x_{n+1}\right| \\ &\le \frac{2M}{n} \to 0, \ (n \to \infty).\end{aligned}$$

So $x \in C_1(\Delta^m)$. For strict inclusion, observe that $(k^m) \in C_1(\Delta^m)$ but $(k^m) \notin \ell_\infty(\Delta^{m-1})$ (If $x_k = k^m$, then $\Delta^m x_k = (-1)^m m!$ and $\Delta^{m-1}x_k = (-1)^{m+1} m!\left[k + \frac{m-1}{2}\right]$ for all $k \in \mathbb{N}$). ■

Theorem 4 *The sequence spaces $C_1(\Delta^{m-1})$ and $c(\Delta^m)$ overlap without containing each other, but $c(\Delta^m) \subset C_1(\Delta^m)$ and the inclusion is strict.*

Proof. Proof is similar to that of Theorem 3. ■

The proof of the following result is easy, so we state without proof.

Theorem 5 *$C_1(\Delta^m)$ is a closed subspace of $C_\infty(\Delta^m)$.*

Theorem 6 *$C_1(\Delta^m)$ is a nowhere dense subset of $C_\infty(\Delta^m)$.*

Proof. Proof follows from the fact that $C_1(\Delta^m)$ is a proper and complete subspace of $C_\infty(\Delta^m)$. ■

Theorem 7 *$C_\infty(\Delta^m)$ is not separable, in general.*

Proof. Proof follows from Theorem 5 of Bhardwaj et al. [5]. ■

Theorem 8 *$C_\infty(\Delta^m)$ does not have Schauder basis.*

Proof. Proof follows from the fact that if a normed space has a Schauder basis, then it is separable. ■

Theorem 9 *$C_1(\Delta^m)$ is separable.*

Proof. Proof follows from Theorem 2.5 of Et and Nuray [12]. ■

Theorem 10 *$C_1(\Delta^m)$ does not have the AK property.*

Proof. Let $x = (x_k) = (k^m) = (1^m, 2^m, 3^m, 4^m, ...) \in C_1(\Delta^m)$. Consider the nth section of the sequence (x_k) as $x^{[n]} = (1^m, 2^m, 3^m, 4^m, ...n^m, 0, 0, ...)$. Then

$$\left\|x - x^{[n]}\right\|_{\Delta} = \left\|\left(0, 0, 0, 0, ...0, n^{m+1}, n^{m+2}, ...\right)\right\|_{\Delta}$$
$$= \frac{1}{n}\left[(-1)^{m+1} m!\left((n+1) + \frac{m-1}{2}\right)\right]$$

which does not tend to 0 as $n \to \infty$. ■

References

[1] B. Altay and F. Basar, On the fine spectrum of the difference operator Δ on c_0 and c, Inform. Sci. **168(1-4)** (2004), 217–224.

[2] Y. Altin, Properties of some sets of sequences defined by a modulus function, Acta Math. Sci. Ser. B Engl. Ed. **29(2)** (2009), 427–434.

[3] Y. Altin ; M. Et and R. Çolak, Lacunary statistical and lacunary strongly convergence of generalized difference sequences of fuzzy numbers, Comput, Math. Appl. **52(6-7)** (2006), 1011–1020.

[4] V. K. Bhardwaj and S. Gupta, Cesàro summable difference sequence space, J. Inequal. Appl. **2013:315** (2013), 9 pp.

[5] V. K. Bhardwaj ; S. Gupta and R. Karan, Köthe-Toeplitz duals and matrix transformations of Cesàro difference sequence spaces of second order, J. Math. Anal. **5(2)** (2014), 1–11.

[6] M. Et, On some generalized Cesàro difference sequence spaces, İstanbul Üniv. Fen Fak. Mat. Derg. **55/56** (1996/97), 221–229.

[7] M. Et ; H. Altinok and Y. Altin, On some generalized sequence spaces, Appl. Math. Comput. **154(1)** (2004), 167–173.

[8] M. Et, Generalized Cesàro difference sequence spaces of non-absolute type involving lacunary sequences, Appl. Math. Comput. **219(17)** (2013), 9372–9376.

[9] M. Et ; M. Mursaleen and M. Işık, On a class of fuzzy sets defined by Orlicz functions, Filomat **27(5)** (2013), 789–796.

[10] M. Et and M. Işık, On α−dual spaces of generalized difference sequence spaces, Appl. Math. Lett. **25(10)** (2012), 1486–1489.

[11] M. Et and R. Colak, On generalized difference sequence spaces, Soochow J. Math. **21(4)** (1995), 377-386.

[12] M. Et and F. Nuray, Δ^m−statistical convergence, Indian J. Pure Appl. Math. **32(6)** (2001), 961–969.

[13] A. A. Jagers, A note on Cesàro sequence spaces, Nieuw Arch. Wisk. **22(3)** (1974), 113–124.

[14] M. Işik, On statistical convergence of generalized difference sequences, Soochow J. Math. **30(2)** (2004), 197–205.

[15] M. Mursaleen ; R. Çolak and M. Et, Some geometric inequalities in a new Banach sequence space, J. Inequal. Appl. (2007), Art. ID 86757, 6 pp.

[16] H. Kızmaz, On certain sequence spaces, Canad. Math. Bull. **24(2)** (1981), 169-176.

[17] P. N. Ng and P. Y. Lee, Cesàro sequence spaces of non-absolute type, Comment Math. **20** (1978), 429-433.

[18] P. N. Ng and P. Y. Lee, On the associate spaces of Cesàro sequence space, Nanta Math. **9(2)** (1976), 168–170.

[19] J. S. Shiue, On the Cesàro sequence space, Tamkang J. Math. **1(1)** (1970), 19-25.

The Numerical Solution of Fractional Bratu-Type Differential Equations

Duygu Dönmez Demir[1*] and *Aylin* Zeybek[2]

*Department of Mathematics, Manisa Celal Bayar University, Manisa, Turkey

Abstract. This paper introduce the differential transform method (DTM) to solve the fractional Bratu-type differential equation modeling a combustion in numerical slab. For the definition of fractional derivative, the Caputo sense is used. The result corresponds to the exact solution when the obtained solution is constructed as power series for some values of fractional order. Finally, some examples are presented to indicate the efficiency of applied method. Comparison of the results obtained by DTM with those obtained by other methods is given.

Keywords: Fractional Bratu-type differential equation; DTM; Caputo sense fractional derivative.

1 Introduction

The mathematical model of many events in nature may be obtained by fractional order differential equations. Fractional differential equations have various applications in many fields such as physics, engineering, fluid and statistical mechanics, biosciences and electrochemistry. Since the exact solutions of fractional differential equations do not exist, many researchers focus on approximate solutions of functional equations of fractional order in particular. There are several numerical methods given for differential equations of integer order which extended to fractional order differential equations. These methods are known as differential transform method (DTM), variational iteration method (VIM), Adomian decomposition method (ADM), homotopy perturbation method (HPM) and etc. [1]. This study focuses on the solution of Bratu-type fractional differential equation [2]

$$\begin{gathered} D^{\alpha} y(x)+\lambda \exp\left[y(x)\right]=0\,,\ 1<\alpha \le 2,\ 0<x<1 \\ y(0)=b_1,\ \ y'(0)=b_2 \end{gathered} \tag{1}$$

where D^{α} represents the derivative operator of fractional order α, $y(x)$ is unknown function on the interval $[0,1]$ and b_1, b_2 and λ are the constants given. The numerical solutions of the mentioned problem have been investigated by ADM [1], HPM [3], VIM [4], RKM [2] before.

We consider DTM in this study. The DTM is an iterative method based on Taylor's series expansion [5]. This technique has been developed by Elsaid [6] as fractional differential transform method (FDTM) and by Odibat [7] et al. as generalized differential transform method (GDTM) for fractional differential equations. In this study, we used FDTM. For exponential nonlinearity, its Adomian polynomials are considered, then, the differential transform of the dependent variable components can be obtained. The numerical results for the different examples having exponential nonlinearity are compared with exact solutions and RKM solutions. It is concluded that the present method gives good results for Bratu-type fractional differential equations.

2 Basic Concepts and Definitions

In this section, we introduce necessary definitions related to fractional calculus which is needed in our study.

2.1 Definition [6].

A real function $f(x),\ x>0$ is said to be in the space $C_{\eta}, \eta \in ¡$, if there exists a real number $p>\eta$, such that $f(x)=x^{p}g(x)$, where $g(x)\in C(0,\infty)$ and it is said to be in the space C_{η}^{n} if $f^{(n)}\in C_{\eta}$, $n\in ¥$.

2.2 Definition [8].

In the fractional calculus, the Caputo's fractional derivative is the definition which the initial conditions give most appropriately to the physical conditions. The Caputo's fractional derivative is defined by

$$ {}_{a}^{C}D_{x}^{\alpha} f(x)=\frac{1}{\Gamma(n-\alpha)}\int_{a}^{x}\frac{f^{(n)}(\xi)}{(x-\xi)^{\alpha+1-n}}d\xi ;\quad n-1<\alpha<n . \tag{2}$$

The Caputo's definition has an advantage since it has the initial conditions in same form with those of the equations of integer order.

2.3 Definition [8].

For $n-1\le\alpha<n,\ n\in Z^{+}, x>a$, the definition of the fractional derivative in the Riemann-Liouville sense is

$$ {}_{a}^{RL}D_{x}^{\alpha} f(x) = \frac{1}{\Gamma(n-\alpha)} \frac{d^n}{dx^n} \int_a^x \frac{f(\xi)}{(x-\xi)^{\alpha+1-n}} d\xi \, . \tag{3} $$

2.4 Definition [6].

The fractional integral of order $\alpha \geq 0$ in the Riemann Liouville sense of a function $f(x) \in C_{\eta}, \eta \geq -1$ is defined as

$$ I_a^{\alpha} f(x) = \frac{1}{\Gamma(\alpha)} \int_a^x (x-\xi)^{\alpha-1} f(\xi) d\xi; \quad \alpha > 0, x > 0 \tag{4} $$

Also, the Caputo fractional derivative [6] can be given by

$$ {}^{C}D^{\alpha} f(x) = I^{n-\alpha} \frac{d^n}{dx^n} f(x); \; n-1 < \alpha < n $$

$$ {}^{C}D^{\alpha} f(x) = \frac{d^n}{dx^n} f(x); \; \alpha = n $$

3 Fractional Differential Transform Method

DTM is considered for the approximate solutions of many differential equations appearing in the fields such as the engineering and sciences. This method produces result without any linearization, perturbation and restriction. DTM is based on Taylor's series expansion constructed in polynomial form. This method has a solution of polynomial series obtained as iterative. Recently, the mentioned technique is effectively used for linear or nonlinear fractional differential equations. We consider analytic function

$$ y(x) = \sum_{k=0}^{\infty} Y(k)(x-x_0)^{\frac{k}{\theta}} \tag{5} $$

where θ denotes the order of fraction to be choosed and $Y(k)$ is the kth fractional differential transform of $y(x)$. Applying the initial conditions by integer-order derivatives for practical applications, then the transformation of initial conditions is obtained as

$$ Y(k) = \begin{cases} \dfrac{1}{(k/\theta)!} \left[\dfrac{\partial^{k/\theta}}{\partial x^{k/\theta}} y(x) \right]_{x=x_0} & , \dfrac{k}{\theta} \in ¢^{+} \\ 0 & , \dfrac{k}{\theta} \notin ¢^{+} \end{cases} \tag{6} $$

where $k = 0, 1, K, (\alpha\theta - 1)$ and α is order of the fractional differential equation [6]. Here, θ should be selected where $\alpha\theta$ is a positive integer. Besides, the fractional differential transform method has some fundamental properties in the following;

We assume that $y(x), z(x)$ and $r(x)$ are the functions of variable x and $Y(k), Z(k)$ and $R(k)$ are their differential transforms with order of fraction θ. Thus, the properties are given as;

- If $y(x)=z(x)\pm r(x)$, then $Y(k)=Z(k)\pm R(k)$.
- If $y(x)=c\,z(x)$, then $Y(k)=c\,Z(k)$ for c is a constant.
- If $y(x)=D_x^{\alpha}z(x)$, then $Y(k)=\dfrac{\Gamma\left(\alpha+1+\frac{k}{\theta}\right)}{\Gamma\left(1+\frac{k}{\theta}\right)}Z(k+\alpha\theta)$.

For the second term which is a nonlinear function in Eq. (1), we consider approximate by the series

$$f(y)=\sum_{n=0}^{\infty}A_n \tag{7}$$

where the Adomian polynomials A_n are calculated by

$$A_n=\frac{1}{n!}\left[\frac{d^n}{d\eta^n}\left[f\left(\sum_{i=0}^{\infty}\eta^i y_i\right)\right]\right]_{\eta=0}, \quad n=0,1,\ldots \tag{8}$$

For the differential transform of the fractional Bratu-type differential equation having the form $D^{\alpha}y=f(y)$ where $f(y)$ represents a nonlinear function, the recurrence scheme is given [6] by

$$\frac{\Gamma\left(\alpha+1+\frac{k}{\theta}\right)}{\Gamma\left(1+\frac{k}{\theta}\right)}Y(k+\alpha\theta)=\tilde{A}_k \tag{9}$$

where $\tilde{A}_k$ is obtained from the Adomian polynomials of $f(y)$.

4 Numerical Results

Example 1

We introduce the following Bratu-type differential equation of fractional order [2]

$$\begin{aligned} &D^{\alpha}y(x)-2e^{y(x)}=0,\ \ 1<\alpha\le 2,\ \ 0<x<1 \\ &y(0)=0,\ \ y'(0)=0 \end{aligned} \tag{10}$$

The exact solution is given by $y(x) = -2\ln(\cos x)$ for $\alpha = 2$. For approximate solution, we apply the fractional differential transform method. Then, the differential transform of $D^{\alpha} y(x)$ is obtained as

$$D^{\alpha} y(x) = \frac{\Gamma(\alpha(k+1)+1)}{\Gamma(1+\alpha k)} Y(k+1) \tag{11}$$

where θ equals to $1/\alpha$. Since $e^{y(x)}$ is the nonlinear term in Eq. (10), we take advantage of the Adomian polynomials. Then, the Eq. (10) become

$$\frac{\Gamma(\alpha(k+1)+1)}{\Gamma(1+\alpha k)} Y(k+1) = 2\tilde{A}_k; \quad k = 0,1,2,\ldots \tag{12}$$

where $\tilde{A}_k$ is the differential transforms of the Adomian polynomials for the exponential nonlinearity e^{y} such that

$$\begin{aligned}
&A_0 = e^{y_0} && \tilde{A}_0 = e^{Y(0)} \\
&A_1 = y_1 e^{y_0} && \tilde{A}_1 = Y(1) e^{Y(0)} \\
&A_2 = \left[y_2 + \frac{y_1^{\,2}}{2!} \right] e^{y_0} && \tilde{A}_2 = \left[Y(2) + \frac{Y(1)^2}{2!} \right] e^{Y(0)} \\
&A_3 = \left[y_3 + y_1 y_2 + \frac{y_1^{\,3}}{3!} \right] e^{y_0} && A_3 = \left[Y(3) + Y(1)Y(2) + \frac{(Y(1))^3}{3!} \right] e^{Y(0)} \\
&\vdots && \vdots
\end{aligned} \tag{13}$$

Applying the differential transform to the initial condition, $Y(0) = 0$ is obtained for $y(0) = 0$. Other differential transform components is found as:

$$Y(1) = \frac{2}{\Gamma(\alpha+1)} \tag{14}$$

$$Y(2) = \frac{\Gamma(1+\alpha)}{\Gamma(2\alpha+1)} \left[2 \frac{2}{\Gamma(\alpha+1)} \right] = \frac{4}{\Gamma(2\alpha+1)} \tag{15}$$

$$Y(3) = \frac{\Gamma(2\alpha+1)}{\Gamma(3\alpha+1)} \left[2 \left(\frac{4}{\Gamma(2\alpha+1)} + \frac{2}{\Gamma^2(\alpha+1)} \right) \right] = \frac{4\left[2\Gamma^2(\alpha+1) + \Gamma(2\alpha+1) \right]}{\Gamma(3\alpha+1)\Gamma^2(\alpha+1)} \tag{16}$$

Considering the inverse differential transform, the series solution is found as

$$y(x) = \frac{2}{\Gamma(\alpha+1)} x^{\alpha} + \frac{4}{\Gamma(2\alpha+1)} x^{2\alpha} + \frac{4\left[2\Gamma^2(\alpha+1) + \Gamma(2\alpha+1) \right]}{\Gamma(3\alpha+1)\Gamma^2(\alpha+1)} x^{3\alpha} + \ldots \tag{17}$$

The comparisons of the exact solution for $\alpha = 2$ and the approximate solutions for the various values of α are demonstrated in Table 1 and Fig. 1.

Table 1. Comparison of the numerical solutions for the some values α for Example 1 (for $x = 0.1$ and $N = n = 5$).

α	DTM	RKM
1.9	1.3814E-2	1.3082E-2
1.8	1.8983E-2	1.7221E-2
1.7	2.5993E-2	2.2504E-2
1.6	3.5472E-2	2.9221E-2
1.5	4.8260E-2	3.7921E-2

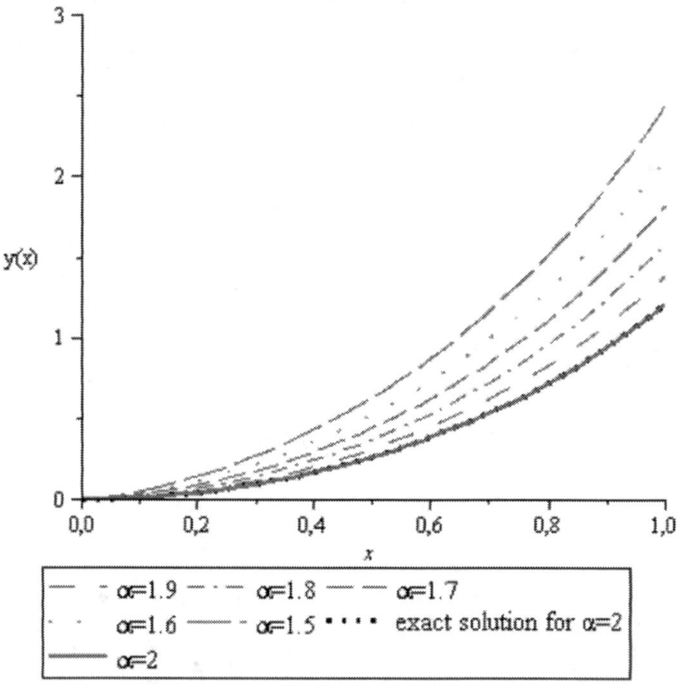

Fig. 1. Comparison the exact solution for $\alpha = 2$ with the approximate solutions for the different values α for Example 1.

Example 2

$$D^{2\alpha} y(x) - \pi^2 e^{y(x)} = 0, \ \ 0 < \alpha \le 1, \ \ 0 < x < 1$$
$$y(0) = 0, y^{\alpha}(0) = \pi \tag{18}$$

For $\alpha = 1$, the exact solution is $y(x) = -\ln(1-\sin(\pi x))$ [1]. Taking the fractional differential transform and choosing $\theta = 1/\alpha$ yields the recurrence scheme

$$\mathrm{Y}(k+1) = \frac{\Gamma(1+\alpha k)}{\Gamma(\alpha(k+1)+1)}\left[\pi^2 \tilde{A}_{k-1}\right], \quad \mathrm{k}=1,2,3,... \tag{19}$$

where $\tilde{A}_{k-1}$ is calculated from of the Adomian polynomials for the nonlinear term e^y as

$$\begin{aligned} \tilde{A}_0 &= e^{Y(0)} \\ \tilde{A}_1 &= Y(1)e^{Y(0)} \\ \tilde{A}_2 &= \left[Y(2) + \frac{Y(1)^2}{2!}\right]e^{Y(0)} \\ &\vdots \end{aligned} \tag{20}$$

Applying the differential transform to the initial condition yields $Y(0) = 0$. Considering the Maclaurin series, we can choose the differential transform of a trial function corresponding to $y_1(x)$ as

$$Y(1) = \frac{\pi}{\Gamma(\alpha+1)} \tag{21}$$

Thus, the differential transform components are obtained as

$$Y(2) = \frac{\Gamma(1+\alpha)}{\Gamma(2\alpha+1)}\pi^2 \tag{22}$$

$$Y(3) = \frac{\pi^3}{\Gamma(\alpha+1)}\frac{\Gamma(2\alpha+1)}{\Gamma(3\alpha+1)} \tag{23}$$

Applying inverse the fractional differential transformation yields the series solution as

$$y(x) = \frac{\pi}{\Gamma(\alpha+1)}x^{\alpha} + \frac{\Gamma(\alpha+1)\pi^2}{\Gamma(2\alpha+1)}x^{2\alpha} + \frac{\pi^3}{\Gamma(\alpha+1)}\frac{\Gamma(2\alpha+1)}{\Gamma(3\alpha+1)}x^{3\alpha} + \cdots \tag{24}$$

Comparing the exact solution for $\alpha = 1$ with the approximate solutions for the various values of α are shown in Fig. 2.

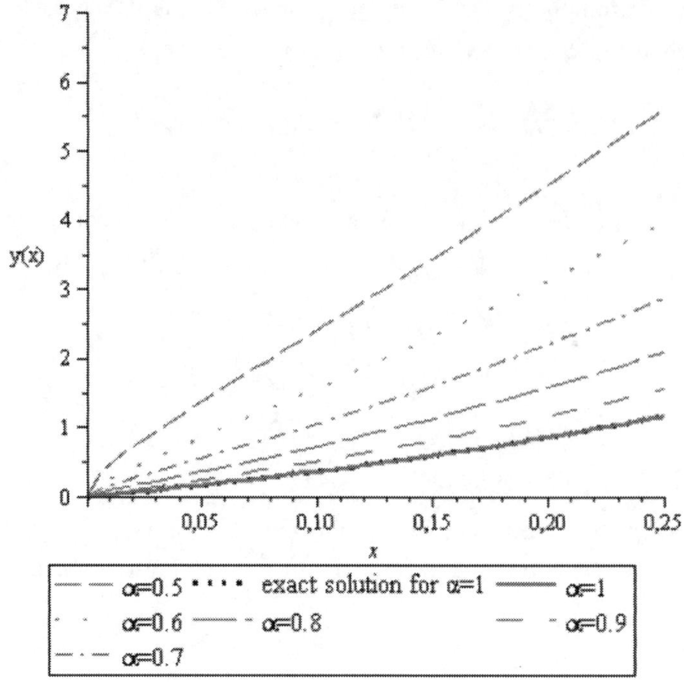

Fig. 2. Comparison the exact solution for $\alpha = 1$ with the approximate solutions for the different values α for Example 2.

Example 3

Consider the initial value problem [1]

$$D^{2\alpha} y(x) + \pi^2 e^{-y(x)} = 0, \quad 0 < \alpha \le 1, \quad 0 < x < 1$$
$$y(0)=0, \; y^{\alpha}(0)=\pi \tag{25}$$

For $\alpha = 1$, the exact solution is $y(x) = \ln(1 - \sin(\pi x))$. Taking the fractional differential transform and choosing $\theta = 1/\alpha$, then, we obtain the recurrence scheme as follows ;

$$\mathrm{Y}(k+1) = \frac{\Gamma(1+\alpha k)}{\Gamma(\alpha(k+1)+1)} \left[-\pi^2 \tilde{A}_{k-1} \right], \quad k=1,2,3,\ldots \tag{26}$$

where $\tilde{A}_{k-1}$ is calculated from of the Adomian polynomials for the nonlinear term, e^{-y}. Then, the approximate solution of Eq. (25) by DTM is

$$y(x) = \frac{\pi}{\Gamma(\alpha+1)} x^{\alpha} - \frac{\Gamma(\alpha+1)\pi^2}{\Gamma(2\alpha+1)} x^{2\alpha} + \frac{\pi^3}{\Gamma(\alpha+1)} \frac{\Gamma(2\alpha+1)}{\Gamma(3\alpha+1)} x^{3\alpha} - \cdots \tag{27}$$

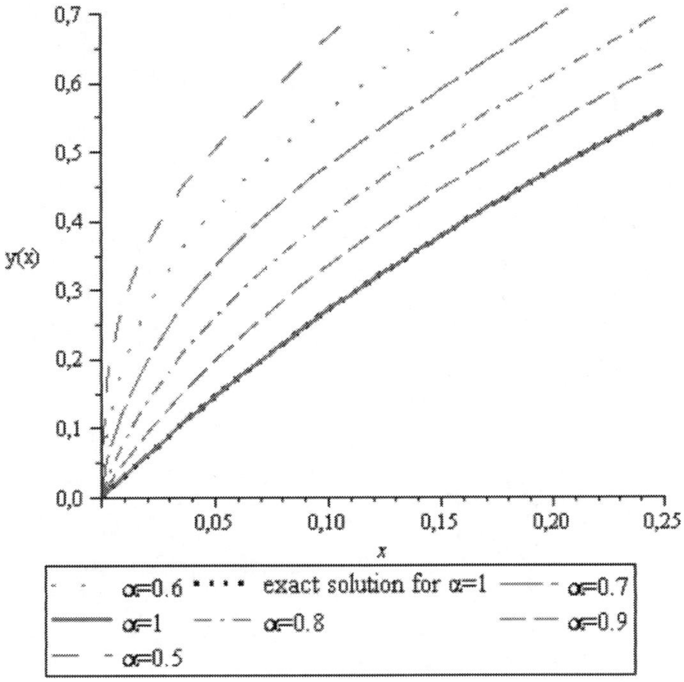

Fig. 3. Comparison the exact solution for $\alpha = 1$ with the approximate solutions for the different values α for Example 3.

Example 4

Consider the following problem [2]

$$\begin{aligned} &D^{\alpha} y(x) - e^{2y(x)} = 0, \quad 1<\alpha \le 2, \quad 0<x<1 \\ &y(0)=0, \; y'(0)=0 \end{aligned} \tag{28}$$

The exact solution is $y(x) = \ln(\sec(x))$ for $\alpha = 2$. The approximate solution is found as

$$y(x) = \frac{1}{\Gamma(\alpha+1)} x^{\alpha} + \frac{2}{\Gamma(2\alpha+1)} x^{2\alpha} + \frac{4\Gamma^2(\alpha+1) + 2\Gamma(2\alpha+1)}{\Gamma(3\alpha+1)\Gamma^2(\alpha+1)} x^{3\alpha} + \ldots \tag{29}$$

The numerical results obtained for $\alpha = 1.5(0.1)1.9$ are compared with the results obtained from the RKM in Table 2. In the Table 3, we present absolute errors between the exact and approximate solutions for $\alpha = 2$.

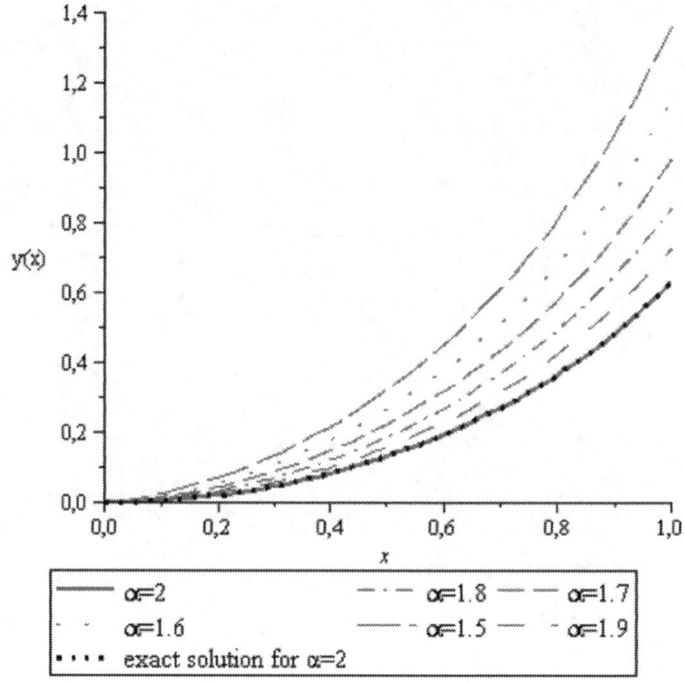

Fig. 4. Comparison the exact solution for $\alpha = 2$ with the approximate solutions for the various values α for Example 4.

Table 2. Comparison of the approximate solutions for the various values, α for Example 4 (for $x = 0.1$ and $N = n = 5$).

α	**DTM**	**RKM**
1.9	6.8982E-3	6.8894E-3
1.8	9.4725E-3	9.4537E-3
1.7	1.2956E-2	1.2917E-2
1.6	1.7652E-2	1.7571E-2
1.5	2.3957E-2	2.3791E-2

Table 3. Absolute errors for $\alpha = 2$, $N = n = 5$.

x	**Example 1**	**Example 2**
0.1	4.4369E-8	4.183E-6
0.2	2.2809E-6	6.775E-5
0.3	3.1482E-5	3.501E-4
0.4	1.7273E-4	1.139E-3
0.5	6.3707E-4	2.893E-3
0.6	1.8167E-3	6.305E-3
0.7	4.3028E-3	1.242E-2
0.8	8.7868E-3	2.286E-2

5 Conclusion

We use the differential transform method for the approximate solutions of fractional Bratu-type differential equations. For calculating the differential transform of the exponential nonlinearity, its Adomian polynomials are used in recurrence schemes. It is seen that the differential transform method is implemented simply and effectively to fractional nonlinear differential equations. It is concluded that the present method gives good results.

References

1. B. Ghazanfari, A. Sepahvandzadeh, J. Math. Computer Sci. **8,** 236-244 (2014)
2. E. Babolian, S. Javadi, E. Moradi, Math. Appl. Sci. **39,** 1548-1557 (2016)
3. B. Ghazanfari, A. Sepahvandzadeh, Journal of Mathematical Modeling, **2,** 143-155 (2015)
4. L. Jin, Int. J.Contemp. Math. Sciences, **5,** 153-158 (2010)
5. N. Bildik, A. Konuralp, F. O. Bek, S. Küçükarslan, Appl. Math. Comput. **172,** 551-567 (2006)
6. A.Elsaid, Appl. Math. Comput. **218,** 6899-6911 (2012)
7. Z. Odibat, S. Momani, V. S. Erturk, Appl. Math. Comput. **197**, 467-477 (2008)
8. I. Podlubny, *Fractional Differential Equations* (Academic Press, 1999)
9. G. C. Wu, E. W. M. Lee, Phys. Lett. A, **374,** 2506-2509 (2010)
10. H. R. Marasi, N. Sharifi, H. Piri, TWMSJ. App. Eng. Math. **5,** 124-131(2015)

DOI: 10.1051/itmconf/20171301009

Cryptanalysis of Application of Laplace Transform for Cryptography

Muharrem Tuncay GENÇOĞLU*

Vocational School of Technical Sciences, Firat University, Elazig, Turkey

Abstract. Although Laplace Transform is a good application field in the design of cryptosystems, many cryptographic algorithm proposals become unsatisfactory for secure communication. In this cryptanalysis study, one of the significant disadvantages of the proposed algorithm is performed with only statistical test of security analysis. In this study, Explaining what should be considered when performing security analysis of Laplace Transform based encryption systems and using basic mathematical rules, password has broken without knowing secret key. Under the skin; This study is a refutation for the article titled Application of Laplace Transform for Cryptography written by Hiwerakar[3].

Keywords: Laplace Transform; Cryptography; Cryptanalysis; A general attack scenario.

* Corresponding author: mtgencoglu23@gmail.com

1 Introduction

Several encryption algorithms were designed by using the dynamics which were presented by Laplace Transform system for cryptographic applications [1-5]. However, security analyses of the proposed algorithms in several designs were shown by using experimental results and statistical tests [6-10]. The resistance of algorithm to brute force attack was correlated to the number of parameters which are only used in key production. Eventually, the weakness of these Laplace Transform based cryptosystem designs is shown by using simple.

The fundamental difficulty of designing a cryptosystem is to express with a mathematical model for structures used in encryption architecture then to prove that these structures are cryptographically secure. Indeed, approaching problem with a cryptanalyst point of view while designing encryption scheme will disappear several possible problems which may exist in further stages. Same situation is valid for Laplace Transform based cryptology, as well. A Laplace Transform based text encryption algorithm was proposed in [3]. Security analyses of the proposed algorithm were done only by using statistical tests and experimental results. In this study, cryptanalysis of the proposed algorithm was performed. Firstly; a general attack scenario was given for cryptanalysis, secondly; how to obtain plaintext from cipher text was shown using this scenario without knowing key parameter. In the last section, obtained results were discussed and some general proposals were presented.

2 Description of the encryption algorithm

Fundamental of the proposed encryption algorithm depends on encryption of the letters with substitution method produced with the help of a Laplace transform. Encryption process is carried out by using of Taylor series. Since the proposed algorithm is a symmetrical encryption algorithm, in the beginning a secret key in between sender and receiver is determined. The encryption algorithm steps are as follows:

Step 1. Before starting encryption process, sender and receiver agree on a key.

Step 2. Laplace Transform which will be used in the algorithm is determined. Hyperbolic functions were used in the proposed encryption algorithm. Standard expansion of Hyperbolic functions were given in Eq. (1). Plaintext is determined by using Eq. (2).

$$\sinh rt = rt + \frac{r^3t^3}{3!} + \frac{r^5t^5}{5!} + \frac{r^7t^7}{7!} + \cdots + \frac{r^{2i+1}t^{2i+1}}{(2i+1)!} + \cdots + \cdots = \sum_{i=0}^{\infty} \frac{(rt)^{2i+1}}{(2i+1)!} \quad (1)$$

Where $r \in N$ is a constant,

$$t^2 \sinh 2t = 2t^3 + \frac{2^3t^5}{3!} + \frac{2^5t^7}{5!} + \frac{2^7t^9}{7!} + \cdots + \frac{2^{2i+1}t^{2i+3}}{(2i+1)!} + \cdots + \cdots = \sum_{i=0}^{\infty} \frac{2^{2i+1}t^{2i+3}}{(2i+1)!} \quad (2)$$

It allocated 0 to A and 1 to B then Z was 25.

Step 3. Given plaintext "FLOWER" was equivalent to

5 11 14 22 4 17.

Recognizing coefficients that

$$G_0 = 5, G_1 = 11, G_2 = 14, G_3 = 22, \qquad G_4 = 4, G_5 = 17,$$

$$G_n = 0 \ \ for \ n \geq 5.$$

Writing these numbers as a coefficients of $t^2 \sinh 2t$, and assuming $f(t) = G\,t^2 \sinh 2t$, He get

$$f(t) = t^2[G_0\,.2t + G_1 \frac{2^3t^3}{3!} + G_2 \frac{2^5t^5}{5!} + G_3 \frac{2^7t^7}{7!} + G_4 \frac{2^9t^9}{9!} + G_5 \frac{2^{11}t^{11}}{11!}]$$

$=\sum_{i=0}^{\infty} \frac{2^{2i+1}t^{2i+3}}{(2i+1)!} G_i$ (3)

$=5\frac{2t^3}{1!} + 11\frac{2^3t^5}{3!} + 14\frac{2^5t^7}{5!} + 22\frac{2^7t^9}{7!} + 4\frac{2^9t^{11}}{9!} + 17\frac{2^{11}t^{13}}{11!}$.

Step 4. Taking Laplace transform on both sides he have

$$L\{f(t)\} = L\{\,Gte^{2t}\} = 5\frac{2t^3}{1!}.\frac{3!}{s^4} + 11\frac{2^3t^5}{3!}.\frac{5!}{s^6} + 14\frac{2^5t^7}{5!}.\frac{7!}{s^8} + 22\frac{2^7t^9}{7!}.\frac{9!}{s^{10}} + 4\frac{2^9t^{11}}{9!}.\frac{11!}{s^{12}} + 17\frac{2^{11}t^{13}}{11!}.\frac{13}{s^{14}}$$

$$\frac{60}{s^4} + \frac{1760}{s^6} + \frac{18816}{s^8} + \frac{202752}{s^{10}} + \frac{225280}{s^{12}}$$

$+\frac{5431296}{s^{14}}$ (4)

Adding 5 to the resultant values

60 1760 18816 202752 225280 5431296 to mod 26,

65=1 mod 26,1765=23 mod 26,

18821=23 mod 26, 202757=9 mod 26, 225285= 21 mod 26,5431301=5 mod 26.

Step 5. Sender sends the values, 2 67 723 7798 8664 208896 as a key.

$G_0{}' = 13, \quad G_1{}' = 23, \quad G_2{}' = 23, \quad G_3{}' = 9, \quad G_4{}' = 21, \quad G_5{}' = 5, \; G_n{}' = 0 \quad (n \geq 6)$

The given plain text was converted to cipher text

13 23 23 9 21 5.

The message "FLOWER" was converted to "NXXJVF".

3 A general attack scenario for Laplace transform based encryption schemes

Below, a general attack scenario which a cryptanalyst can use while analyzing any Laplace transform based encryption schemes was briefly summarized.

Case 1. The structures used in encryption scheme must be expressed with a mathematical model. It must be investigated if the model can be expressed with simpler equations or cannot and, if exist, algebraic dependencies must be revealed.

Case 2. Encryption system must be shown to be resistant to known attacks. According to Taylor series expansion(Laplace Transform) and modular arithmetic of principle; it should be assumed that the attacker knows everything except secret key in encryption scheme and what kind of things can be obtained with specifically chosen plain text/cipher text pairs about encryption scheme should be investigated.

Case 3. Since the security of encryption algorithm is dependent on chosen key space, the specifications which can be done on key space must be investigated. Key design algorithm must be

DOI: 10.1051/itmconf/20171301009

expressed mathematically; the existence of poor keys caused by design must be investigated.

Case 4. Topological properties of Laplace Transform systems used in encryption architecture should be investigated in detail. It must not be forgotten that the required confusion and diffusion properties which cryptographic systems need to be secure are provided by Laplace Transform used in encryption scheme.

Case 5. The problems which can occur because of divide rules when Laplace Transform systems and mode are carried out on digital computers must be investigated. Although very strong structures are used, special attacks to the design must be investigated by taking into account that the tiniest opening can affect entire system.

4 Cryptanalysis

In this section, how a cryptanalysis is carried out by applying the attack scenario given in previous section onto the proposed Laplace Transform based text encryption algorithm [3] step by step was demonstrated. Encryption architecture was expressed with a simple mathematical model as shown in Eq. (5). Ih the proposed algorithm, it was stated that a relationship between numbers correspond to ciphertext and modular arithmetic exists. . It is not necessary to know the secret key since cihpher is solved according to modular arithmetic principle. The existence of dependencies in between numbers correspond to ciphertext and modular arithmetic is one of the drawbacks in algorithm.

$$L\left\{\sum_{i=0}^{\infty} \frac{2^{2i+1} t^{2i+3}}{(2i+1)!} G_i\right\} = \sum_{i=0}^{\infty} \frac{2^{2i+1}.(2i+3)!}{(2i+1)!.s^{2i+4}} G_i \tag{5}$$

Encrypted text is converted to numbers with the method used by the author;

"NXXJVF"→ 13 23 23 9 21 5.

These numbers were obtained in Eq. (5). Then,We get

$$G_0.\frac{2.3!}{1!.s^4} + G_1.\frac{2^3.5!}{3!.s^6} + G_2.\frac{2^5.7!}{5!.s^8} + G_3.\frac{2^7.9!}{7!.s^{10}} + G_4.\frac{2^9.11!}{9!.s^{12}} + G_5.\frac{2^{11}.13!}{11!.s^{14}}. \tag{6}$$

Since $G_i \leq 25$ and numbers have used equivalents in mode 26 we get

$$G_0.12 + 5 = 26.K_0 + 13 \Rightarrow G_0 = \frac{26.K_0 + 8}{12} \Rightarrow \begin{cases} G_{0,1} = \mathbf{5}\ for\ K_0 = \mathbf{2} \\ G_{0,2} = 18\ for\ K_0 = 8 \end{cases}$$

$$G_1.160 + 5 = 26.K_1 + 23 \Rightarrow G_1 = \frac{26.K_1 + 18}{160} \Rightarrow \begin{cases} G_{1,1} = \mathbf{11}\ for\ K_1 = \mathbf{67} \\ G_{1,2} = 24\ for\ K_1 = 147 \end{cases}$$

$$G_2.1344 + 5 = 26.K_2 + 23 \Rightarrow$$

DOI: 10.1051/itmconf/20171301009

$$G_2 = \frac{26.K_2 + 18}{1344} \Rightarrow \left\{ \quad K_2 = \mathbf{723}\ i\c{c}in\ G_2 = \mathbf{14} \right.$$

$$G_3.9216 + 5 = 26.K_3 + 9 \Rightarrow G_3 = \frac{26.K_3 + 4}{9216} \Rightarrow \left\{ G_3 = \mathbf{22}\ for\ K_3 = \mathbf{7798} \right.$$

$$G_4.56320 + 5 = 26.K_4 + 21 \Rightarrow G_4 = \frac{26.K_4 + 16}{56320} \Rightarrow \left\{ \begin{array}{l} G_4 = \mathbf{4}\ for\ K_4 = \mathbf{8664} \\ G_4 = 17\ for\ K_4 = 36824 \end{array} \right.$$

$$G_5.319488 + 5 = 26.K_5 + 5 \Rightarrow G_5 = \frac{26.K_5}{319488} \Rightarrow \left\{ G_5 = \mathbf{17}\ for\ K_5 = \mathbf{208896} \right.$$

5 11 14 22 4 17 → "FLOWER".

5 Conclusion

A symmetrical encryption algorithm was proposed by Hiwarekar[3]. In the proposed algorithm, by using modular arithmetic the secret key detected between sender and receiver and cipher text solved. Namely; proposed encryption algorithm without knowing the key is broken only by seeing encrypted text.Therefore, claimed by author "It is very difficult for an eyedroper to trace the key by any attack." is dasabled also the password is broken without a computer with simple divisibility and module theory.

References

1. Bodkhe D.S, Panchal S.K. (2015). Use of Sumudu Transform in Cryptography, Bulletin of the Marathwada Mathematical society, **16/2:** 1-6.
2. Hiwarekar A.P. (2012). A new method of cryptography using Laplace transform, International Journal of Mathematical Archive, **3/3 :** 1193-1197.
3. Hiwarekar A.P. (2015). Application of Laplace Transform for Cryptography, International Journal of Engineering & Science, **5/4 :** 129-135.
4. Lakshmi G.N, Kumar B.R, Sekhar A.C. (2011). A cryptographic scheme of Laplace transforms, International Journal of Mathematical Archive, **2/12:** 2515-2519.
5. Gençoğlu M.T. (2016). Use of Integral Transform in Cryptology, Science and Eng.J of Fırat Univ., **28/2:** 217-220.
6. Ge X, Liu F, Lu B, Yang C. (2010). Improvement of Rhouma's attacks on Gao algorithm, Physics Letters A, **374:** 1362-1367.
7. Sakallı M. T, Aslan B. (2014). On the algebraic construction of cryptographically good 32 × 32 binary linear transformations, Journal of Computational and Applied Mathematics, **259 :** 485–494.

Combining Cryptography with Steganography

Muharrem Tuncay GENÇOĞLU*

Vocational School of Technical Sciences, Firat University, Elazig, Turkey

Abstract. In this paper a different cryptographic method is introduced by using Power series transform, codes of ASCII and science of steganography. Here,we produce a new algorithm for cryptology,we use Expanded Laplace transformation of the exponential function for encrypting the plain text and we use codes of ASCII for support to the confidentiality of the chipertext. After, Chipertext have embedded by steganographic method in another plaintext to hide the existence of chipertext. We show corresponding inverse of Power Series transform for decryption.

Key words: Cryptology, Encryption, Decryption, Laplace Transform, ASCII, Steganography.

* Corresponding author: mtgencoglu23@gmail.com

1 Introduction

Foundation based on ancient the confidential communication,with the technological progress ıt has varied in terms of form and methods, have maintaned continuous its importance. To be very important of privacy in applications; protected information before hand of third parties were aimed to sending related destination and studies in this direction were made [2-6-7]. Network security problem has become very important in recent years. E-banking, e-commerce, e-government, e-mail, SMS services, security of ATMs, the existence of financial information has become indispensable in our lives. In these environments the processed and transferred to the protection of information or to ensure safety is of great importance. The Digital environment while providing data communication, from the sender to the recipient data unauthorized access, damage, prevent as there are many threats. These threats for the elimination of many encryption technique improved[2,6-9]. Cryptography is the all of mathematical technical studies related to information security. Cryptology is cipher science and ensures security of information.

The main goal of cryptography is to allow two people to communicate through non-secure channels. Encryption is the process of blocking information to make it unreadable without special knowledge.These operations are expressed using an algorithm. In general this is called the symmetric algorithms. For encryption and decryption must be used the same secret key in the symmetric algorithms[5]. The converse is also true.The security of this algorithms is associated with key[2]. The original information is known as plain text and cryptic text is encrypted format of this text. Encrypted text message contains all of the information in plain text message but It is not a readable format by a human or a computer without a suitable mechanism to decryption. The cipher is expressed by the parameters often called as the key by part of the external information. Encryption procedure is changed to vary of details of the algorithm operation based on the key. Without an appropriate key decryption is almost impossible. Advanced Encryption Standard (AES) method is the most used. Figure-1 also shows a symmetrical crypto system [2,7,8,9,10,11]. Steganography is come into question to hide a text as a complementary security solution at this point. Steganography as word meaning means hidden text or covered text.It is art of storing information which can not be detected the presence[10]. The objective of the Steganography is hide the presence of a message and is create a channel to the implicit [11].

In this study; steganography and cryptrography using together is intented to increase security for confidential data.Power series are used for cryptography. This process has been supported with 8 bit ASCII code and a high security application has been implemented for confidential data with steganography combined. In the second section of the study;Respectively definations and some standard results are given for the proposed method.In the third section,flow diagrams are given together with recommended method and practice. In the fourth section,the evaluation of the results from the study are situated.

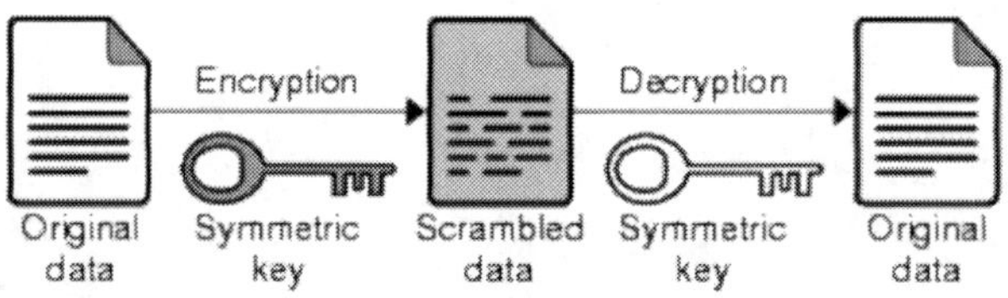

Figure 1.

2 Preliminaries

2.1 Definition

Let f(t) be defined for t>0. We say f is of exponential order if there exist numbers α, M>0 so that

$$|f(t)| \leq Me^{\alpha t} \tag{2.1}$$

If f(t) is exponential function then we have f(t)=∞ for t→∞ [1].

2.2 Definition

Let f(t) be given for t≥0 and assume the function satisfy the property of α exponential order and t,s∈R. The Laplace transform of f(t) is defined by [1]

$$F(s) = \int_0^\infty e^{-st} f(t)dt. \qquad (2.2)$$

Lets define a new transformation function by expanding the Laplace transformation using Definitions 1 and 2.

2.3 Definition

Transformation of f(t) for every t≥0 is defined as:

$$F(h) = T[f(t)] = \int_0^\infty \frac{1}{h} e^{\frac{-t}{h}} f(t)dt. \qquad (2.3)$$

(Extended Power Series Transformation) We present f(t)=T⁻¹[F(h)] to define the inverse transformation of f(t). Obtained extended power series transformation has the following standard results [3,4]

$$1. T\{t^n\} = \frac{n!}{s^{n+1}} \Rightarrow T^{-1}\left\{\frac{1}{s^{n+1}}\right\} = \frac{t^n}{n!}$$

$$2. T\{t^n e^{st}\} = \frac{n!.h^n}{(1-sh)^{n+1}} \Rightarrow \quad T^{-1}\left\{\frac{h^n}{(1-sh)^{n+1}}\right\} = \frac{t^n.e^{st}}{n!} \quad (t\geq 0) \quad (2.4)$$

2.4 Definition

In order to keep the text information in the computer memory computer system assigns a numerical value to each letter or symbol. This process depends on the encoding system. By setting the numerical value of symbols, in order to represent non-numeric or alphabetic type of information on the computer the most commonly used as the coding system is used in ASCII coding system [4].

2.5 Definition

The process of fitting a data or message into another object is called steganography. The goal is to conceal the existence of the message [6].

3 Application

By combining methods of cryptography and steganography application stages of this hybrid model that increases data security and privacy are as follows;

3.1 Encryption

Assume that we want to send the message "FIRAT". Firstly we consider extended Taylor series with e^t:

$$f(x) = f(a) + \frac{f'(a)}{1!}(x-a) + \frac{f''(a)}{2!}(x-a)^2 + \cdots + \frac{f^n(a)}{n!}(x-a)^n + \cdots$$

$$= \sum_{n=0}^{\infty} \frac{f^n(a)}{n!}(x-a)^n. \quad (3.1)$$

Then, if we expand;

$$e^t = 1 + \frac{t}{1!} + \frac{t^2}{2!} + \frac{t^3}{3!} + \cdots = \sum_{n=0}^{\infty} \frac{t^n}{n!} \tag{3.2}$$

with t^3, then we get:

$$t^3 e^t = t^3 + \frac{t^4}{1!} + \frac{t^5}{2!} + \frac{t^6}{3!} + \cdots = \sum_{n=0}^{\infty} \frac{t^{n+3}}{n!} \tag{3.3}$$

Therefore, we obtain:

$$f(t) = \sum_{n=0}^{\infty} K_n \frac{t^{n+3}}{n!}. \tag{3.4}$$

If we enumarate letters of the alphabet from scratch "FIRAT" plain text be equal 6,9,19,0,22. If we write $K_0=6, K_1=9, K_2=19, K_3=0, K_4=22$ in to (3.4), we get

$$f(t) = \sum_{n=0}^{\infty} K_n \frac{t^{n+3}}{n!}$$

$$= K_0 \frac{t^3}{0!} + K_1 \frac{t^4}{1!} + K_2 \frac{t^5}{2!} + K_3 \frac{t^6}{3!} + K_4 \frac{t^7}{4!} \tag{3.5}$$

If we apply extended power series transformation to both sides of (3.5), we get

$$T[f(t)](h) = T[\sum_{n=0}^{\infty} K_n \frac{t^{n+3}}{n!}](h)$$

$$= T[K_0 \frac{t^3}{0!} + K_1 \frac{t^4}{1!} + K_2 \frac{t^5}{2!} + K_3 \frac{t^6}{3!} + K_4 \frac{t^7}{4!}](h)$$

$$= 6.3!\, h^3 + 9.4!\, h^4 + 19.5! \frac{h^5}{2!} + 0.6! \frac{h^6}{3!} + 22.7! \frac{h^7}{4!}$$

$$\sum_{n=0}^{\infty} K_n (n+3)! \frac{h^{n+3}}{n!} = 36h^3 + 216h^4 + 1140 \frac{h^5}{2!} + 0 \frac{h^6}{3!} + 4620 \frac{h^7}{4!}. \tag{3.6}$$

The provisions of 36,216,1140,0,4620 in the modes (28) are (K_n) 8,20,20,0,0. If we write quotient in mode operation instead of these numbers, we obtain the key (K_n') 1,7,40,0,165. "FIRAT" plain text converts "HSSAA" by (3.3).

If we convert "HSSAA" encrypted text to 8-bit characters in the ASCII code we obtain 72,83,83,65,65. If these codes are written in binary system we get the keys $(1001000)_2$, $(1010001)_2$, $(1010001)_2$, $(1000001)_2$, $(1000001)_2$. ASCII 8 bit keys are in the text as follows: A provision giving the binary number system in the space between each word of the text namely if the number between two words 1 then we get $(1)_2$ and we define 2 with $(0)_2$.

Kriptoloji,_haberleşen__iki__veya_daha__fazla__ tarafın__bilgi_alışverişini__emniyetli_olarak__yapmasını__ sağlayan_ve_gizli_ya__da_özel__bilgiyi__istenmeyen_ kişilerin_anlamayacağı_hale__getirerek__korumayı__ esas__alan,__temeli_matematiksel_yöntemlere__dayalı__ uygulamaların__ve__tekniklerin__bir_bütünüdür.	Cryptology is a set of practices and techniques based on the basis of mathematical methods providing safety of the exchange of information communicating two or more the parties and bringing protection to make people to not understand unsolicited confidential or proprietary information.

Table 1. Embedded Text

Sender also send this text clearly with (1,7,40,0,165) secret key.

Hence theorem can be following

3.1.1 Theorem

The given plain text in terms of (K_n) ,under Laplace transform of $K_n \frac{t^{n+3}}{n!}(h)$, can be converted to cipher text ,

$$(K_n') = (K_n) - 28q_n \quad \text{(n=0,1,2,...)} \tag{3.7}$$

Where a key

$$q_n = \frac{K_n - K_n'}{28} \quad \text{(n=0,1,2,...)} \tag{3.8}$$

3.2 Decryption

The recipient receives a text message and by reading the spaces between words with software that will get the data buried create the necessary numerical equivalents. If these numbers are divided into 8-bits groups then ASCII provision of the data buried has been obtained. We can see the hidden data buried "HSSAA" in the Table 2.

Kriptoloji,_haberleşen__iki__veya_daha__fazla__ tarafin__bilgi_alışverişini__emniyetli_olarak__yapmasını__ sağlayan_ve_gizli_ya__da_özel__bilgiyi__istenmeyen_ kişilerin_anlamayacağı_hale__getirerek__korumayı__ esas__alan,__temeli_matematiksel_yöntemlere__dayalı__ uygulamaların__ve__tekniklerin__bir_bütünüdür.					
Text Bits	100 1000	101 0001	101 0001	100 0001	100 0001
Hidden Data	72	83	83	65	65
Encryption Message	H	S	S	A	A

Table 2. Embedded text and solution

If we write H,S,S,A,A→8,20,20,0,0 and secret key values (1,7,40,0,0,165) into

$$A_n = \frac{K_n - K_n'}{28}$$

$\mathbf{36} = 28x1 + 8$

$\mathbf{216} = 28x7 + 20$

$\mathbf{1140} = 28x40 + 20$

$\mathbf{0} = 28x0 + 0$

$\mathbf{4620} = 28x165 + 0$ are obtained.

If we apply these values 36,216,1140,0,4620 to the

$$\sum_{n=0}^{\infty} K_n (n+3)! \frac{h^{n+3}}{n!}$$

then, we get

$$\sum_{n=0}^{\infty} K_n (n+3)! \frac{h^{n+3}}{n!} = 36h^3 + 216h^4 + 1140\frac{h^5}{2!} + 0\frac{h^6}{3!} + 4620\frac{h^7}{4!}$$

$$= 6.3!\,h^3 + 9.4!\,h^4 + 19.5!\frac{h^5}{2!} + 0.6!\frac{h^6}{3!} + 22.7!\frac{h^7}{4!} \quad (3.7)$$

If we apply inverse Extended Power Series Transformation to both sides of the (3.7), then we get

$$T^{-1}\left[\sum_{n=0}^{\infty} K_n (n+3)! \frac{h^{n+3}}{n!}\right] = T^{-1}\left[6.3!\,h^3 + 9.4!\,h^4 + 19.5!\frac{h^5}{2!} + 0.6!\frac{h^6}{3!} + 22.7!\frac{h^7}{4!}\right]$$

$$\sum_{n=0}^{\infty} K_n \frac{t^{n+3}}{n!} = 6.t^3 + 9.t^4 + 19.\frac{t^5}{2!} + 0.\frac{t^6}{3!} + 22.\frac{t^7}{4!}.$$

If we convert the $K_n$coefficients we will get the first plain text 6,9,19,0,22→F,I,R,A,T.

Hence theorem can be following

3.2.1 Theorem

The given cipher text in terms of (K_n'),with a given key q_n, can be converted top lain text (K_n) under the inverse Laplace transform of

$$T^{-1}\left[\sum_{n=0}^{\infty} K_n (n+3)! \frac{h^{n+3}}{n!}\right] = \sum_{n=0}^{\infty} K_n \frac{t^{n+3}}{n!} \,,$$

Where

$K_n = 28q_n + K_n'$ (n=0,1,2,...).

Operations performed in this section is shown in Figure 2 and Figure 3.

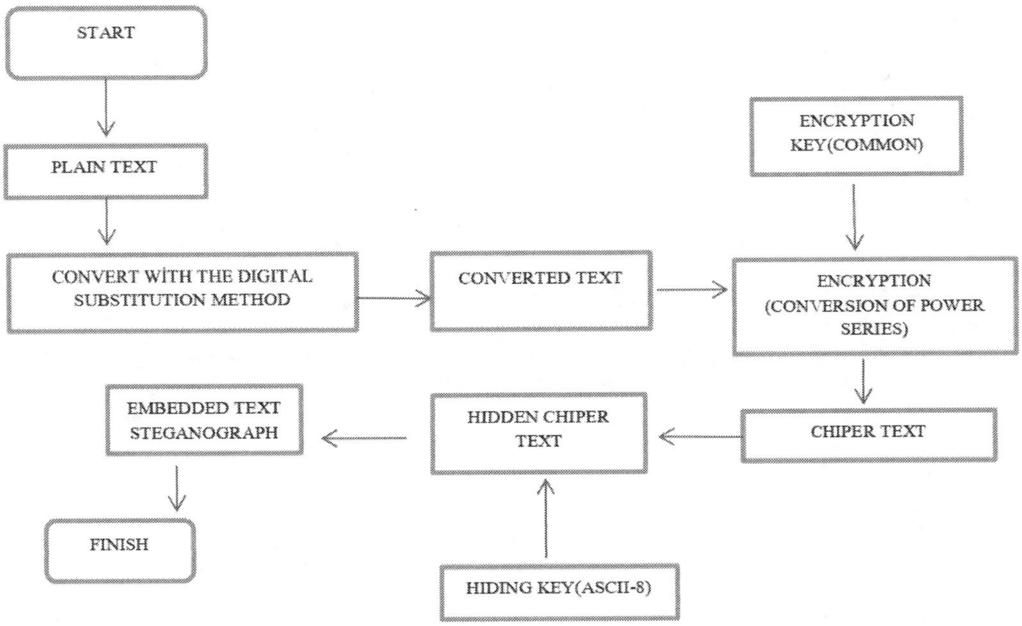

Fig. 2. Flow Diagram of Encryption System

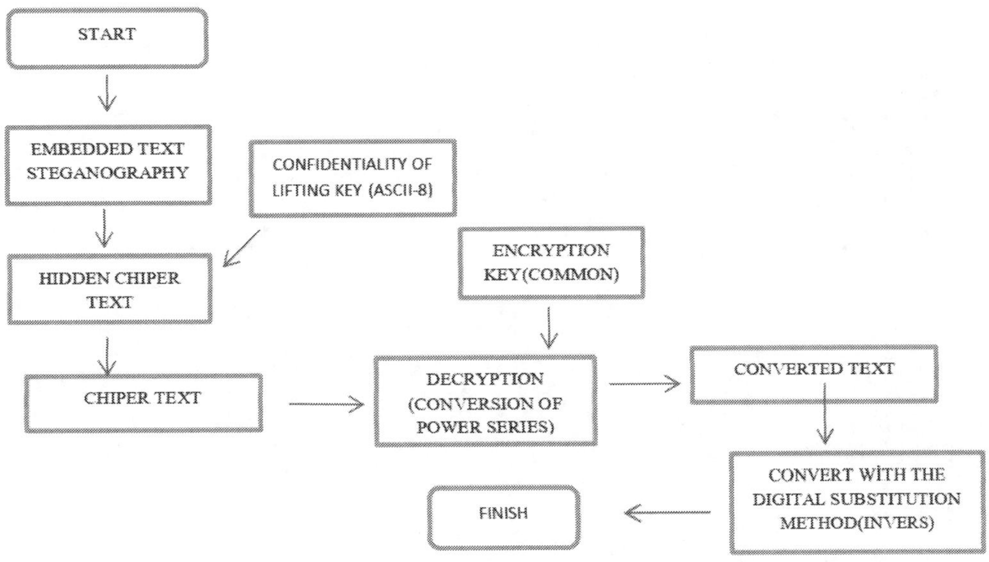

Fig. 3. Flow Diagram of Decryption System

4 Remark

If we expand the Laplace transformation with different values $t, t^2, \ldots$ we will get different encrypted texts.

5 Conclusion

By using a conversion which we called the power series conversion algorithm has been created.The keys generated using this algorithm is similar to the method known as substitution method in literatüre.But this method has been obtained by the method emerged as a result of digitization.In order to provide greater security developed a hybrid model using steganography and explained the details of it.Through this proposed hybrid model user has hidden this message with keys instead of (K_n') coefficient of q_n coefficient obtained by taking and has hidden existence of this message with ASCII code.Then,using another a password is hidden encrypted text into a text.In this way, higher safety feature is provided to data using steganography approach.

References

1. M.Aydın, G. Gökmen,B. Kuryel, G. Gündüz, *Diferansiyel Denklemler ve Uygulamaları*, Barış Yayınları, (1990)
2. Koç,Ç.K., *Cryptographic Engineering*, Springer, (2009),
3. Belgacem,F.B.M.,Karaballi,A.A.,Kalla,L.S. Analytical Investigations of the Sumudu Transform and Applications to Integral Productions Equations,Mathematical Problems in Engineering, **3**, 103-118 (2003)
4. Gençoğlu,M.T., Use of Integral Transform in Cryptology.Science and Eng. J of Fırat Univ. **28(2),** 217-220 (2016)
5. https://tr.wikipedia.org/wiki/ASCII
6. Martin,K.M., *Everyday Cryptography Fundemental Principles and Applications*,Oxford University Press. (2012)
7. Delfs,H.,Knebl,H.,**Introduction to Cryptography Principles and Applications**,Springer, (2007)
8. Yalman,Y.,Ertürk,İ. Kişisel Bilgi Güvenliğnin Sağlanmasında Steganografi Bilminin Kullanımı.*ÜNAK* (2009)
9. Paar,C.,Pelzl,J., *Understanding Cryptography* ,Springer, (2010)
10. S. Usha, G. A. Sathish Kumal, K. Boopathybagan., A. Secure Triple Level Encryption Method UsingCryptography and Steganography. *International Conference on Computer Science and NetworkTechnology, IEEE.* (2011)
11. Johnson, N.F. ve Jajodia, S.,Exploring steganography: Seeing the unseen, Computer, **31(2)**, 26-34 (1998)

*The Creativity Dimension of Instructional Materials Designed by Prospective Teachers: The Comparison across Domains

Esen Ersoy[1]*, *Emine Şendurur*[2] *and İsmail Çetin*[2]

[1]Ondokuz Mayıs University Education Faculty, Department of Mathematics Education, Samsun/Turkey
[2]Ondokuz Mayıs University Education Faculty, Department of Computer Education and Instructional Technologies, Samsun/Turkey

Abstract. This study aims at revealing the creativity dimension of the materials designed and developed by the second year students studying at the department of Computer Education and Instructional Technology. A part of the participants has completed the process by designing materials in their own field, information technologies; while some of them have done so by designing materials in the field of mathematics. The data have been retrieved from an experimental study of 13 weeks. "Teaching Material Creativity Rubric" developed by the researchers, has been used as the data collection tool. The rubric has been developed in order to evaluate the creativity dimensions of products. While developing the rubric, the creative product evaluation dimensions of [14] have been a source of inspiration. The products developed by the students have been evaluated through the retrieved data, in terms of their creativity. The rubric developed includes Originality, Practicality & Sensibility, Productivity & Flexibility, Feasibility, Inclusiveness, and Insightfulness dimensions. In this research, the data of the aforementioned dimensions and sub-dimensions have been evaluated. The results present that the creativity level of the products on teaching information technologies, which have been developed by Computer Education and Instructional Technology (CEIT) students, is high. It has been argued that the creativity of domain-specifically developed materials is higher, through literature. Keywords: Material Design in Computers, Mathematics Teaching, Originality, Applicability, Creativity, Creativity and Domainswords.

1. Introduction

Technology supported teaching has a positive impact on students' creativity [2,3,43]. Enabling students to reveal their creativity through thinking in classroom settings [37] is a part of the process of education. What is important in this process is to make use of the products in the most efficient way. Positive results could be often obtained at times when students participate in the process as designers [28].

* Corresponding Author: esene@omu.edu.tr

Although government agency, curriculum developer, teacher, pupil, and researcher have impacts on creativity [5], focusing only on the pupil's creativity might be very hard to change. In the process of education, producing as many ideas as possible and obtaining original products could bring creative products out. Thereby, creating settings, in which students can comfortably present their ideas, could contribute to increasing the number of creative products.

According to [17], for creativity in mathematical education, students' needs and learning settings should be taken into consideration. Permanent learning should be enabled through effectively using knowledge in mathematical education [10], in addition to creative products. The literature suggests that the creativity of the students, who have high performance, is also high; however, getting high grades does not always guarantee having highly creative designs [4]. Creativity is a cognitive skill owned by each and every individual. What is important is to bring that skill out. On the other hand, creativity can be improved through collaboration and technology [24]. The previously conducted studies [3,17,20,39,42] suggest that creativity can be brought out throughout the process of education and training. Technology might sometimes affect the enhancement of creativity [8,19,23,33], however, this effect can be either limited to such skills as writing [23] or not exist at all [31]. In order to achieve effectiveness, technology does not necessarily have to be directly used the process of design, i.e. it can have an indirect impact on the overall creativity. For example, the study of [21] has demonstrated that those who play more video games display a greater creativity.

People can mentally synthesize novel combinations of objects and their component parts, which often results in creative insights and discoveries [15,16]. Moreover, extensive brainstorming can also result in creativity [9]. In the study, the students have obtained new products by creating new combinations while developing their materials. In the settings, where synthesizing occurs, student produce creative products by combining their previous knowledge with new knowledge they acquire. Creative products are obtained through the production of original ideas. This research is of importance in terms of students' ability to produce many ideas through the materials they developed; to comfortably present their ideas; and to produce original materials, in other words, in terms of revealing the creativity of the students.

2. Method

The research is built upon experimental pattern. 13 week long application has been conducted with experimental group and control group students. In the experimental group, there were a mathematics expert and an IT expert; whereas in the control group, there were two IT experts. In the experimental group, materials for mathematical curriculum were developed; while in the control group, materials for IT curriculum were developed. In both groups, educational materials were designed and developed on a weekly basis. The content planned in the experimental group was prepared by the researchers respectively as following: "curriculum examination, concept map, puzzle, presentation, handbook, digital story (1 week for scenario), wiki, animation (1 week for storyboard), and worksheet".

2.1. Study Group

The research group consists of second year students studying at the department of Computer Education and Instructional Technology in the faculty of Education of a university located in the Black Sea region of Turkey. The study has been conducted with 76 students. In the beginning of the semester, students were randomly assigned to one of the groups, either the experimental (Math) group or control (IT) group.

2.2. Data Collection Tools

"Teaching Material Creativity Rubric" was developed as the data collection tool by the researchers. The rubric has been developed in order to evaluate the creativity dimensions of products. While developing the rubric, the creative product evaluation dimensions of [14] have been a source of inspiration. The rubric developed includes Originality, Practicality & Sensibility, Productivity & Flexibility, Feasibility, Inclusiveness, and Insightfulness sub-dimensions. The originality dimension has been evaluated under the categories of Well-crafted; No copy/paste; Uncommon; No imitation/replication of something and Novel. Practicality & Sensibility dimension is evaluated under the categories of Logical, Useful, Valuable, Understandable, Appropriate to integrate real settings, Sensitive to the needs of target learners/user friendly, Focus on solution, Implementation and Purpose. Productivity & Flexibility dimension is examined under the categories of Time (duration of material development), Number of ideas/Idea fluency, Amount of discussion during the process, Interdisciplinary uses of material, Materials with plan-B, such as medium-free alternative uses, /producing alternative plans and Economical. Feasibility dimension is evaluated under the categories of Applicable to real content, Workability and Updatability/Sustainability. Inclusiveness dimension is examined under the categories of Elimination of bad parts of previous materials, reaching the basic standards, including the previous good features of materials, and Thoroughness. Insightfulness dimension is evaluated under Consider design principles, Consider pedagogical issues, Consider usability, Consider budget, Consider learning outcomes and Consider learners' readiness categories. In the total creativity of the rubric and the evaluations of sub-dimensions, 1:low - 5: high coding were paid regard to. The highest score to be achieved by rubric is 130; whereas the lowest score has been calculated as 25. Total creativity score of the rubric and creativity scores of sub-dimensions have been calculated in the research. In this study, in order to understand the differences among creativity scores, each developed instructional material has been evaluated by researchers, and then the average has been taken as a creativity score.

So as to ensure the validity of the rubric, expert opinions have been taken. After receiving feedbacks from 6 experts in field of creativity and 4 experts in the field of mathematical education, necessary adjustments have been applied and the content validity has been ensured. In the phase of reliability, a pilot group has been chosen before the application and rubric has been applied to it. Moreover, "observer agreement" has been studied in order to ensure reliability. Computed inter-observer agreement is 71% for experiment group; and 73% for control group, which are at the acceptable limits [34].

2.3. Procedures

In the beginning of the semester, students were randomly assigned to one of the groups, either the experimental group or control group. Since the students worked on projects, they were assigned to groups of 4 or 5. During all laboratory sessions they all worked in groups. All students attended the same theoretical hours of the course, then they continued with 2 hours of practice, which required designing and developing an instructional material. In the first week, they were allowed to examine the curriculum: mathematical curriculum for the experimental group; Information Technologies curriculum for the control group. Throughout the semester, students created materials about the goals and objectives selected from the curriculum. In the following weeks, they created the materials, which can be relatively stated as: concept map, puzzle, presentation, handbook, digital story (1 week for scenario), wiki, animation (1 week for storyboard), and worksheet. They spent a week for scenario of the digital story and storyboarding of the animation. Although those were the basis of the digital story and animations, we rated them separately in terms of creativity, because their technical skills might have affected their final product.

We uploaded weekly lab sheet on a learning management system. Students also uploaded their products or links of products to that system. There was a time restriction for all lab sessions, which took about 2x45 minutes. The products produced by the experimental group were rated by both a Math expert and an Information Technology (IT) expert. On the other hand, two IT experts evaluated the creativity of the products of the control group.

2.4. Problem Statement

"What is the creativity dimension of the materials developed by Computer Education and Instructional Technology (CEIT) students?"

2.4.1. Sub-problems

1. Is there a significant statistical difference between the total creativity scores of the materials, which were developed by Computer Education and Instructional Technology (CEIT) students for mathematical education and information technologies education?

2. Is there a significant statistical difference among sub-dimensions of the creativity of the materials developed by Computer Education and Instructional Technology (CEIT) students for mathematical education and information technologies education

   a. Originality

   b. Practicality & Sensibility

   c. Productivity & Flexibility

   d. Feasibility

   e. Inclusiveness

   f. Insightfulness?

3. Findings Related to the First Sub Problem

In order to determine whether there is a significant statistical difference among the total creativity scores of the materials developed for mathematical education; Two-Sample Kolmogorov-Smirnov Z test has been applied. The reason why we chose this test is due to the fact that the study has a small sampling size, i.e., and this test is more powerful in samples with less than 25 observations per group [13].The null and the alternative hypothesis are as follows:

H0= There are no differences between distributions of two groups.

Ha1= There are significant differences between distribution of two groups.

Ha2= The cumulative distribution function (cdf) of the distribution of group 1 (IT material development) is larger than the cdf of the distribution of group 2 (math material development), in terms of creativity scores.

The results of the test indicate that there is a single maximum difference between two distributions, which are IT instructional material development group and math instructional material development group. In other words, two-sample Kolmogorov-Smirnov Z test followed the two groups' distributions for creativity ($D_{creativity} = 1.00$, $N = 8$ for each group, and $p < .05$). Considering critical value for equal size at .05 alpha level, we can confirm that the absolute value ($D = 1.00$) of the test statistic exceeds the critical two-tailed value ($D_{critical} = .63$), which means rejecting the equality of distributions. Moreover, the computed absolute value ($D = 1.00$) is greater than the one-tailed critical value ($D_{critical} = .50$), which means cumulative distribution function (cdf) of IT group distribution is larger than cdf of math group distribution, in terms of total creativity scores. The output of the test is summarized in Table 1.

 DOI: 10.1051/itmconf/20171301011

Table 1. Two-sample kolmogorov-smirnov Z test for total creativity scores

		Creativity
Most Extreme Differences	Absolute Positive Negative	1.00 1.00 .000
Kolmogorov-Smirnov Z Asymp. Sig. (2-tailed)		2.00 .001

3.1. Findings Related to the Second Sub Problem

The findings belonging to the second sub-problem are stated below according to the sub-dimensions of the materials developed. For each sub-dimension of creativity, the following hypothesis are tested with Two-Sample Kolmogorov-Smirnov Z test.

H0= There is no differences between distribution of two groups.

Ha1= There are significant differences between distribution of two groups.

Ha2= The cdf of the distribution of group 1 is larger than the cdf of the distribution of group 2 in terms of originality, practicality-sensibility, productivity-flexibility, feasibility, inclusiveness, and insightfulness.

a. In order to determine whether there is a significant statistical difference among the originality scores of the materials developed for mathematical education; Two-Sample Kolmogorov-Smirnov Z test has been applied. The test shows that the maximum absolute difference in originality distribution of two groups is significant ($D_{originality} = .75$, $D_{critical} = .63$, $N = 8$ for each group, and $p < .05$). Since the computed absolute value is greater than the one-tailed critical value ($D_{originality} = .75$, $D_{critical} = .50$, $N = 8$ for each group, and $p < .05$), it can be concluded that the cdf of IT group distribution is larger than cdf of math group distribution in terms of total originality scores. The output of the test is summarized in Table 2.

Table 2. Two-sample kolmogorov-smirnov Z test for originality scores

		Creativity
Most Extreme Differences	Absolute	.750
	Positive	.750
	Negative	.000
Kolmogorov-Smirnov Z		1.50
Asymp. Sig. (2-tailed)		.022

b. In order to determine whether there is a significant statistical difference among the practicality and sensibility scores of the materials developed for mathematical education; Two-Sample Kolmogorov-Smirnov Z test has been applied. The test shows that the maximum absolute difference in practicality and sensibility distribution of two groups is significant ($D_{Prac\&Sens} = .88$, $D_{critical} = .63$, $N = 8$ for each group, and $p < .05$). Since the computed absolute value is greater than the one-tailed critical value ($D_{Prac\&Sens} = .88$, $D_{critical} = .50$, $N = 8$ for each group, and $p < .05$), it can be concluded that the cdf of IT group distribution is larger than cdf of math group distribution in terms of total practicality and sensibility scores. The output of the test is summarized in Table 3.

Table 3. *Two-sample kolmogorov-smirnov z test for practicality & sensibility scores*

		Creativity
Most Extreme Differences	Absolute	.875
	Positive	.875
	Negative	.000
Kolmogorov-Smirnov Z		1.75
Asymp. Sig. (2-tailed)		.004

c. In order to determine whether there is a significant statistical difference among the productivity and flexibility scores of the materials developed for mathematical education; Two-Sample Kolmogorov-Smirnov Z test has been applied. The test shows that the maximum absolute difference in productivity and flexibility distribution of two groups is significant (DProd&Flex = .88, Dcritical = .63, N = 8 for each group, and $p < .05$). Since the computed absolute value is greater than the one-tailed critical value (DProd&Flex = .88, Dcritical = .50, N = 8 for each group, and $p < .05$), it can be concluded that the cdf of IT group distribution is larger than cdf of math group distribution in terms of total productivity and flexibility scores. The output of the test is summarized in Table 4.

Table 4. *Two-sample kolmogorov-smirnov z test for productivity & flexibility scores*

		Creativity
Most Extreme Differences	Absolute	.875
	Positive	.875
	Negative	.000
Kolmogorov-Smirnov Z		1.75
Asymp. Sig. (2-tailed)		.004

d. In order to determine whether there is a significant statistical difference among the feasibility mscores of the materials developed for mathematical education; Two-Sample Kolmogorov-Smirnov Z test has been applied. The test shows that the maximum absolute difference in feasibility distribution of two groups is significant (DFeasibility = .88, Dcritical = .63, N = 8 for each group, and $p < .05$). Since the computed absolute value is greater than the one-tailed critical value (DFeasibility = .88, Dcritical = .50, N = 8 for each group, and $p < .05$), it can be concluded that the cdf of IT group distribution is larger than cdf of math group distribution in terms of total feasibility scores. The output of the test is summarized in Table 5.

Table 5. Two-sample kolmogorov-smirnov z test for feasibility scores

		Creativity
Most Extreme Differences	Absolute	.875
	Positive	.875
	Negative	.000
Kolmogorov-Smirnov Z		1.75
Asymp. Sig. (2-tailed)		.004

e. In order to determine whether there is a significant statistical difference among the inclusiveness scores of the materials developed for mathematical education; Two-Sample Kolmogorov-Smirnov Z test has been applied. The test shows that the maximum absolute difference in inclusiveness distribution of two groups is significant (Dinclusiveness = .88, Dcritical = .63, N = 8 for each group, and p < .05). Since the computed absolute value is greater than the one-tailed critical value (Dinclusiveness = .88, Dcritical = .50, N = 8 for each group, and p < .05), it can be concluded that the cdf of IT group distribution is larger than cdf of math group distribution in terms of total inclusiveness scores. The output of the test is summarized in Table 6.

Table 6. *Two-sample kolmogorov-smirnov z test for inclusiveness scores*

		Creativity
Most Extreme Differences	Absolute	.875
	Positive	.875
	Negative	.000
Kolmogorov-Smirnov Z		1.75
Asymp. Sig. (2-tailed)		.004

f. In order to determine whether there is a significant statistical difference among the insightfulness scores of the materials developed for mathematical education; Two-Sample Kolmogorov-Smirnov Z test has been applied. The test shows that the maximum absolute difference in **insightfulness** distribution of two groups is significant ($D_{insightfulness}$ = 1.00, $D_{critical}$ = .63, N = 8 for each group, and p < .05). Since the computed absolute value is greater than the one-tailed critical value ($D_{insightfulness}$ = 1.00, $D_{critical}$ = .50, N = 8 for each group, and p < .05), it can be concluded that the cdf of IT group distribution is larger than cdf of math group distribution in terms of total **insightfulness** scores. The output of the test is summarized in Table 7.

Table 7. *Two-sample kolmogorov-smirnov z test for insightfulness scores*

		Creativity
Most Extreme Differences	Absolute	1.00
	Positive	1.00
	Negative	.000
Kolmogorov-Smirnov Z		2.00
Asymp. Sig. (2-tailed)		.001

4. Discussion and Conclusion

Technology is, in essence, a manifestation of human creativity. Thus, an important way, in which students can come to understand it, would be by engaging in acts of technological creation. Technology as a context for creativity is an important area of research [26]. Although there are many ways to evaluate creativity, in this study we followed the dimensions offered by [14]. The reason why we used those criteria is because of its focus on evaluation of product development process. Basically, there are six dimensions including originality, practicality and sensibility, productivity and flexibility, feasibility, inclusiveness, and insightfulness.

In this study, it has been aimed to reveal prospective IT teachers' creativity in their domain-specific and domain-general material designs. So as to reveal creativity, it is a necessity to determine individual and social skills; to allow for self-induced ideas; to classify similarities and differences in a versatile manner; and to allow for flexible thinking (Meissner, 2006). The results suggest that the creativity scores of the products developed by CEIT students for IT education are higher than those of the products developed for mathematical education. This might be due to the fact that CEIT students have a better command of information technologies. Even though, there are studies in the literature claiming that creativity is not domain-specific [30], there also exist studies, which propose that interdisciplinary approaches along with deep field knowledge contribute to creativity [32]. In this study, the fact that CEIT students have developed more creative materials in their own field might be because they do not find themselves adequate in the field of mathematics. It is possible to say that the aforementioned finding is contrary to those of [30] and partially supports the point of view of [32].

Originality dimension refers to uncommon products that are somehow surprising, well-crafted, novel or at least innovative. An original instructional material should not be the replication or imitation of some other material; therefore copying and pasting any content are not welcome, neither. The results of the study demonstrate that the products developed by CEIT students for IT education are original, in other words; well-designed, innovative, and creative. Although, the creativity can be enhanced through the use of computer-aided instruction in the field of mathematics [2], the involvement of computers in the design and development process does not seem to contribute to the creativity of mathematical materials. This may be result from the fact that CEIT students have examined or used many educational materials about their own field. In today's world, on-hand materials about coding, graphic design, animation etc. can be easily accessed through platforms such as YouTube and Facebook by students, who wish to develop themselves in those fields. On the other hand, the number of digital materials about mathematics at higher education level are less than that of the IT materials. For that reason, the students might have tended to develop materials different from those they had seen before, while developing cliché materials for mathematics since they were unable to make comparisons.

Practicality and sensibility dimension refers to useful or valuable products based upon a logical or understandable background. Such a product is assumed to have high quality. Offered solutions of the product should be sensitive to the needs of target learners and appropriate to integrate real settings. The results indicate that the products developed by CEIT students about IT education is more applicable and understandable when compared to the other group. This might be due to the fact that the students have a command of their own field. For example, since mathematics is a more abstract field, CEIT students' interdisciplinary approach to the subject and materialization and ability to reflect it on the material could prevent the emergence of practical and sensible products.

Productivity and flexibility dimension emphasizes on idea fluency as well as the amount of discussion. In addition to the time spent for material development, interdisciplinary uses of developed material and alternative plans are also important in this dimension. The results of the study suggest that the products developed by CEIT students about IT education is more productive when compared to those of the other group. According to the previous studies [22,29]; it is asserted that the productivity and flexibility dimensions of the materials developed will come to the light within a certain period of time. In the study conducted, a 13-week application has been used. Throughout this process, the students have worked in groups. Their idea fluency, discussion amounts, and the time they spared for preparing materials have changed in time. The difference between two groups might be due to the fact that they had to spend extra time in order to comprehend the conceptions about mathematics and to reflect them in the best possible way.

Feasibility dimension focuses on how applicable the product to real content. It is also important to create a product that is workable and sustainable. Especially in the field of information technology, developed instructional materials should be easily kept up-to-date. The results indicate that the products developed by CEIT students about IT education is more applicable when compared to the other group. The previous studies [1] support the research results of [40].

Inclusiveness dimension refers to thoroughness of the product. It should at least reach the basic standards of the previously available materials by eliminating the bad features. Thereby, it has been observed that the integration dimension of the products developed by CEIT students about IT education are more comprehensive. This, again, might result from the fact that IT materials are more easily and comfortably accessed by the students.

Insightfulness dimension refers to consideration of certain issues related to instructional material development. In our study, these consist of design principles, pedagogical issues, usability, budget, learning outcomes, and learners' readiness level. The results indicate that in the process of developing educational materials, the products developed by CEIT students about IT education are more understandable. Some of the previous studies [27] support this research.

The experimental researches conducted in the field of technological education [7,8,11,19,25,35,38]; have also demonstrated the significance of the creativity dimension. Since evaluating the creativity of students at an informational dimension can only be done through the planned course contents [11];the educational process should be designed by the experts in the field. In this study, the lesson plans prepared for IT education are also created in order to determine creativity. The plans prepared contain curriculum examination, concept map, puzzle, presentation, handbook, digital story (1 week for scenario), wiki, animation (1 week for storyboard), and worksheet. The application have been relatively applied to both groups. According to the retrieved results, the creativity, originality, practicality & sensibility, productivity & flexibility, feasibility, inclusiveness, and insightfulness dimensions of the products prepared for IT education are better than those of the other products. Although both groups were provided with equal laboratory opportunities, their creativity scores displayed differences since the fields, in which they developed materials, were also different. The students did not often spend extra time for understanding/discovering the conceptions or terms while developing materials in their own field. However, they spent extra effort for understanding the conceptions or terms and presenting them in the right manner, before stepping up to the phase of developing materials in a field, which is different from their own field and can be considered to be quite an abstract field. This might have made them feel limited or inadequate at times. The significant points that can be recommended based on the results of this study are as follows:

*the prospective teachers should be provided with adequate support in their fields throughout the process of designing and developing materials in interdisciplinary studies,

*having proper technical skills does not always guarantee for students to develop creative materials,

*for the materials to be developed in interdisciplinary studies, good examples of the field should be brought together and presented,

*important points regarding the subject should be elaborated before developing materials for abstract subjects,

The scope of the study has been determined as CEIT students; and one of the groups has been expected to develop materials in a field different from their own field. Different departments and fields could be included in the study in order to make it more comprehensive and to evaluate the results in other fields. One of the restrictions of the study has been the fact that the students have conducted a group work. More detailed observations could be conducted through creating smaller groups. Besides that, the technical skills of the students and their reflections on creativity could also be examined.

The creativity dimensions of the products developed in creative learning settings are often high [6,18,12,36,41]. These results display parallelism with the research. As [17] discusses student needs and setting are required for creativity. In the research, a learning setting and plan, which meet the needs of the students for learning, have been prepared. The lesson plans, which reveal the creativity of students, have been applied to the students for 13 weeks. It is possible to say that the study has yielded positive results in terms revealing the creativity of the students and the effectiveness of the education.

References

1. Alpert, D., Bitzer, D.L. Advances in Computer-based Education. *American Association for the Advancement of Science.* **167,** 1582-1590. (1970).
2. Aqda, M.F., Hamidi, F., Rahimi, M. The comparative effect of computer-aided instruction and traditional teaching on student's creativity in math classes. *Procedia Computer Science,* **3,** 266-270. (2011).
3. Aslan, A.E., Puccio, G.J. Developing and Testing a Turkish Version of Torrance's Test of Creative Thinking: A Study of Adults. *USA: Journal of Creative Behavior.* **40** (3), 163-178. (2006).
4. Atkinson, S. Does the need for high levels of performance curtail the development of creativity in design and technology project work? *International Journal of Design and Education.* **10,** 255-281 (2000).
5. Barlex, D. Creativity in school design and technology in England: a discussion of influences. *International Journal of Technology Design Education,* **17**, 149-162 (2007).
6. Besemer, Susan P., Donald J. Treffinger. Analysis of Creative Products: Review and Synthesis *The Journal of Creative Behavior* **15** (3), 158-178 (1981),
7. Besemer, S.P., O'Quin, K. The development, reliability and validity of the revised creative product semantic scale. *Creativity Research Journal,* **2,** 268-279 (1989).
8. Betz, J.A. Computer games: Increase learning in an interactive multidisciplinary environment. *Journal of Technology Systems*, **24** (2), 195-205 (1996).
9. Bryant, C. A 21st-century art room: the remix of creativity and technology. *Art Education,* **63** (2), 43-48 (2010).
10. English, L.D., Halford, G.S. Mathematics Education Models and Processes, Models and processes. Mahwah, NJ : Lawrence Erlbaum, (1995).
11. Eyadat, W.M., Eyadat, Y.A. Instructional technology and creativity among university students: the missing link. *World Journal on Educational Technology*, **2** (2), 87-99 (2010)
12. Feldhusen, John F., Pamela R. Clinkenbeard. Creativity instructional materials: A review of research. *The Journal of Creative Behavior* **20** (3) 153-182, (1986).
13. Field, A. *Discovering Statistics: Using SPSS* (2nd Ed.). London: Sage (2005).
14. Finke, R.A., Ward, T.B., Smith, S.M. *Creative Cognition: Theory, Research, and Applications,* Cambridge, MA: MIT Press, (1996).
15. Finke, R.A. Imagery, Creativity and Emergent Structure. Consciousness and Congnition **5**, 381-393. Article No. 0024. (1996).
16. Finke, R.A., Slayton, K. Explorations of creative visual synthesis in mental imagery. *Memory & Cognition*, **16** (3), 252-257, (1988).
17. Higginston, W. Creativity in Mathematics Education: The Role of the Teacher. *9th International Congress on Mathematical Education*, Tokyo, (2000).
18. Hill, A. *Community-based Projects in Technology Education: An Approach for Relevant Learning*. Paper presented at the International Workshop Seminar for Scholars in Technology Education, Washington, DC, (1998).
19. Howe, R. (1992). Uncovering the creative dimensions of computer-graphic design products. *Creativity Research Journal*, **5** (3), 233-243.
20. Isaksen, S.G., Puccio, G.J. Adaption-Innovation And The Torrance Tests of Creative Thinking: The Level-Style Issue Revisited. *Psychological Reports* **63**, 659-670 (1988).

21. Jackson, L.A., Witt, E.A., Games, A.I., Fitzgerald, H.E., Van Eye, A., Zhao, Y. Information technology use and creativity: findings from the children and technology Project. *Computers in Human Behavior,* **28**(2), 370-376, (2012).
22. Jang, S. Exploration of secondary students' creativity by integrating web-based technology into an innovative science curriculum. *Computers & Education*. **52,** 247-255. (2008).
23. Joram, E., Woodruff, E., Bryson, M., Lindsay, P. The effects of revising with a Word processor on writing composition. *Research in the Teaching of English,* **26**(2), 167-192 (1992).
24. Karakaya, A.F., Demirkan, H. Collaborative digital environments to enhance the creativity of designers. *Computers in Human Behavior,* **42,** 176-186 (2015).
25. Lewis, T. Creativity: A Framework for the Design/problem Solving Discourse in Technology Education. *Journal of Technology Education*, 17(1), 35-52 (2006).
26. Lewis, T. Research in technology education: Some areas of need. Journal of Technology Education, **10** (2), 41-56, (1999).
27. Lewis, T., Petrina, S., Hill, A.M. Problem Posing-Adding a Creative Increment to Technological Problem Solving. *Journal of Industrial Teacher Education.* **36** (1) (1998).
28. Liu, M., Hsiao, Y.P. Middle School Students as Multimedia Designers: A Project-Based Learning Approach. *Journal of Interactive Learning Research,* **13**(4), 311-337 (2002).
29. Lubart, T.I., Sternberg, R.J. An investment approach to creativity: theory and data. In S. M. Smith, T. B. Ward and R. A. Finke (Eds). *The creative cognition approach* (pp. 271-302). San Diego: Academic Press (1995).
30. Lucas, B. Creative teaching, teaching creativity and creative learning. *In* A. Craft, B. Jeffrey, M. Liebling (Eds.) *Creativity in Education.* London: Continuum. (2001).
31. Michael, K.Y. The Effect of a Computer Simulation Activity versus a Hands-on Activity on Product Creativity in Technology Education. Journal of Technology Education. **13** (1), 31-43. (2001).
32. Mishra, P., Henriksen, D., The Deep-Play Research Group Rethinking technology and creativity in the 21st century: on being in-disciplined. *TechTrends,* **56**(6), 18-21. (2012).
33. Neo, M., Neo, T.-K. Exploring students' creativity and design skills through multimedia project: a constructivist approach in a Malaysian classroom. *Design and Technology: An International Journal,* **18**(3), 48-59. (2013).
34. Neuendorf, K. A. The content analysis guidebook. *Thousand Oaks, California: Sage Publications.* (2002). - See more at: http://digital-activism.org/2013/05/picking-the-best-intercoder-reliability-statistic-for-your-digital-activism-content-analysis/#sthash.9Hb1j1RG.dpuf
35. Ottenbreit-Leftwich, A.T., Glazewski, K.D., Newby, T. J., Ertmer, P.A. Teacher value beliefs associated with using technology: Addressing Professional and student needs. *Computers & Education* **55,** 1321-1335, (2010).
36. Parnes, S.J. *The Creative Studies Project.* In Isaksen, S.G. (Ed.), *Frontiers of Creativity Research: Beyond the basic* (Pp. 156-188). Buffalo, NY: Bearly Limited. (1987).
37. Quitadamo, I.J. *Critical Thinking in Higher Education: The Influence of Teaching Styles and Peer Collaboration on Science and Math Learning.* Washington State University. (2002).
38. Shuell, T. J., Farber, S. L. Students' perceptions of technology use in college courses. Journal of Educational Computing Research, **24**, 21-30. (2001).
39. Sosik, J.J., Kahai, S.S., Avolio, B.J. Transformational Leadership and Dimension of Creativity: Motivating Idea Generation in Computer-Mediated Groups. *Creativity Research Journal.* **11**(2), 111-121, (1998).
40. Sun, p., Tsai, R.J., Finger, G., Chen, Y., Yeh, D. What drives o successful e-Learning? An empirical investigation of the critical factors influencing learner satisfaction. *Computers & Education.* **50**(4). 1183-1202, (2008).
41. Şendurur, E., Ersoy, E., Çetin, İ. The design and development of creative instructional materials: the role of domain familiarity for creative solutions. Int J Technol Des Educ. DOI 10.1007/s10798-016-9391-y (2016).
42. Törnkvist, S. Creativity: Can It Be Taught? The Case of Engineering Education. European Journal of Engineering Education. **23**(1), 5-12, (1998).

THE EVALUATION OF THE PROBLEM SOLVING IN MATHEMATICS COURSE ACCORDING TO STUDENT VIEWS

Esen ERSOY[1*], *Belgin Bal-Incebacak*[2]

[1]Education Faculty, Mathematics Department, 55139 Samsun, Turkey
[2]Education Faculty, Primary School Department, 55139 Samsun, Turkey

Abstract. This study was conducted to determine the problem solving skills of the third grade students studying at the department of elementary school mathematics teaching. The study was conducted in the second semester of the academic year of 2015-2016. The study group consists of 47 third year student who study at Ondokuz Mayıs University, Faculty of Education Elementary School Mathematics Teaching ad take the selective course of Problem Solving in Mathematics. Within the scope of this course, the researchers explained subjects related to problem and problem solving, problem solving skills and solved problems during the first 4 weeks of the course. For the rest of the weeks, the students were divided into groups. They have solved two non-routine problems each week for 8 weeks. The method of study is the interview method, which is one of the qualitative research methods. In light of the retrieved findings, the answers given by the students have been thematized as the stages of problem solving, understanding the problem, implementing the problem, evaluation of the problem, reasons for taking the courses, association problems, ways of finding different solutions, development of procedural skills, creating formulas, mathematical thinking, use of mathematical language, suitability of the course, views on problem solving, and the contribution of the course. Keywords: Problem, Problem solving, Mathematics, Polya

1. Introduction

The expectations from individuals change with the lifestyles, improving technology and requirements. It is thought that the individuals who will adapt to the environment lived in should be trained as individuals who know problem solving, self and environment aware, solution oriented and practical. [42] states that the way of raising an individual is through the adoption of an understanding of an educational system that solves the events, able to see the relationships between the events, and establish cause-effect relationship, gaining problem-solving and reasoning. The objective in education should be to train individuals who use old information, conceptualization, analysis and synthesis with creativity to obtain new information. [32] in the traditional mathematics teaching, teachers were teaching students by giving information in pieces, but the new understanding is to teach mathematics

* Corresponding author: esene@omu.edu.tr

by doing mathematics. [26] stated that operative knowledge is not enough in problem solving in accordance with new understanding, students knowing this are not sufficient to be good problem solvers. In order to be a problem solver, it is necessary to understand and identify the problem well.

Problem can be defined as mind confusing [15], characterized as obstacle and difficulty [23] the effort made to reach a certain objective [9, 41] and what is done in the process of overcoming the obstacle. To solve the existing problems, problem solving must be known. Problem solving was reported by [16] and [24] as a complex process containing affective and behavioral prowess as well as cognitive skills, and besides, it can be observed due to being a mental process. Problem solving can be defined for mathematics lesson as the process of perceiving real life, adapting and using mathematical skills in daily life, recognizing numbers [46] being open-minded, creative [44] classification, making comparisons, perceiving mathematical concepts [19].

When the literature is examined, the most accepted four-step process in the problem solving process is as follows. Understanding the problem, strategy selection, application of selected strategy and evaluating the solution [2,8,20,35,36,37]. It is very important that these steps are well known by teacher candidates. These stages are in fact the basic conditions for the student to become a good problem solver. It is necessary to structure the step of understanding the problem well. In the step of understanding the problem, [8] stated that writing the given and required in the problem, explaining the problem with the student's own expression, and drawing a shape suitable for the problem is necessary. The student should also know the strategy on how a problem is solved to become a good problem solver. When the literature is examined, the most accepted four-step process in the problem solving process is as follows. Creating flow chart, utilizing solutions of similar simple problems, trial and error, using and writing equations, eliminating and drawing diagrams, do what is needed, eliminating unnecessary data, work retrospectively, solving by drawing graphs selecting operation, using materials, modeling, model making, reasoning reasoning, search for pattern, relation, finding, expressing the problem in different way, simplifying the problem, simplification, solving a part of a problem, creating a systematic list, using figures and images solving by creating at able guessing, guesswork and checking, using assumptions, finding insufficient data [2]
[8,17,20,21,27,35,36] These strategies can be taught only by solving the problems according to these strategies and making practices. For this purpose, in this study, problem situations that students can solve with the strategies listed above are shown to them every week, solved by applying one by one, and solutions are presented in the process.

When the literature on cognitive field studies is examined, it is known that students who actively participate in the process better understand the process [49]. That is why students must be taught how to reach the source of information and the knowledge on how to solve them must be given. [45] stated that problem-solving behaviors should be gained to students. Most students develop new problem solving strategies by observing problem solving examples of other people [11]. Therefore, it is known that leaving the students alone with the problem situations that can be solved with different strategies will be good for the development of this skill. [31] stated that if the teacher constantly solves the problem that the students should solve, they will prevent the students from developing their problem solving behaviors. In this respect, it is important for the teacher to be a good problem solver, teach how to explain problem solving in terms of teaching this to younger age groups. Therefore, it is thought that it will be beneficial for the students to be educated in this way while being a teacher candidate during the undergraduate period. It is important to do this study in terms of seeing whether the students think this way before and after the process.

The problem of this research aims to reveal the opinions of 3rd grade students studying at Ondokuz Mayıs University Faculty of Education Primary Education Mathematics Teaching about the lesson, problem situations, problem solving and problem solving skills before and after taking the elective lesson of problem solving in Mathematics. In this way, the feelings and thoughts of students towards the lesson and the ideas against the problems will be learned. While contributing to the course to be given in the next semester, it also has a qualification as a study to be a guideline for other researchers who give this course.

2. Metot

In this section, information on the pattern, study group, process, data collection tool and analysis of study are given.

2.1. Working Design

The research method of this study is interview, one of the case study patterns in accordance with the nature of the qualitative study. [48] stated that the method used mainly is interviewing from the qualitative research methods. [18] stated that the interview was a method used to reveal the opinions, thoughts, interests, attitudes and behaviors of students about various subjects and their causes. The interview is a method of gathering information by taking the answers of the closed-ended to open-ended questions asked to related persons [29]. [38] stated that interview is more than a method where only answers to questions were found and hypotheses were tried to be evaluated. Because with the interview, he stated that a person's experiences with other people, the meanings from their experience, symbolize the feeling of curiosity. [25] stated that the view is a careful inquiry and listening process.

2.2. Study Group

The study was conducted in the second semester of the academic year of 2015-2016. The study group consists of 47 third year student who study at Ondokuz Mayıs University, Faculty of Education Elementary School Mathematics Teaching ad take the selective course of Problem Solving in Mathematics.

2.3. Process

Within the scope of this course, the researchers explained subjects related to problem and problem solving, problem solving skills and solved problems during the first 4 weeks of the course. For the rest of the weeks, the students were divided into groups. They have solved two non-routine problems each week for 8 weeks. At the end of each course the problems solved were discussed and the problem situations were dealt within the scope of Polya's problem solving stages.

2.4. Data Acquisition Tool and Analysis

The data collection tool of this work is a semi-structured interview form. The semi-structured interview form was prepared by researchers. The interview form consists of 20 questions. Two interviews were conducted before and after the implementation. The data obtained with the interview form data collection tool were analyzed by inductive analysis method. Codes were developed for content analysis and inductive analysis of probable categories and themes were made [34]. This method is also called open coding [40]. After establishing categories and themes in inductive analysis, it is appropriate to perform deductive analysis [40] in the validation phase, the last stage of the qualitative research, and in the phase of data not matching the categories.

3. Findings

In the light of the findings, the answers of the students were themed as the stages of problem solving, understanding the problem, application of the problem, evaluation of the problem, problem solving stages, reasons for taking the lessons, forming relationship difficulties, finding different solution paths, proactive inhibition, reasoning skills, formula creating, mathematical thinking, solution ways, development in procedural skills, using mathematics language, suitability of the course, thoughts on problem solving, the contribution of the lesson. The categories and codes belonging to the theme are listed below.

3.1. Problem solving stages

Table 1. Analysis of data according to the stages of problem solving

	Pre-Implementation				Post-Implementation		
Theme	Category	Code	f	%	Code	f	%
Problem solving stages	The areas in which students feel insufficient in terms of their problem soling skills	Directly solving by reading	16	34,0			
		I did not use problem solving stages	18	38,2			
		I solved them in a solution-oriented way	8	17,0			
		From my mind, through multiple-choice logic	13	27,6			
		With no rules	4	8,5			
		By roughly working on it, not in a systematic way	6	12,7			
	The areas in which students feel sufficient in terms of their problem solving skills	Logical reasoning	4	8,5	Understanding the problem	43	91,4
		Drawing figures	3	6,3	Writing down the information	43	91,4
		Creating an equation	3	6,3	Writing down what is asked	43	91,4
		Listing	1	2,1	Strategy selection	43	91,4
		Drawing a Diagram	1	2,1	Implementation of the strategy	43	91,4
		Creating a list	1	2,1	Evaluation of the solution	43	91,4
		Writing down the given information	1	2,1	I use all stages	42	89,3
		Writing down what is asked	1	2,1	I think of methods convenient for the solution when I read the question	5	10,6
					Writing down an equation/inequation	2	4,2
					Estimation control	4	8,5
					Drawing a table	2	4,2
					Drawing a Diagram	2	4,2
					Making a systematic list	1	2,1
					Logical reasoning	1	2,1
					Drawing figures	1	2,1
					I write down all methods and check them	1	2,1

The table demonstrates that there are two categories for students' views for the pre-implementation of problem solving stages and one category for the post-implementation views of students. The pre-implementation views are examined in two categories including the areas in which students feel insufficient in terms of their problem solving skills and the areas in which students feel sufficient in terms of their problem solving skills. It was observed that in the pre-implementation phase, the majority of them only read and solved the problem and did not use the problem solving stages, whereas; in the post-implementation phase, the majority of the students solved the problems by paying attention to the problem solving stages of Polya. Furthermore, it was detected that in both the pre-implementation and the post-implementation phases, they solved the problems by employing strategies such as drawing a figure, creating a diagram, and making a list.

3.2. Data analysis based on the stage of understanding the problem

Data analysis based on the stage of understanding the problem and the categories and codes belonging to the theme are listed below.

Table 2. Data analysis based on the stage of understanding the problem

	Pre-Implementation				Post-Implementation		
Theme	Category	Code	f	%	Code	f	%
Understanding the problem	Making use of the data	Considering what is asked	16	34,0	Finding what is given in the question	42	89,3
		Considering the data	11	23,4	Writing down what is asked	42	89,3
					Distinguishing between the necessary and unnecessary information	8	17,0
	Problem types / Subject knowledge	By drawing	5	10,6			
		By establishing a connection among data	4	8,5			
		Reasoning	1	2,1			
	Deficiencies in Understanding/ Errors	By repeatedly reading	15	31,9			
		By mentally making operations	14	29,7			
		I would solve them only in a solution-oriented way	7	14,8			
		I had difficulty with understanding the problem	4	8,5			

Table 2 suggests that students primarily made use of the data at the stage of comprehending the problem. This only involved looking at what is given and asked in the question at the pre-implementation phase, however, at the post-implementation phase, it also involved writing down what is given and asked in the question as well as distinguishing between the necessary and unnecessary information. In the pre-implementation phase, the students also tried to draw what is given in the question and create a link between them. The errors or insufficiencies observed in the pre-implementation phase include solving the problem by reading it over and over again, mentally making mathematical operations, trying to solve the problem by looking at the solution in a solution-oriented way, not solving the question as a result of having difficulty with understanding the problem, yet, in the post-implementation phase these errors or insufficiencies were overcome.

3.3. Problem implementation

Problem implementation and the categories and codes belonging to the theme are listed below.

Table 3. Data analysis based on the stage of problem implementation

	Pre-Implementation				Post-Implementation		
Theme	Category	Code	f	%	Code	f	%

Problem Implementation	Planning the solution	By thinking of how to solve the question through the use of the given information	19	40,4	Strategy selection according to the given information and what is asked in the question	45	95,7
		Logical reasoning	2	4,2	Solution plan that is convenient for the strategy	12	25,5
		I solved them in a solution-oriented way	13	27,6	By evaluating all 8 strategies	1	2,1
		I solved them based on certain patterns	3	6,3			
	Deficiencies in Understandin g/ Errors	I did not implement the strategy	24	51,0	Level of knowledge about the selected strategy	3	6,3
		I start making operations before understanding	9	19,1			
		Without properly reading	4	8,5			
		I solved them through multiple-choice logic	4	8,5			
		By roughly drawing	2	4,2			
		Without paying attention to Polya's stages	1	2,1			
	Understandin g the problem	By writing an equation	4	8,5	Finding a connection	1	2,1
		By drawing a diagram	3		Making a systematic list	1	2,1
		By drawing figures	2	4,2	Drawing a table	1	2,1

Looking at the Table 3, there are three categories in the problem implementation stage including planning the solution, deficiencies in understanding/errors, and understanding the problem. It was found out that they implemented the solutions to the problems by thinking of how to solve the question through the use of the given information, through logical reasoning, by solving the problem in a solution-oriented way, and solving them based on certain patterns. In the post-implementation phase, they approached the questions by selecting the suitable strategy by looking at what is given and expected in the question, developing a solution plan convenient with the strategy, and evaluating the eight different strategies they learned. In terms of the errors and deficiencies, in the pre-implementation phase, it was observed that they did not implement strategies, started operating without understanding, roughly drawing, and solved via multiple choice logic without paying attention to Polya's stages. In the post-implementation phase, they expressed that they approached the question after selecting the strategy based on how competent they are with that strategy. In the implementation phase, it was observed that they tried to write equations, to draw diagrams and figures so as to understand the problem. Similarly, they expressed that in the post-implementation phase, they implemented the solution to the problem by finding a link, making a systematic list, and drawing a table.

3.4. Evaluation of the Problem

Evaluation of the problem and the categories and codes belonging to the theme are listed below.

Table 4. Data analysis based on the stage of evaluation of the problem

	Pre-Implementation				Post-Implementation		
Theme	Category	Code	f	%	Code	f	%
Problem Evaluation	Positive	Crosschecking	27	57,4	Crosschecking	25	53,1
		Reaching retroactively	3	6,3	Evaluation of the solution	19	40,4

		Is it convenient for the solution to the problem?	1	2,1	I evaluate them by solving other problems of the same kind	15	31,9
					I can give assignments	13	27,6
					Studying retroactively	7	14,8
					Solving the problem through different strategies	6	12,7
					Verbal explanation	6	12,7
	Negative/ False	I do not evaluate	16	34,0	The easiest part is the evaluation	2	4,2
		By looking at the answer key	4	8,5	Generalization	4	8,5
		I evaluate if I am not sure of the accuracy of the solution	4	8,5			
		I crosscheck if I am not sure	2	4,2			
		I do it if I have time	1	2,1			
	Evaluating a different person	By giving classic questions	2	4,2	Written evaluation	2	4,2
		Giving a grade	2	4,2			
		By conducting an exam	1	2,1			

Examining the Table 4, it becomes apparent that there are three categories analyzed at the problem evaluation phase including positive, negative, and evaluating a different person. Primarily, it is demonstrated in the table that evaluations were conducted at the pre-implementation and post-implementation phases. Crosschecking, working retroactively, and evaluation of the solution were carried out both prior to and after the implementation. At the post-implementation phase, it was found out that other methods such as solving other problems of the same kind, giving assignments, and verbally explaining were also used. In the negative and false section, the majority of the students used expressions such as "I do not evaluate", "I evaluate by looking at the answer key", "I evaluate if I am not sure of the accuracy of the solution', "I crosscheck if I am not sure", and "I do it if I have time". In the post-implementation phase, they uttered that the evaluation part was the easiest and simplest part and they evaluated by making generalizations. The students, who perceived problem evaluation as evaluating a person, mentioned that they would use techniques such as giving classic questions, giving a grade or conducting an exam, however, they changed their perspective in the post-implementation phase and stated that they may make a written evaluation in future.

3.5. Problem solving stages

Problem solving stages and the categories and codes belonging to the theme are listed below.

Table 5. Data analysis based on the stage of problem solving stages

	Pre-Implementation				Post-Implementation		
Theme	Category	Code	f	%	Code	f	%
Problem solving stages	Right approach	Detecting the given information	15	31,9	Understanding the problem	45	95,7
		Detecting what is asked in the question	14	29,7	Strategy selection	45	95,7

		Writing an equation	6	12,7	Implementation of the strategy	45	95,7
		Creating an equation	6	12,7	Evaluation of the solution	45	95,7
		Logical reasoning	5	10,6	Polya's stages	1	2,1
		Solving through a scheme	5	10,6			
		Mentally solving	4	8,5			
		Drawing a table	4	8,5			
		Drawing figures	4	8,5			
		Drawing a Diagram	3	6,3			
		Estimation control	2	4,2			
		Making a systematic list	2	4,2			
		Finding a connection	2	4,2			
		Using variables	2	4,2			
	False approach	I did not have these stages	9	19,1			
		My approach would be solution-oriented	8	17,0			
		I implement it without selecting a strategy	7	14,8			
		I would solve it by marking x as the unknown	6	12,7			
		Using the evaluation part	4	8,5			
		Reading the question for many times	3	6,3			

Looking at the Table 5, it can be observed that there are two approaches towards problem solving including the right approach and the false approach. The right approach involves those, who solved the question by paying attention to Polya's stages. The false approach includes not using Polya's stages, solving the problem in a solution-oriented manner, solving the questions by using similar solutions without selecting a strategy, solving by marking x as the unknown in an equation, not using the evaluation stage, and wasting time on repeatedly reading the question. In fact, they have various false approaches here. In the post-implementation phase, it was observed that almost all of the student realized how to solve the questions by using right approaches.

3.6. Reasons why they take the course

Reasons why they take the course and the categories and codes belonging to the theme are listed below.

Table 6. Data analysis based on the stage of reasons why they take the course

Theme	Category	Code	f	%
Reasons Why They Take the Course	Personal Development	Solving the problems in a better way	40	85,1
		For learning the problem solving methods	11	23,4
		Because I believe problem solving will be a guide for me	7	14,8
		Learning the methods	2	4,2
		How do I solve a problem more easily	2	4,2
		Failing at solving mathematical problems	1	2,1

		For learning a subject I dislike	1	2,1
	Liking	Becauase of the teacher	5	10,6
		I thought the course would be enjoyable	3	6,3
		Because it greatly attracted my attention	1	2,1
	Professional Development	For being useful for students throughout the problem solving process	18	38,2
	Obligation	Because other elective courses are too irrelevant to me	3	6,3
		The system appointed me	1	2,1
		I had no reason to take this course, I understood its value once I started taking it	1	2,1

Taking a look at the reasons why students take the elective course, it is seen that they generally select it because it is related to personal development. Moreover, it was detected that some of the reasons were due to liking, professional development, and obligation. In the personal growth section, the reasons why students take this course can be listed as solving the problems in a better way, learning the problem solving methods, believing problem solving will be a guide for them, viewing themselves as unsuccessful at this course, and learning this course which is regarded as an disliked subject. Moreover, some other reasons were believing the course would be enjoyable, liking the teacher, and considering the course as an attention-grabbing one. Also, the course's being the best alternative available, and the fact that the system appointed this course to them, and taking this course for no reason are some of the other reasons why students took this course.

3.7. Finding a different solution

Finding a different solution and the categories and codes belonging to the theme are listed below.

Table 7. Data Analysis Based on the Stage of Finding a different solution

Theme	Category	Code	f	%
Finding a different solution	Found		17	36,1
		I usually found a different solution	9	19,1
		I find different solutions by using reasoning skills	2	4,2
		It depends on the question	2	4,2
		I find different solutions by drawing tables	1	2,1
		I find different solutions by creating a list	1	2,1
	Requires time	It usually is the same	17	36,1
		I make trials to find different solutions	7	14,8
		I cannot find different solutions yet	2	4,2
		I need to practice	1	2,1

Examining the data from the perspective of finding different solutions, it is seen that there are students who found it and who required time. The expression of the majority of them stated that they sometimes found a different solution, again an equal number of students mentioned that they found the same solution with the rest of their classmates. A small number of students never found a different solution, however, one of them uttered that he/she would require practice and may find a different solution afterwards.

3.8. Data analysis based on the level of using reasoning skills

Data analysis based on the level of using reasoning skills and the categories and codes belonging to the theme are listed below.

Table 8. Data analysis based on the stage of data analysis based on the level of using reasoning skills

Theme	Category	Code	f	%
Reasoning skills	Positive	I do not experience difficulty	27	57,4
		I can use reasoning skills	13	27,6
		Reasoning is my job	3	6,3
		I am a "thinking" person	3	6,3
		I like handling problems	1	2,1
	Negative	Sometimes	7	14,8
		I have concerns about expressing these stages	1	2,1
		I use them in accordance with multiple-choice logic	1	2,1

Evaluating themselves in terms of their reasoning skills, the majority of the students stated that they did not experience difficulty with reasoning and they used their reasoning skills. Additionally, some of the students mentioned that they had concerns with expressing the reasoning stages and used them in accordance with multiple-choice logic without thinking about it.

3.9. Data analysis based on the level of using formula developing skills

Data Analysis Based on the Level of Using Formula Developing Skills and the categories and codes belonging to the theme are listed below.

Table 9. Data analysis based on the data analysis based on the level of using formula developing skills

Theme	Category	Code	f	%
Developing a formula	Positive	I develop them by establishing a connection	31	65,9
		I develop formulas	13	27,6
		I figure out the general formula through the given information	8	17,0
		I create formulas by looking at basic sample problems	3	6,3
		I develop formulas by generalizing	3	6,3
		I develop formulas by making a systematic list	2	4,2
		I can develop the formulas I had forgotten	1	2.1
		I can formulize it if there is a convenient pattern	1	2.1
	Negative	I use what is available	3	6,3
		I cannot develop formulas	1	2.1

Analyzing the data in terms of the level of using formula developing skills, the majority of the students are able to develop formula by establishing a connection, using the given information or generalizing and making a systematic list. They stated that if there is a pattern they could develop a formula through the solution of a simple problem. 4 students uttered negative views on the situation. These students said that they used the existing methods and did not develop new formulas.

3.10. Data analysis based on the level of using mathematical thinking skills

Data analysis based on the level of using mathematical thinking Skills and the categories and codes belonging to the theme are listed below.

Table 10. Data analysis based on the level of using mathematical thinking skills

Theme	Category	Code	f	%

Mathematical thinking	Positive Impact		47	100
		Solving step by step improved my mathematical thinking	13	27,6
		I realized how to use the mathematical language	2	4,2
		It enabled me to discover the existence of language	1	2,1
			1	2,1

Analyzing the data in terms of the level of using mathematical thinking skills, they mentioned that the process had a positive impact on them and solving the problem stage by stage improved their mathematical thinking. Furthermore, they stated that during this process they realized how to use the mathematical language.

3.11. Data analysis based on the level of using solutions

Data analysis based on the level of using solutions and the categories and codes belonging to the theme are listed below.

Table 11. Data analysis based on the level of using solutions

Theme	Category	Code	f	%
Solutions	Those who thought of all solutions	I think of some of them all solutions	27	57,4
		I think of some of them at the same time	3	6,3
	Trial-Error	I try one of them, if it fails, I try another one	6	12,7
		I implement the first solution I think of	4	8,5
	Deficient/false behavior	I do not try them all	4	8,5
		I try to solve them by using the strategies I like	2	4,2
		I do not think of all of them	2	4,2

Evaluating the data in terms of their level of using solutions, it becomes apparent that there are students, who considered all of the solutions available in the process, as well as students, who still used the trial-error method and demonstrated false of insufficient behaviors.

3.12. Data analysis based on the development level of operation skills

Data analysis based on the development level of operation skills and the categories and codes belonging to the theme are listed below.

Table 12. Data analysis based on the development level of operation skills

Theme	Category	Code	f	%
Development Level of Operation Skills	Those who think their operation skill has improved	My operation skills improved	42	89,3
		My skills improved while using the given information and what is asked in the question	4	8,5
		They improved because the process goes step by step	3	6,3
		They improved because we used different strategies	2	4,2
		Our practical operation strength increased	1	2,1
		It improved our ability to explain	1	2,1
		Rather than our operational skills, our methodological approach improved	1	2,1
		Our interpretation strength increased	1	2,1
	Those who think their operation skills did not improve	I am ineligible	1	2,1

Examining the students in terms of their operation skill levels, it is observed that the majority of them believe that their operation skills improved. Moreover, there are some students whose methodological skills improved besides their operational skills. One student still does not feel sufficient, thus believes his/her skills did not improve.

3.13. Data analysis based on the level of using mathematical language

Data analysis based on the level of using mathematical language and the categories and codes belonging to the theme are listed below.

Table 13. Data analysis based on the level of using mathematical language

Theme	Category	Code	f	%
Using mathematical language	Those who can use mathematical language	I use mathematical language	42	89,3
		I used the given information while writing mathematically	6	12,7
	Those who do not use mathematical language	It is not something that can be immediately learned	2	4,2
		I find myself ineligible	2	4,2
		I did not make any progress about this subject in this course	1	2,1
	Those who tried using it	We tried to use it	2	4,2
		We paid attention to using it	3	6,3

Analyzing the data based on the level of using mathematical language, the majority of them in fact uses mathematical language. However, some of the students argued that this process cannot be immediately learned, thus they did not see any progress. There were students, who tried to use the mathematical language during this process.

3.14. Data analysis based on the competency level for the first mathematics course

Data analysis based on the competency level for the first mathematics course and the categories and codes belonging to the theme are listed below.

Table 14. Data analysis based on the competency level for the first mathematics course

Theme	Category	Code	f	%
Competency for the first maths course	Competency	Very competent	47	100
		Thanks to this course, it is possible to solve this problems in a much more accurate way	1	2,1
	Suggestions	It would be more convenient if given at the same semester as material design and origami	1	2,1

Analyzing the data relevant to determining the class level for the mathematics course, it is observed that there is consensus among students that the course is given at the right time. In addition to that, they suggested that the course could be given at the same semester with the material design and origami courses, so that they can support one another.

3.15. Analysis of the data relevant to determining the class level

Analysis of the data relevant to determining the class level and the categories and codes belonging to the theme are listed below.

Table 15. Analysis of the data relevant to determining the class level

Theme	Category	Code	f	%
Class Level	University	From the 2nd grade	14	29,7
		From the 3rd grade	12	25,5
		From the 1st grade	10	21,2
		It should be given for one semester each year	7	14,8
		It should be given at the 4th grade, so that it can be permanent	3	6,3
		For two semester at the 2nd semester	1	2,1
		It should be gradually given in four years	1	2,1
	Secondary School	It should be given at the elementary school level	3	6,3
		It should be given at the 5th and 6th grades	3	6,3
		From the elementary school 8th grade	2	4,2
		From the elementary school 7th grade	2	4,2

Analyzing the answers given to the question about why this course should be given at a certain class level, it is observed that the majority of them thinks this course should be given at the university level, even though there are also some students, who believe it should be given starting from secondary school level. Generally, the students suggested that it should be given in the first years of university education, whereas; some of the students suggested that it should be given at different class levels. Another observation is that some students believe it should be given starting from the fifth grade.

3.16. Three concepts occurring to them about problem solving

Three concepts occurring to them about problem solving and the categories and codes belonging to the theme are listed below.

Table 16. Three concepts occurring to them about problem solving

Theme	Category	Code	f	%	Category	Code	f	%
Problem çözme denince aklına gelen üç kavram	Problem solving	Strategy	24	51,0	Personal opinion	Esen teacher	2	4,2
		Problem	18	38,2		It is an effective course	1	2,1
		What is given/what is expected	10	21,2		A nice semester	1	2,1
		Understanding the problem	6	17,7		My success in the group work	1	2,1
		Solution	5	10,6		Relaxation	1	2,1
		Problem solving	5	10,6		When I had a hard time solving the problem in the first lesson	1	2,1
		Stages	5	10,6		The fact that I am not afraid of problem solving anymore	1	2,1
		Evaluation	5	10,6		Taking my problem solving skills to an advanced level	1	2,1
		Group work	5	10,6		Learning how to scrutinize a problem	1	2,1
		Diagram-connection	4	8,5		Effective solution and understanding	1	2,1
		Gradual progress	1	2,1		The fact that strategy selection is sometimes difficult	1	2,1
		Systematical thinking	1	2,1		The fact that selecting a false strategy creates chaos	1	2,1
		Implementation of the strategy	2	4,2		Going through the books to find the solution to the problem	1	2,1
		Reasoning	1	2,1		Solving the problem on the board	1	2,1
	Mathematic	Mathematics	4	8,5		Control	1	2,1
		Applied Mathematics Teaching	2	4,2		Description	1	2,1
		Different opinions	2	4,2		How I used to solve problems	1	2,1
		Logic	2	4,2		How they solved the problems for us	1	2,1
		Mathematics Curriculum	1	2,1		Did I understand the problem?	1	2,1
		Mathematical Language	1	2,1		Was I able to solve it?	1	2,1
		Equation	1	2,1		Is my solution correct?	1	2,1
						Occupation	1	2,1
						Relation	1	2,1
						There are no problem without a solution	1	2,1
						Definitely write down the given information	1	2,1

Examining the answers of the students to the question about expressing problem solving in three concepts, the categories can be listed as problem solving, mathematics, and personal opinion. In the problem solving category, they often discussed the problem solving stages. In the mathematics category, they used expressions such as mathematical language and logic. In the personal opinion category, it is observed that they evaluated by considering how the course makes them feel emotionally.

3.17. Analysis of the data relevant to determining the level of contribution and productivity of the course

Analysis of the data relevant to determining the level of contribution and productivity of the course and codes belonging to the theme are listed below.

Table 17. Analysis of the data relevant to determining the level of contribution and productivity of the course

Theme	Category	Code	f	%
Productivity and Contribution of the Course	Positive	Very efficient	40	85,1
		I can solve problems more comfortably	36	76,5
		It enabled me to pay attention to the necessary steps	20	42,5
		I can teach my students more systematically in my internship	18	38,2
		We comprehended how an educator must solve a problem	16	34,0
		I can find what to implement	13	27,6
		Comprehending the strategies efficiently	13	27,6
		I will not have difficulty with explaining a problem to a student in my professional life	13	27,6
		I can solve problems more easily	12	25,5
		We solve problems in a more knowledgeable and conscious way	11	23,4
		It enabled me to extract new problem situations and create solutions	11	23,4
		Deciding what to do before starting with the problem	11	23,4
		I enjoyed it throughout the semester	10	21,2
		It strengthened our frienships	7	14,8
		Seeing new solutions	6	12,7
		Encountering different types of questions	8	17,0
		I do not look at problems with prejudice	4	8,5
		We learned different methods and techniques	4	8,5
		I am not concerned with understanding the problem correctly	7	14,8
		I took on another level in problem solving	3	6,3
		Making practices in the classroom made the lesson more enjoyable	3	6,3
		It made think logically	2	4,2
		Our ability to analyze and synthesize improved	1	2,1
		It is a very serious subject, which should be given for another semester, so that success will be tripled	1	2,1
		I wish we had solved more problems	1	2,1
	Negative	I did not know what to do at the beginning of the semester	7	14,8
		The theorical knowledge given at the beginning of the semester was boring	2	4,2
		The fact that the course was at afternoon decreased our productivity	1	2,1
		There were disruptions during the presentation	1	2,1

Examining the impact of the course on the student, there occurs two categories: positive views and negative views. There are positive views such as approaching problems more comfortably thanks to this course, positive contribution to their instruction of the course as well as their professional development, and learning how to proceed while teaching students. The negative views include believing that the timing is not good and the four-week course is boring, thus their efficiency decreases. Moreover, some of the negative comments were due to the various disruptions during the presentation of the course.

4. Discussion and Conclusions

Examining the responses of the students, it is observed that they greatly enjoyed the course, learned many things, their outlook on the mathematics course changed, and got the chance to implement the problem solving stages. Moreover, they mentioned that this course will become very beneficial for them once they start teaching in various provinces and serves as a major step towards becoming a good teacher. They stated that there has been major changes in their pre-implementation and post-implementation approaches towards problems and that they now approach a problem in a more systematic and conscious manner. They stated that before the implementation, they usually solved the problems through trial and error method or the methods they had known before. Furthermore, they added that after the implementation they realized that there is a strategy for each solution and they learned how to use those strategies

[1,3,4,5,6,7,10,12,13,22,28,33,39,43,47] They stated that they had many false approaches inn the problem solving process before the implementation. These errors were found out through the content analysis in the classroom. Before the implementation, they associated problem solving with concepts such as chaos and difficulty, however, after the implementtion they started associating problem solving with strategy, solution, gradual progress, logic, information given and asked. It was stated that students' skills of solving non-routine problems were low. Examining the results of the interviews, it is observed that the students argue that the strategies they had used to approach problems were deficient and false.

[26] states that mathematical education continuously becomes less successful. He argues that this is because students do not learn how to build mathematics in their elementary school years. The interviews suggest that argument. Because students believe that they will become better teachers after this course. Because they believed they will be able to explain that to the students in a better manner. Some students mentioned that they wished they had learned this method sooner, which indicates that they associate their lack of success with not learning about this subject during their education.

[30] utters that students develop their academic self-perception as they find alternative solutions to the current and possible problems. In this study, it was observed that the more knowledge students obtained about this subject, the more their academic self-confidence increased. Most of the statements received from students supports this assertion. Their statements include expressions such as "I am not afraid of problems, because I know how to approach each problem".

References

1. Altıparmak, K., Öziş, T. Matematiksel İspat ve Matematiksel Muhakemenin Gelişimi Üzerine bir inceleme. *Ege Eğitim Dergisi,* **6***(1),* 25-37, (2005).
2. Altun, M. *Matematik öğretimi.* Bursa: Erkam Matbaası, (2002).
3. Altun, M. *İlköğretim ikinci kademe (6, 7 ve 8. sınıflarda) matematik öğretimi* (5. Baskı). Bursa: Aktüel (2008).
4. Arcavi, A. The role of visual representations in the learning of mathematics. *Educational Studies in Mathematics,* **52,** 215-241, (2003).
5. Arslan, Ç., Altun, M. Learning to solve non-routine mathematical problems. *İlköğretim Online,* **6** (1), 50-61, (2007).

6. Bal-İncebacak, B., Ersoy, E. 7. sınıf öğrencilerinin matematiksel muhakeme becerilerinin TIMSS'e göre analizi. *Uluslararası Sosyal Araştırmalar Dergisi*, **9**(46), 474- 481, (2016a).
7. Bal-İncebacak, B., Ersoy, E. Problem solving skills of secondary school students, *China-USA Business Review*, **15**(6), 275-285 doi: 10.17265/1537-1514/2016.06.002, (2016b).
8. Baykul, Y. *İlköğretimde matematik öğretimi*. Ankara: Pegem Akademi, (2012).
9. Bingham, A. *Çocuklarda problem çözme yeteneklerinin geliştirilmesi.* (çev. F. Oğuzkan). İstanbul: Milli Eğitim Basımevi, (1983).
10. Cramer, K., Post, T. Connecting research to teaching proportional reasoning. *Mathematics Teacher*, **86**(5), 404- 407, (1993).
11. Crowley, K., Siegler, R. S. Explanation and generalization in young children's strategy learning. *Child Development*, **70** (2), 304-316, (1999).
12. Delice, A. *A Comparative Study of Students' Understanding of Trigonometry in The United Kingdom and The Turkish Republic*. Unpublished doctora thesis, İngiltere: University of Leeds, İnstitute of Education Sciences, (2003).
13. Delice, A., Sevimli, E. Geometri Problemlerinin Çözüm Süreçlerinde Görselleme Becerilerinin İncelenmesi: Ek Çizimler, *M.Ü. Atatürk Eğitim Fakültesi Eğitim Bilimleri Dergisi*, **31**, 83-102, (2010).
14. Duatepe, A., Akkuş-Çıkla, O., Kayhan, M. Orantısal akıl yürütme gerektiren sorularda öğrencilerin kullandıkları çözüm stratejilerinin soru türlerine göre değişiminin incelenmesi. *Hacettepe Üniversitesi Eğitim Fakültesi Dergisi*, **28**,73-81, (2005).
15. Duman, B. *Neden beyin temelli öğrenme*., 2. baskı, Ankara: Pegem Akademi Yayıncılık, (2009).
16. Erden, M. İlkokulların birinci devresine devam eden öğrencilerin dört işleme dayalı problemleri çözerken gösterdikleri davranışlar. *Hacettepe Üniversitesi Eğitim Fakültesi Dergisi*, **1**, 105-113, (1986).
17. Erden, M., Akman, Y. *Eğitim psikolojisi.* Ankara: Arkadaş Yayınevi (1998).
18. Ersoy, E. *Eğitimde ölçme ve değerlendirme* (Savaş Baştürk, Edt.). *Görüşme*, (s. 277-283). Ankara: Nobel Akademik Yayıncılık, (2014).
19. Gönen, M.,Dalkılıç, N. *Çocuk eğitiminde drama.* İstanbul : Epsilon Yayıncılık, (2000).
20. Hacısalihoğlu, H. H., Mirasyrdioğlu, Ş., Akpınar, A. (2003). *Matematik öğretimi.* Ankara: Asil Yayın Dağıtım.
21. Hatfield, M. M., Edwards, N. T., Bitter, G. G. *Mathematics methods for elementary and middle school teachers*, Third Edition, Boston: Allyn/Bacon, (1997).
22. Işık, C., Kar, T. İlköğretim 6, 7 ve 8. sınıf öğrencilerinin sayı algılama ve rutin olmayan problem çözme becerilerinin incelenmesi, *Ahi Evran Üniversitesi Eğitim Fakültesi Dergisi*, **12** (1), 57-72, (2011).
23. Kalaycı, N. *Problem çözme. Öğretim ilke ve yöntemleri.* (F. Dilek Gözütok). Ankara: Ekinoks Eğitim Danışmanlık Hizmetleri ve Basın Yayın Dağıtım San. Ltd. Şti, (2006).
24. Korkut, F. Lise öğrencilerinin problem çözme becerileri. *Hacettepe Üniversitesi Eğitim Fakültesi Dergisi*, **22**, 177-184, (2002).
25. Kvale, S., Brinkmann, S. *Interviews*. Thousand Oaks: Sage, (2009).

26. McCarthy, D.S. *A Teaching Experiment Using Problem Based Learning at the Elementary Level to Develop Decimal Concepts,* Doctoral Program, Buffalo The State University of New York, (2009).
27. MEB. *İlköğretim Matematik Dersi Öğretim Programı ve Kılavuzu 6-8. Sınıflar.* Ankara: Devlet Kitapları Basımevi, (2005).
28. Nemirovsky, R., Noble, T. On mathematical visualization and the place where we live. *Educational Studies in Mathematics* **33**(2), 99–131, (1997).
29. Nevra-Seggie, F., Bayyurt, Y. *Nitel araştırma yöntem, tekniki analiz ve yaklaşımları,* Ankara:Anı Yayıncılık, (2015).
30. Oğuz, V. *Proje Yaklaşımının Anasınıfina Devam Eden Çocukların Problem Çözme Becerilerine Etkisinin İncelenmesi.* Yayımlanmış doktora tezi. Ankara Üniversitesi Fen Bilimleri Enstitüsü, Ankara, (2012).
31. Oğuz, V., Köksal-Akyol, A. *Aile ve Çocuk.* (Editör: Neriman Aral), *Çocuklarda Problem Çözme Becerisi* (s. 217-237), Ankara. Ankara: Üniversitesi Basımevi, (2012).
32. Olkun, S., Toluk, Z. *İlköğretimde etkinlik temelli matematik öğretimi* (3. Baskı). Ankara: Anı Yayıncılık, (2004).
33. Olkun, S., Aydoğdu, T. Üçüncü uluslararası matematik ve fen araştırması (TIMSS) nedir? Neyi sorgular? Örnek geometri soruları ve etkinlikler. *İlköğretim-Online* **2***(1). [Online]:* http://ilkogretim-online.org.tr (2003).
34. Patton, M. Q. *How to use qualitative methods ın evaluations.* Newbury Park, California: Sage Publications, (1987).
35. Polya, G. *How to solve it,* Princeton NJ: Princeton U. Press, (1945).
36. Polya, G. *How to solve ıt.* New Jersey, NJ: Princeton University Pres, (1988).
37. Polya, G. *Nasıl çözmeli*? (çev.) Feryal Halatçı, İstanbul: Sistem Yayıncılık. (ss.168-169), (1997).
38. Seidman, I. *Interviewing as qualitative research.* New York: Teachers college Press, (2006).
39. Slovin, H. Moving to proportional reasoning. *Mathematics Teaching in the Middle School,* **6**(1), 58-60, (2000).
40. Strauss, A., Corbin, J. *Basics of qualitative research: Techniques and procedures for developing grounded theory.* 2d ed. Thousand Oaks, CA: Sage, (1998).
41. Troutman, A. P., Lichtenberg, B. K. *Mathematics a good beginning: strategies for teaching children.* United States of America: Brooks/Cole Publishing Company, (1994).
42. Umay, A. Matematiksel Muhakeme Yeteneği. *Hacettepe Eğitim Fakültesi Dergisi,* **24**, 232- 243, (2003).
43. Umay, A., Kaf, Y. Matematikte kusurlu akıl yürütme üzerine bir çalışma. *Hacettepe Üniversitesi Eğitim Fakültesi Dergisi,***28,** 188-195, (2005).
44. Üstündağ, T. *Yaratıcılığa yolculuk.* Ankara: Pegem A Yayıncılık, (2002).
45. Van Tıl, C. T., Van Der Vleuten, C. P. M., Van Berkel, H. J. M. *Influence of Regularity of Studying in a Problem-based Curriculum on Students Achievement' [Invloed van regelmaat van studeren binnen PGO op de studieresultaten van studenten]* , in J. A. Smal (ed.) GOC-Proceedings. Houten, The Netherlands: Bohn Stafleu Van Loghum, (1997).

46. Wagner, J. B. *Dorothy Heathcote drama as a learning medium*. Washington D. C.: National Education Association, (1976).
47. Yakimanskaya, I .S. *The development of spatial thinking in schoolchildren.* NCTM: Reston, USA, (1991).
48. Yin, R.K. *Qualitative research from start to finish.* New York: The Guilford Press, (2011).
49. Harris, K., Marcus, R., McLaren, K., & Fey, J. (2001). Curriculum materials supporting problem-based teaching, *School Science & Mathematics*, Sayı: 20, 191-192

ITM Web of Conferences **13**, 01013 (2017) DOI: 10.1051/itmconf/20171301013
CMES2017

Modeling Routing Constructs to Represent Distributed Workflow Processes Using Extended Petri Nets

Mehmet Karay*
Final International University, Business Administrative, Toroslar Avenue, No:6, 99370, Catalkoy, Kyrenia, North Cyprus

Abstract.The contribution introduces explaining and modelling routing construct to represent distributed workflow processes using extended Petri nets with the new construction which is the interruption routing. Petri nets notation is used for representation of the four main routing constructs as well as for the workflow process. In this paper, four main routing constructs and the new construct interruption explained and modeledby using extended Petri nets.

1 Introduction

Workflow refers to a set of sequential tasks performed to achieve a pre-set goal in an organisation. In other words, workflow processes enable the flow of work among multiple participants. For contemporary business environments, Workflow is an essential component with its ability to build a well-equipped infra-structure for business processes by enhancing them with the supportive information and corresponding technology.

In order to manage workflow effectively, organizations started to employ technology in terms of computer software which is called Workflow Management System (WFMS). WFMS is defined as "A system that defines, creates and manages the execution of workflows through the use of software, running on one or more workflow engines, which is able to interpret the process definition, interact with workflow participants and, where required, invoke the use of IT tools and applications" by Workflow Management Coalition [1]. WFMSs have recently been an important issue to investigate workflow processes in information systems of organizations. As a result of the challenges of globalization and the changes in economy along with the rapid improvements in technology, a growing competition has emerged within the market. So as to survive in this competitive environment, today's organizations must be time efficient, and cost effective. To address these requirements, it has been unavoidable for organizations to reconsider the way they organize workflow processes they use for workflow management systems.

* Corresponding author: mehmet.karay@final.edu.tr

As the need and importance of workflow systems has significantly increased for organizations over the last decades, it has been crucial to put serious efforts for their enhancement. It is a widely accepted point that Modelling is an efficient technique used in developing systems. Instead of working on the real phenomenon, it is more practical, cost effective and convenient to work on a model. There are a number of modeling approaches each coming with their weaknesses and strengths[11].

Among these approaches, Petri nets have been in use as a graphical modelling language for more than 50 years. They were first introduced by Dr. Carl Adam Petri in his Ph.D. thesis in 1962. Since then, they have been a successful modeling tool in computer science. Beneath the success of Petri nets lies their well-defined mathematical theory together with a well-organized bipartite graphical structure to represent active performance of systems. The theory of mathematics provides detailed modeling and analysis of the performance of system and the graphical structure makes it possible to visualize the performance of system.

W.M.P. van der Aalst, in his paper, provided three good reasons for using a Petri net based Workflow Management Systems (WFMS). The first reason was stated as the fact that business logic can be represented by a formal but also graphical language. The following reason was that in contrast with many other process modelling techniques, the state of case can be modelled explicitly in a Petri net. As the third reason, he stated that Petri nets are marked by the availability of many analysis techniques and this is a great asset in favour of a Petri net based WFMS.

However, Petri nets, in its ordinary form, are sometimes insufficient in some cases. That is because classical Petri nets do not represent a universal algorithmic system which means they cannot be used to create on arbitrary Turing machine. Therefore, they cannot be used to implement arbitrary algorithms of information processing, which is essential for the study of behaviour of dynamic system. This can be done only with the use of extended Petri nets that allow to express all fundamental properties of processes. These properties include transfer of data (in the form of move of tokens), time delay, control of move of tokens and data transformation. These properties are associated with firing of transitions in the extended Petri net. But, classical Petri nets express only move of abstract tokens from place to another places as a result of transition firing, and no time delay is associated with firing of transitions.For this reason, in this paper, extended Petri nets are used to model workflow processes.

2 A survey of a class of extended petri nets

As was stated in the introduction, general Petri nets, oriented from works of K. Petri in 1962, are not suitable for simulation of discrete-event systems. We give here a brief survey of a class of extended Petri nets that represent a complete algorithmic system and were used in this paper for simulation modeling. These nets are timed Petri nets, with attributed tokens and the abilities to control transfer of information, carried with token attributes, and to process this information during transition firing.

The structural elements of extended Petri nets are places, transitions and directed arcs. There are two types of places in these nets-simple and queue places.

A simple place or S-place (represented graphically as a circle) can hold one token at most, while a queue place or Q-place (depicted in the form of an oval) can accumulate an arbitrary number of tokens.

The minimal building blocks of extended Petri nets are elementary nets. Each elementary net consists of only one transition and a few incident input and output places. In the used extended Petri nets, there are five basic types of elementary nets, denoted by T, X, Y, I and G. Each type has its own structure and functionality. By connecting elementary nets with each other, one can create Petri nets of arbitrary size and complexity.

For the purpose of this paper, elementary nets of types T, Y, X and I were used. The following paragraphs contain brief descriptions of these types.

Elementary net of type T is shown in Figure 1. When its transition fires, the net merges tokens from all input places and creates new tokens in all output places. The transition can fire only if each input place contains a token and all simple output places (S-places) are empty. At the end of transition firing, the default data processing function, associated with this net, copies attributes of the token of place x_1 and assigns the values of these attributes to each new token in all output places y_1, y_2,...,y_n. If necessary, any desired data processing function can be performed on attributes of new tokens in output places. The net of this type actually implements logical operation AND on input places. Note that, in general, any non-zero time delay can be associated with the transition. This is true also for all other types of elementary nets. In addition, this elementary net, without input places, can be used as a token generator.

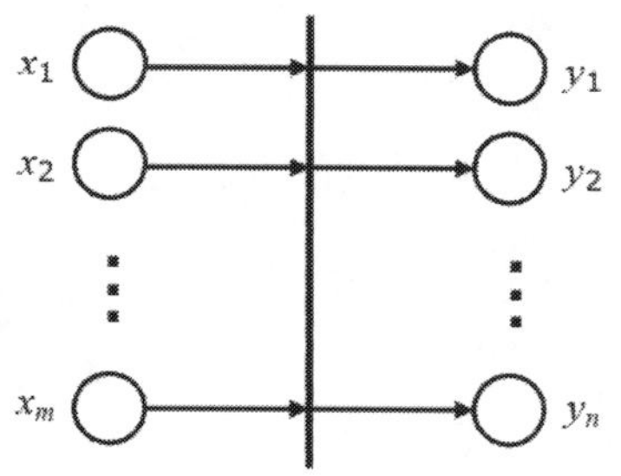

Fig. 1. Elementary net of type T.

Elementary net of type Y is given in Figure 2. As it is shown, the transition bar of this net has a short line segment directed to input places. Logically the net of this type implements the conditional multiplexing operation by selecting one token from one of the input places when the transition fires. As a result of transition firing, one token will be created in each output place. Values of attributes of the token, selected from an input place, are copied and assigned to corresponding attributes of new tokens created in all output places. On default, the token from the first non – empty input place is selected. But with an explicitly assigned input control function, the selection can be done from any other non – empty input place. Note that tokens in the remaining input places are not affected. Without input places, this elementary net behaves like a net of type T without input places.

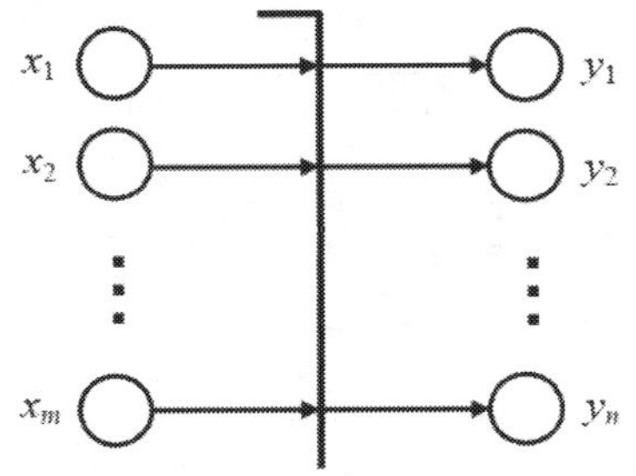

Fig. 2. Elementary net of type Y.

Elementary net of type X is presented in Figure 3. It provides routing of tokens from input places to one of the output places. The transition bar of this net has a short line segment directed to the output places. For firing of the transition here, it is necessary that each input place holds a token and the selected output place is empty. When the transition fires then, on default, attributes of input place x_1 are copied and their values are assigned to attributes of a new token created in one of the output places.

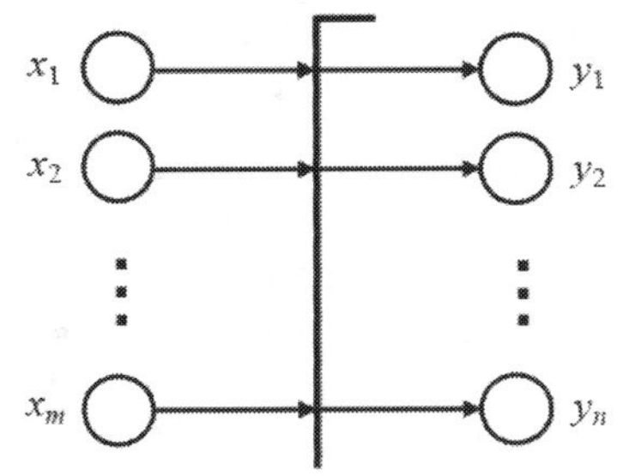

Fig. 3. Elementary net of type X.

Elementary net of type I, shown in figure 4, provides a possibility to interrupt the firing interval of transition by an external event. As shown in the figure, this net has two input places x1 and x2 and two output places y1 and y2. Place x1 is the main input, while place x2 serves as an interrupting place. Correspondingly, y1 and y2 are outputs for the main input and for the interrupting input, respectively. The arrow symbol on the transition bar is directed from the interrupting input to the main input.

There are two main scenarios in the operation of this elementary net. First, if there is a token in place x1, but place x2 is empty, the net works as usual net of type T with two incident places x1 and y1. Second, if, during started firing of transition, a token appears in place x2, then the ongoing firing instantly stops, the token from place x1 goes to place y1 with its attributes, and the interrupting token from place x2 goes, with its attributes and zero delay, to place y2. After the interruption, the tokens in place y1 and y2 can be treated separately in subsequent parts of the model. Usually, the interruptingtoken will initiate some interruption procedure by this same net, while the interrupted token will wait until the net completes processing the interruption event.

DOI: 10.1051/itmconf/20171301013

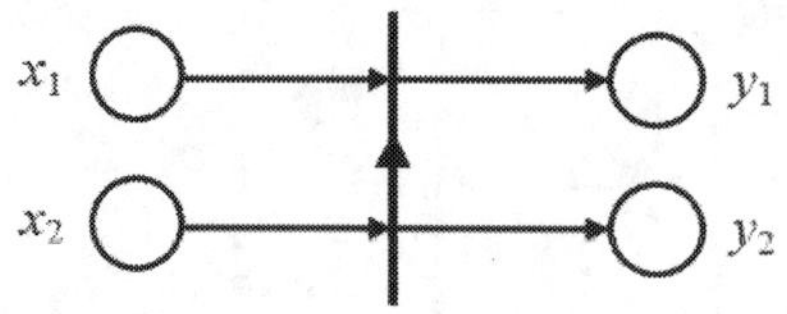

Fig. 4. Elementary net of type I.

Detailed description of the elementary nets can be found in [5].

For the description of models in terms of extended Petri nets there is the Model Description Language (MDL). The MDL is an extension of object Pascal, with added features for the description of elementary nets, token attributes and initialized data. There is also Modelling Control Language (MCL) that is used to parameterize the compiled models and control their runs on the computer. All these capabilities are implemented in the simulation system Winsim. Its software, user guide and many examples of solved models are available on the CD attachment of [5].

3 Concepts of a workflow process

A workflow process can be examined based on three main concepts. These are case, task, and condition.

The case is the situation in which the need for a workflow management is emerged. Workflows are founded by different cases. The execution of each individual work is performed for a specific case. To illustrate, a healthcare data analysis or an insurance claim both represent different cases. Dealing with cases in the most successful way is the primary objective of workflow management. A workflow process always handles similar cases.

Throughout a workflow process, the case is handled by executing sequential tasks. A task is a piece of work to be done in a certain time interval. The workflow process definition includes the specification of tasks to be carried out and also their order of execution depending on a set of conditions. Therefore the order of the tasks is pre-determined and directed by the set of conditions. A condition holds (true) or does not hold (false). Each task has pre- and post-conditions: the preconditions should hold before the task is executed, and the post conditions should hold after the task is executed [Aalst, 2004].

When Petri nets are used to model a workflow process, tasks are modeled by transitions, conditions are modeled by places, and cases are modeled by tokens.

4 Basic types of workflow constructs

A case is handled throughout a process. The paths that regulate the flow of a process are called routes. The routing of cases is one of the main issues of a workflow process as it determines which process should be done first and which one should go next. These routes are constructed in five different ways. These are sequential, parallel, conditional, iterative, and interruption construct. The first four are applied in the previous studies of modeling workflow processes. In this thesis, as an innovation, interruption construct is also employed for modeling of a workflow process for the first time.

Sequential construct is used when the tasks need to be performed one after another. The tasks which are ordered sequentially are interrelated to each other. One task has to be completed in order to move on the next one. An example of a sequential construct is shown in Figure 5(a).

Differing from sequential routing, parallel construct enables the process to cope with more than one tasks at the same time or in any order. Figure 5(b) shows an example of parallel construct.

Some cases are subject to relevance and they might need to be handled in different ways under different circumstances. The alternative route to be selected varies depending on a specific condition. To model this type of process, conditional routing is used. An example of a conditional construct can be found in Figure 5(c).

In some cases, it is required to repeat the execution of a task. Multiple execution of a particular task is managed by iterative routing. For instance, this type of routing is used when a task requires to be repeated until a subsequent test gives positive results. Figure 5(d) illustrates an example of iterative construct.

The last and the most important form of workflow construct in this thesis is interruption. Interrupts are widely used in information systems to temporarily stop a running process and allow a process with a higher priority to run instead. This type of workflow construction is needed to enable the process to perform a more urgent service. In this case, the interrupted process will continue running after the termination of the process with the higher priority. Figure 5(e) shows an example of interruption construct.

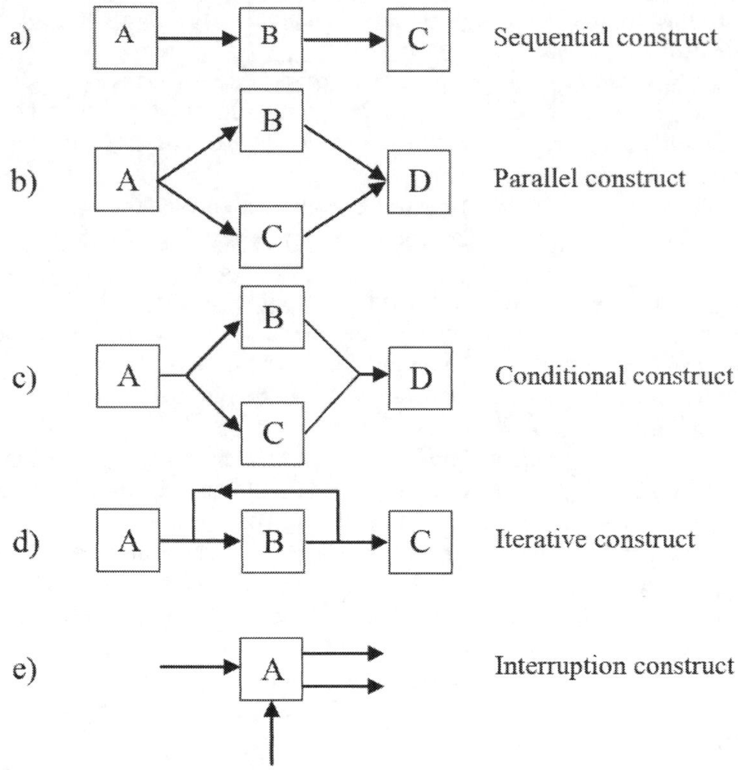

Fig. 5. Workflow constructs

5 The use of basic types of workflow routing constructs to describe a set of interacting distributed workflow information systems

In this section, an example set of interacting distributed workflow information systems is described by using basic types of workflow routing constructs. The example considers a travel agency organizing a trip (Aalst, [2000]). In this case, there are a number of tasks for the travel agency to execute.

In the first place is the registration of the customer. After that, an employee looks for chances to contact the customer. The next step is to reach the customer to learn if she or he is still interested in the trip or if she or he would like to get information about some other alternatives. So far these tasks should be modeled by using sequential routing construct as all the tasks are executed one after another.

From this point on, the process has three possibilities. The first one is when the customer is not interested anymore. The second is when the customer would like to see more alternatives, and the last one is when customer selects an opportunity. If the customer isn't interested in the trip at all, the process comes to an end as a consequence. If the customer wants to be informed of other alternatives, then the employee starts to search for the required information to contact the customer again later. So, the process is repeated and this is modeled by iterative routing construction. On the other hand, if the customer decides to book a trip, then the system both books the trip and starts to prepare one or two types of insurance on request at the same time. As these processes are done simultaneously, they are modeled by parallel routing construction.

There are two types of insurance provided by the agency; trip cancellation insurance and baggage loss insurance. The customer can only choose to get the trip cancellation or only the baggage loss as well as she or he can choose both. Another option for the customer is not to choose any types of insurance. This decision making process is modeled by conditional routing construction.

With the above example, the use of four workflow routing constructs is described in process. The fifth routing construct which is the most significant one for this thesis, named as interruption, is investigated by the researchers in the literature [10].

6 Petri nets as an efficient formal tool to represent and investigate systems of distributed workflow processes

In this section, it is briefly explained why Petri nets are considered to be a desirable theoretical framework for the formal description, modeling and analysis of distributed systems.

To start with, it is convenient to design models in graphical form with Pet nets as Petri nets themselves are graphs consisting of a limited set of types of structural and dynamic elements. Also, several software packages along with their broad range of analyzing procedures can be found to support the design of Petri nets.

Next, using Petri nets gives the advantage of being able to represent various resources of distributed systems such as queues and data bases. The availability and the quantity of the corresponding resource is modeled by the existence of one or more tokens in a place.

In Petri nets, firing of transitions is used to model events related to these transitions. A process considered as a sequence of events is represented by the corresponding sequence of firing transitions of the net. Many transitions are allowed to fire in parallel which enables modeling the existence of many parallel processes in a system together with representing different types of process synchronization.

Another benefit of Petri nets is that they enable detecting deadlocks, proving the properties of safeness, boundedness and liveness, and determining the possibility of reachability of some *target state* from a given *initial state*. The reachability analysis represents one of the most important tasks in the study of discrete-event systems and, it is possible to solve that formally with the help of Petri-net-based models.

Petri nets are also practical to use with large and complex distributed systems as it is possible to use them for modular and hierarchical modeling.

The last but not the least, Petri nets have a number of extensions such as Petri nets with inhibitor arcs, timed Petri nets, stochastic Petri nets, colored Petri nets, predicate/transition Petri nets which enable qualitative and quantitative analyses. These are called extended Petri nets. In some types of extended Petri nets, tokens have attributes which controls the movements of token in the net and in the data transformation during a transition firing process. These types of extended Petri nets are universal algorithmic systems which are suitable to be used for simulation modeling of information systems.

The formal definition and different characteristics of the formal analysis of Petri-net based models are given in more detail in the [10].

7 Mapping of basic types of workflow routing constructs into extended petri nets

The basic type of workflow routing constructs include sequential, parallel, conditional, iterative, and interruption routings which are previously described in Chapter 2. In this section, mapping of these workflow routing constructs into extended Petri nets is described. Building blocks are used to model workflow routing constructs.

Aallst (2004) explained and applied the first four constructs in his paper.

Sequential routing: Causal relationships among transitions (tasks) are dealt with sequential routing. Considering two tasks *A* and *B* and assuming that task *B* is executed after the end of task *A*, then it can be said that *A* and *B* are executed sequentially. Table 1(a) illustrates how the sequential routing can be modeled by adding places to model causal relations among transitions. S2models causal relation between task *A* and task *B*.

Parallel routing: This routing construct is applied in cases where the order of execution is less strict. When two tasks B and C need to be executed but the order of execution is arbitrary, elementary net of type T is used as a building block. Table 1(b) involves an example of such a parallel routing where tasks *B* and *C* are executed in parallel.

Conditional routing: This routing construct is utilized to let routing vary between cases. This makes the routing of a case depended on the workflow attributes of a case, the behaviour of the environment, or the workload of the organization. For the modeling of two or more alternatives, two building blocks are used. These are elementary nets of Type X and Y. This kind of routing is shown in Table 1(c) where task A is has the chance to be followed by either task B or task C. The choice is made between task B and task C. Followed by the execution of one of these two tasks, task D is executed.

Iteration routing: This routing construct is employed to allow repeated execution of a particular task.As building blocks elementary nets of type X and Y are used for modeling. As shown in Table 1(d), task *B* may be executed one or more times where task *C* is the control task.

Interruption routing: In addition to those pre-applied constructs by Aallst(2004), this study also employs interruption routing which provides a very important possibility for an external event to interrupt an activity during its firing transition. To model this type of routing construct, elementary net of type I is used as shown in Table 1(e).

Aallst(2004) modeled workflow management systems with classical Petri nets by using sequential, parallel, conditional and iterative routing constructs. This study takes another step with extended petri nets by adding the *interruption* feature which elementary net of type I makes possible.

Table 1. Basic types of workflow processes as given by Aallst(2004) and in terms of extended Petri nets

Types of the workflow process	By Aallst (2004)	With the extended Petri nets
a)Sequential	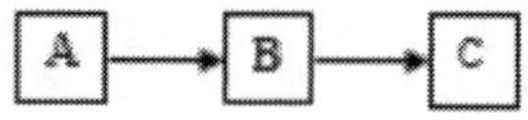	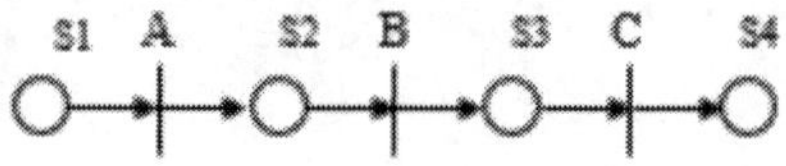
b)Parallel	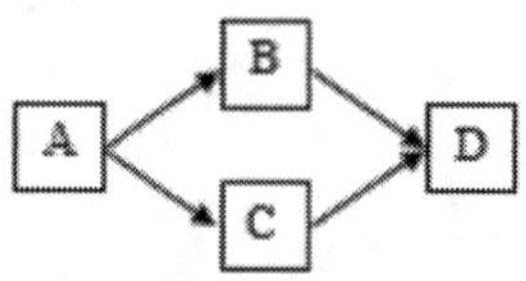	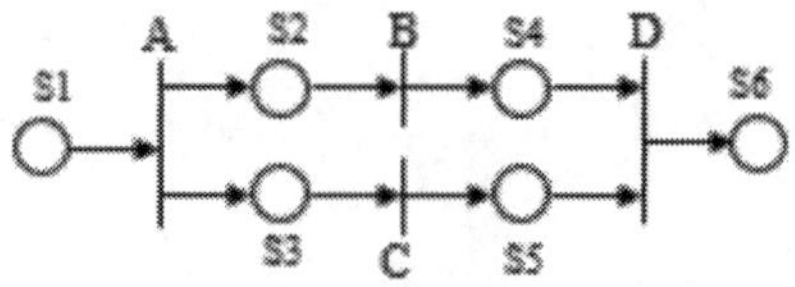
c)Conditional	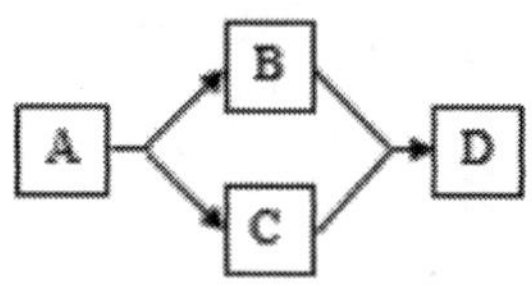	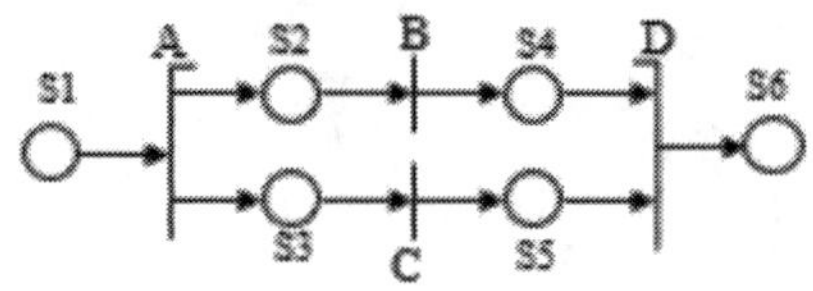
d)Iterative	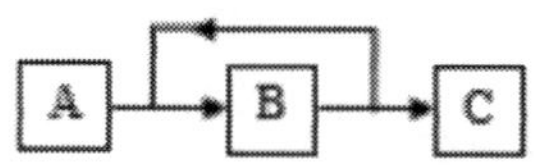	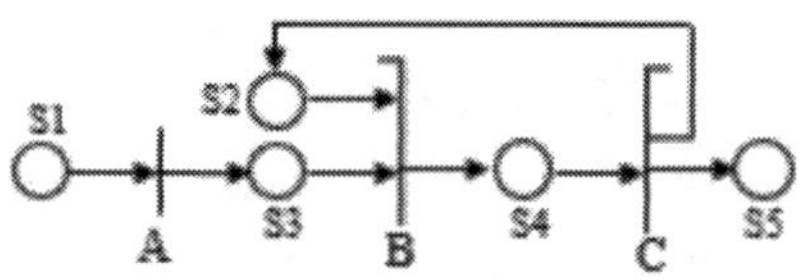
e)Interruption	N/a	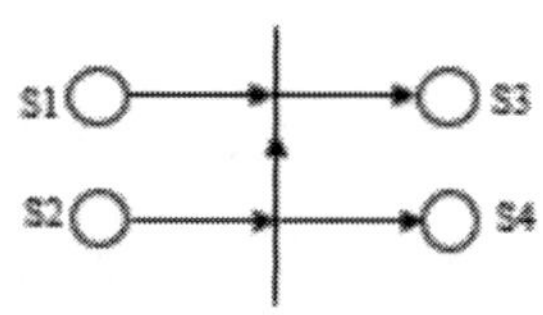

8 Conclusion

Petri nets are very powerful modeling tool with a convenient graphical representation. Their ability to represent complex systems makes them useful for performance analysis. Extended Petri nets refer to a more featured version of Petri nets which allow transfer of data, time delay, control of move of tokens and data transformation. These properties are associated with firing of transitions in the extended Petri net. Classical Petri nets express only the move of abstract tokens from place to another places as a result of transition firing, and no time delay is associated with firing of transitions.

Modeling workflow processes using extended Petri nets allows application of different analysis methods that can be used to examine the behaviour of the process and to compute its performance measures. Transforming Extended Petri net model into graph model allows to express all aspects of control and data transformation in the models of information systems. The extended Petri nets include all basic constructs of work-flow processes proposed by W. Aalst and a new construct for dealing with the interruption of processes. There is a simulation system Winsim for implementation and running simulation models in terms of extended Petri nets. Developed workflow routing constructs in terms of extended Petri nets was used and investigated theoretically and in simulation experiments the model of a queuing system with relative and absolute priorities and developed in terms of extended Petri nets and investigated theoretically and in simulation experiments the model of a cell of flexible manufacturing cell.

References

1. Workflow Management Coalition: Workflow Management Coalition – Terminology &Glossary. Technical Report. Document Number WFMC-TC-1011 (1999).
2. W. M. P. van der Aalst, K. M van Hee: Workflow Management – Models, Methods, and Systems, The MIT Press, Cambridge, Massachusetts, London, England, 2002.
3. Hollingsworth, D.: Workflow Reference Model. Technical Report. The WorkflowManagement Coalition, Document Number WFMC-TC-1003 (1995).
4. Desrochers, A., and R. Ai-Jaar. 1995. Applications of Petri Nets in Manufacturing Systems: Modeling, Control and Performance Analysis. IEEE Press.
5. A. Kostin and Ilushechkina L, "Modeling and simulation of distributed systems", World Scientific Publ. Co., 2010.
6. T. Murata, Petri nets: Properties, Analysis and Applications, Proc. IEEE, vol. 77, no. 4, pp. 541 – 580, 1989.
7. Kostin, A.E., Models and Algorithms for Organization of Distributed Data Processing in Information Systems, Diss. DSc , Moscow Institute of Electronic Technology (Technical University), 1989 (in Russian).
8. Y. Fanaeian and A. Kostin, "Simulated Study of an Anycast – Based Rotuing Method for Wireless Sensor Networks with the use of Petri nets", International Journal of Science and Advanced Technology, vol. 3, no. 4, pp.18 – 27, 2013.
9. Lu, R., Sadiq, S.: A Survey of Comparative Business Process Modeling Approaches. BIS 2007, 82-94 (2007).
10. M. Karay and A. Kostin, "Using Extended Petri Nets for Modeling and Simulation of Queuing Systems with Priorities", International Journal of Science and Advanced Technology, vol. 4, no. 7, pp. 1-6, 2014.

ITM Web of Conferences **13**, 01014 (2017) DOI: 10.1051/itmconf/20171301014
CMES2017

Some Wave Simulation Properties of the (2+1) Dimensional Breaking Soliton Equation

Emine Nesligül Aksan[1,*], *Hasan* Bulut[2] , and *Miraç* Kayhan [1]

[1]Department of Mathematics, University of İnonu, Malatya, Turkey

[2] Department of Mathematics, University of Fırat, Elazığ, Turkey

Abstract. In this paper, we apply an efficient method which is improved Bernoulli sub-equation function method (IBSEFM) to (2+1) dimensional Breaking Soliton equation. It gives some new wave simulations like complex and exponential structures. We test whether all structures verify the (2+1) dimensional Breaking Soliton model. Then, we draw three and two dimensional plane by using Wolfram Mathematica 9.

1 Introduction

Every branch of non-linear partial differential equations (NLPDEs) in applied science plays an major role such as chemical physics, chemical kinematics, optical fibers, fluid mechanics so on. The only wave solutions to the non-linear partial differential equation have been used to describe some physical phenomena. In the literature, plenty of effective methods have been improved to discover the only wave solutions of NLPDEs. Some of those methods are Sine-Gordon expansion method, Homotopy analysis method, The modified simple equation method, generalized Bernoulli sub-equation method, improved Bernoulli sub-equation method. In this article we handle (2+1) dimensional Breaking Soliton Equation [2] defined as:

$$u_{xt}-u_{xxxy}-2u_{xx}u_{y}-4u_{x}u_{xy}=0, \qquad (1)$$

Xie gained non-travelling wave solutions by the developed tanh function method, that they introduced a generalized Riccati equation and obtained different solutions.In this article, we built new non-travelling wave solutions of partial differential equations. Last of all, new non-travelling wave solutions including solution and cyclical solution of PDE are gained. A generalized flexible-coefficent algebraic method with computerized emblamatic calculation is improved to handle (2+1) dimensional Breaking Soliton equation with flexible parameter. Chen al. work the (2+1) Breaking Soliton equation by using the new generalized transformation in homogeneous balance method.

* Corresponding author: mirackayhan@yandex.com

2 Fundamental Properties of the Method

IBSEFM created with extending BSEFM [5,6] will be offered in this sub-division. Hence, we take account of the follows.

Step 1. We think following PDE;

$$P\left(u,u_x,u_y,u_z,u_t,u_{tt},u_{xt},u_{xx},u_{xy},u_{yy},u_{yt},u_{zz},\ldots\right)=0, \tag{2}$$

and get the wave conversion

$$u(x,y,z,t)=U(\eta),\ \ \eta=k_1x+k_2y+\lambda t. \tag{3}$$

If equation (3) is written in equation (2) transform a (NODE) as following;

$$N(U,U',U'',U''',\cdots)=0. \tag{4}$$

Step 2. Thinking test equation of results in equation (2), it can be written as following;

$$U(\eta)=\frac{\sum_{i=0}^{n}a_iF^i}{\sum_{j=0}^{m}b_jF^j}=\frac{a_0+a_1F+a_2F^2+\cdots+a_nF^n}{b_0+b_1F+b_2F^2+\cdots+b_mF^m}, \tag{5}$$

and with respect to the Bernoulli theorem,

$$F'=\sigma F+dF^M,\ w\neq 0,\ d\neq 0,\ M\in\mathbb{R}-\{0,1,2\}, \tag{6}$$

where $F=F(\eta)$ is Bernoulli differential polynom. Take place (6) connection in equation(4), it provide us an equation of polynom $\Omega(F)$ of F;

$$\Omega(F)=\rho_sF^s+\cdots+\rho_1F+\rho_0=0. \tag{7}$$

With respect to the balance essence, we able to gain the values of n,m and M.

Step 3. Let parameter of $\Omega(F)$ must have zero will provide equations system;

$$\rho_i=0,\ i=0,\cdots,s. \tag{8}$$

Solve the system, we going to define the values of $a_0,\cdots,a_n$ and $b_0,\cdots,b_m$.

Step 4. When we solving non-linear Bernoulli differential equation Eq.(6), we gain two cases with respect to σ and d conditions as following;

$$F(\eta)=\left[\frac{-d}{\sigma}+\frac{\tau}{e^{\sigma(M-1)\eta}}\right]^{\frac{1}{1-M}},\ \ \sigma\neq d, \tag{9}$$

$$F(\eta)=\left[\frac{(\tau-1)+(\tau+1)\tanh\left(\frac{\sigma(1-M)\eta}{2}\right)}{1-\tanh\left(\frac{\sigma(1-M)\eta}{2}\right)}\right]^{\frac{1}{1-M}},\ \ \sigma=d,\ \ \tau\in\mathbb{R}. \tag{10}$$

Using a whole distinction system for polynom of F, we analyse equation (8) by using of Mathematica 9 and partition the exact solutions to equation (4). To interpretations better results gained in this wise, we are able to draw 3D and 2D shape of the results gained by taking suitable coefficients.

3 Application

In this subdivision of manuscript, we supply some experimental solutions to describe the demonstration of the analytic algorithm suggested.

If transformation from Eq.(3) to Eq.(1), is applied;

$$\lambda U'' - k_1^2 k_2 U^{(4)} - 6k_1 k_2 U'' U' \tag{11}$$

Integrating and arrangement the values of integration to zero and $V = U'$ written , we gain

$$\lambda V - k_1^2 k_2 V'' - 3k_1 k_2 V^2 = 0 \tag{12}$$

Balancing U'' with U^2 in (12) gives

$$2M + m = n + 2. \tag{13}$$

If we take $M = 3$, $m = 1$, $n = 5$ for Eq. (13), after that, we can write the following equations;

$$U(\eta) = \frac{a_0 + a_1 F(\eta) + a_2 F^2(\eta) + a_3 F^3(\eta) + a_4 F^4(\eta) + a_5 F^5(\eta)}{b_0 + b_1 F(\eta)} = \frac{\Upsilon(\eta)}{\Psi(\eta)} \tag{14}$$

$$U'(\eta) = \frac{\Upsilon'(\eta)\Psi(\eta) - \Upsilon(\eta)\Psi'(\eta)}{\Psi^2(\eta)} \tag{15}$$

and

$$U''(\eta) = \frac{\Upsilon''(\eta)\Psi(\eta) - \Upsilon(\eta)\Psi'(\eta)}{\Psi^2(\eta)} - \frac{\left[\Upsilon(\eta)\Psi'(\eta)\right]'\Psi^2(\eta) - 2\Upsilon(\eta)\left[\Psi'(\eta)\right]^2\Psi(\eta)}{\Psi^4(\eta)},$$

$$\vdots$$

(16)

where $F' = wF + dF^3$, $w \neq 0$, $d \neq 0$. When we substitutes Eqs.(14,16) in the Eq.(12), we obtain a system of equations for Eq.(12). Hence, we achieve an equation from the parameters of polynom of F . By solving equation by using of Wolfram Mathematica 9, it yields the following coefficients;

Case 1. For $\sigma \neq d$, we can find following parameters:
When we substitute these parameters with Eq.(3) in Eq.(14), we gain some new exponential, complex and hyperbolic function solutions to the Eq.(1) as following, respectively;

Case 1. 1. For $b \neq d$, it imaginable following parameters;

$$a_2 = \frac{4d\sqrt{\lambda}b_0}{\sqrt{k_1}},\ a_3 = \frac{4d\sqrt{\lambda}b_1}{\sqrt{k_1}},\ a_4 = -8d^2 b_0 k_1,\ a_5 = 8d^2 b_1 k_1,\ \sigma = \frac{-\sqrt{\lambda}}{2k_1\sqrt{k_2}} \tag{17}$$

Substituting these parameters in Eq.(14) with Eq.(9), we find the following exponential function solution for Eq.(1);

$$u_1(x,y,t) = -\frac{4d\sqrt{\lambda}k_1}{e^{f(x,y,t)}\tau\sqrt{\lambda} + 2dk_1\sqrt{k_2}}. \tag{18}$$

Where $f\left(x,y,t\right) = \dfrac{\sqrt{\lambda}\left(t\lambda + xk_1 + yk_2\right)}{k_1\sqrt{k_2}}$.

When we consider the suitable values of parameters, we can find new wave simulations for Eq.(1) as following figures:

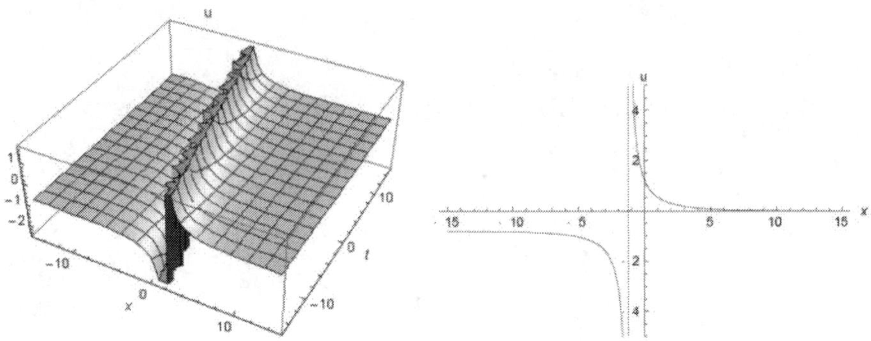

Fig.1. The three dimensional and two dimensional planes of the travelling wave results Eq.(18) by considering the values
$\lambda = 0.1,\ d = 0.3,\ y = 0.4,\ k_1 = 0.5,\ k_2 = 0.6,\ \tau = -1\ \ -15 < x < 15,\ -5 < t < 5,\ and\ t = 0.02$

Case 1. 2. For $b \neq d$, it imaginable following parameters;

$$a_0 = \frac{\lambda b_0}{3k_1k_2},\ a_1 = \frac{\lambda b_1}{3k_1k_2},\ a_2 = \frac{4id\sqrt{\lambda}b_0}{\sqrt{k_2}},\ a_3 = \frac{4id\sqrt{\lambda}b_1}{\sqrt{k_2}},\ a_4 = -8d^2b_0k_1$$
$$a_5 = -8d^2b_1k_1,\ \sigma = -\frac{i\sqrt{\lambda}}{2k_1k_2}. \tag{19}$$

Substituting these parameters in Eq.(14) with Eq.(9), we reach the following exponential function solution for Eq.(1);

$$u_2 = \frac{4d\sqrt{\lambda}k_1}{-e^{\frac{i\sqrt{\lambda}f(x,y,t)}{k_1\sqrt{k_2}}}\tau\sqrt{\lambda} + 2idk_1\sqrt{k_2}} + \frac{\lambda f\left(x,y,t\right)}{3k_1k_2}, \tag{20}$$

where a $f\left(x,y,t\right) = t\lambda + xk_1 + yk_2$. When we consider the suitable values of parameters, we can find new wave simulations for Eq.(1) as following figures:

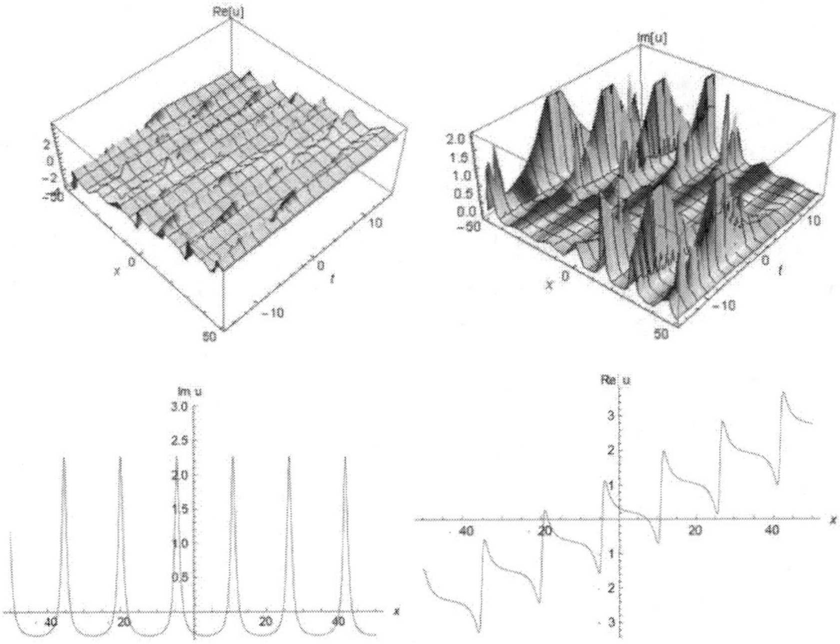

Fig.2. The three dimensional and two dimensional planes of the travelling wave results Eq.(20) by considering the values
$\lambda = 0.1,\ d = 0.3,\ y = 0.4,\ k_1 = 0.5,\ k_2 = 0.6,\ \tau = -1\ -50 < x < 50, -0.4 < t < 3\ and\ t = 0.7$

4 Conclusion

In this paper, IBSEFM has been applied to (2+1) dimensional Breaking Soliton Equation. We have gained some new traveling wave results like complex and exponential function solutions. According to new results and all figures, it has been observed that this way method is a efficient tool for solving such mathematical models.

References

1. M. T. Darvishi, M. Najafi, World Ac. of Sci. **55**, 919–922 (2011)
2. M. T. Darvishi, M. Najafi, Int. J. of Appl. Mat, **1**, 1-7 (2012)
3. Giese, R. Bjerkving, M. Westphal, J. Clin. Oncol, **21(8)**, 1624(2003)
4. H. Bhrawy, M. A. Abdelkawy, A. Biswas. Indian J. Phys, **87**,1125–1131 (2013)
5. H.M. Baskonus, H. Bulut, Waves in Rnd. and Comp. Med, **26**, 189–196 (2016)
6. H.M. Baskonus, 1st Int. Sym. on Co.Mat. Sci. Morocco(2016)
7. F. Özpinar, H. M. Baskonus and H. Bulut, Entropy, **17(12)**, 8267–8277 (2015)
8. H.M. Baskonus, H. Bulut, Waves in Rnd. and Comp. Med, **25(4)**, 720-728 (2015)
9. Bekir, F. Uygun, Arab J. of Mat. Sci. **18**, 73-85 (2012)
10. A.J.M. Jawad, M.D. Petkovic and A. Biswas, Appl. Math. Comput.,**217**, 869-877 (2010)
11. F.B.M., Belgacem, 2010. Appl. Mat. Sci. **4 (74)**, 3665–3686.
12. F. Belgacem, A.A. Karaballi, and S.L. Kalla, Math. Prob. Eng., **3-4**,103 (2003)

Novel Behaviors to the Nonlinear Evolution Equation Describing the Dynamics of Ionic Currents along Microtubules

Haci Mehmet Baskonus[1,*], *Fevzi* Erdogan[2,**], *Arif* Ozkul[3,***], and *Ilham* Asmouh[4,****]

[1] *Department of Computer Engineering, Munzur University, Tunceli, Turkey*
[2] *Department of Mathematics, Yuzuncu Yil University, Van, Turkey*
[3] *Department of Mathematics, Firat University, Elazig, Turkey*
[4] *Department of Mathematics and Applications, Abdelmalek Essaadi University, Fas*

Abstract. In this work, we consider the Bernoulli sub-equation function method for obtaining novel behaviors to the nonlinear evolution equation describing the dynamics of ionic currents along Microtubules. We obtain new results by using this technique. We plot two- and three-dimensional surfaces of the results by using Wolfram Mathematica 9. At the end of this manuscript, we submit a conclusion in the comprehensive manner.

1 Introduction

For the past two decades, searching for the solutions to the various nonlinear evolution equations have became a great activities between scientists in the world. Nonlinear evolution equations arise in the various fields of nonlinear sciences such as optical fibers, chemistry, biology, fluid mechanics etc, and their solutions play a vital roles in our real life situations. Various powerful techniques have been developed for the study of the behavior to the different nonlinear evolution equations such as the (G'/G)-expansion method [1], the new extended (G'/G)-expansion method [2], the sine-cosine method [3], the sine-Gordon expansion method [4, 5], the tanh method [6, 7], the extended tanh method [8], the Hirota bilinear method [9], the modified simple equation method [10], the modified tanh-coth method [11], the Riccati-Bernoulli sub-ODE method [12] and so on. In general, various powerful techniques have been developed by various researchers to tackle different class of nonlinear evolution equations [13–17].

However, in this article, we explore the search for the new solutions to the transmission line models of nano-ionic currents along microtubules [18, 19] by using the Bernoulli sub-equation function method (BSEFM) [20].

The transmission line models of nano-ionic currents along microtubules is given by [18, 19]:

$$wL^2u_{xxt} + L^2u_{xx} + 2\kappa\delta uu_t - \kappa u_t = 0. \tag{1}$$

* e-mail: hmbaskonus@gmail.com
** e-mail: fevzier@gmail.com
*** e-mail: arifozkul@outlook.com
**** e-mail: ilham.fst@gmail.com

2 Analysis of the Method

In this section, we present the steps to be followed for the Bernoulli sub-equation function method in searching new solutions to various nonlinear evolution equations.

Step-1. We consider the partial differential equation in two variables such as x, t and a dependent variable u;

$$P(u_x, u_t, u_{xx}, u_{xt}, \ldots), \tag{2}$$

and take the wave transformation

$$u(x,t) = U(\zeta),\ \zeta = x - kt, \tag{3}$$

where $k \neq 0$. Substituting Eq. (3) into Eq. (2), it gives us the following nonlinear ordinary differential equation;

$$N(U, U^{'}, U^{''}, U^{'''}, \ldots). \tag{4}$$

textbfStep-2. Take trial equation as follows

$$U(\zeta) = \sum_{i=0}^{n} a_i F^i(\zeta) = a_0 + a_1 F(\zeta) + a_2 F^2(\zeta) + \ldots + a_n F^n(\zeta), \tag{5}$$

in which

$$F^{'} = bF(\zeta) + dF^M(\zeta), \tag{6}$$

where $b \neq 0,\ d \neq 0,\ M \in \mathbb{R} \setminus \{0, 1, 2\}$, and $F(\zeta)$ is Bernoulli differential polynomial. Substituting above relations into Eq. (4), we obtain an equation of polynomial $\Omega(F(\zeta))$ of $F(\zeta)$;

$$\Omega(F(\zeta)) = \rho_s F^s(\zeta) + \ldots + \rho_1 F(\zeta) + \rho_0 = 0. \tag{7}$$

According to the balance principle, we can get values of and M and n.

Step-3. Let us consider the coefficients of $\Omega(F(\zeta))$ all be zero, we will obtain an algebraic equations system:

$$\rho_i = 0,\ \ i = 0, \ldots, s. \tag{8}$$

Solving this system, we will determine the values of $a_0, \ldots a_n$.

Step-4. When we solve nonlinear Bernoulli differential equation that is Eq. (6), we obtain the following two situations according to b and d;

$$F(\zeta) = \left[\frac{-d}{b} + \frac{c}{e^{b(M-1)\zeta}}\right]^{\frac{1}{1-M}},\quad b \neq d \tag{9}$$

$$F(\zeta) = \left[\frac{(c-1) + (c+1)tanh\left(\frac{b(1-M)\zeta}{2}\right)}{1 - tanh\left(\frac{b(1-M)\zeta}{2}\right)}\right]^{\frac{1}{1-M}},\quad b = d, c \in \mathbb{R}. \tag{10}$$

Using the complete discrimination of the polynomial of F, we solve the system of equations by using the Wolfram Mathematica 9 and obtain the various cases for the results of the coefficients. We obtain the exact solutions to Eq. (2) by putting the obtained values of the coefficients into the trial solution that is Eq. (5).

3 Application

In this section, we apply the BSEFM to the transmission line models of nano-ionic currents along microtubules [18, 19] given in Eq. (1).

Utilizing the wave transformation $u = U(\zeta)\ \zeta = kx - ct$ on Eq. (1), yields;

$$-ck^2wL^2U'' + L^2k^2U' - c\kappa\delta U^2 + c\kappa U = 0, \tag{11}$$

applying the homogeneous balancing technique on Eq. (11) by using the highest derivative U'' and the highest power nonlinear term U^2, yields the following relation between n and M;

$$n = M + 1 \quad n, M \in \mathbb{Z}^+. \tag{12}$$

Choosing $M = 3$, gives $n = 4$. Using $n = 4$ along with Eq. (5), yields the following;

$$U(\zeta) = a_0 + a_1F + a_2F^2 + a_3F^3 + a_4F^4, \tag{13}$$

differentiating Eq. (13) twice, produces the following equations;

$$\begin{aligned} U' = {} & a_1bF + a_1dF^M + 2a_2bF^2 + 2a_2dF^{M+1} + 3a_3F^3 \\ & +3a_3dF^{M+2} + 4a_4bF^4 + 4a_4dF^{M+3}, \end{aligned} \tag{14}$$

$$\begin{aligned} U'' = {} & a_1bF' + a_1dF^{M-1}F' + 4a_2bFF' + 16a_4dF^3F' + 2a_2d(M+1)F^MF' \\ & +9a_3bF^2F' + 3(M+2)a_3F^{M+1}F' + 4a_4d(M+3)F^{M+2}F', \end{aligned} \tag{15}$$

where $F' = bF + dF^3$, $b \neq 0$, $d \neq 0$. Substituting Eq. (14) and (15) into Eq. (11) and simplifying, we get an equation that involves the polynomials of F. We therefore collect the system of equations from that polynomial there by equating the summation of each coefficients of F that are having the same power to zero. We simplify the system of equations with the help of Wolfram Mathematica 9 and get the following cases of solutions;

Case-1. when $b \neq d$
$a_0 = \frac{a_4b^2}{d^2}, a_1 = 0, a_2 = \frac{2a_4b}{d}, a_3 = 0, w = -\frac{1}{10bc}, \delta = \frac{d^2}{a_4b^2}, \kappa = \frac{12bk^2L^2}{5c},$

substituting these values into Eq. (13), we get the following exponential function solution;

$$u_1(x,t) = \frac{a_4b^4E^2}{d^2\left(de^{-2bct+2bkx-bE}\right)^2}. \tag{16}$$

 DOI: 10.1051/itmconf/20171301015

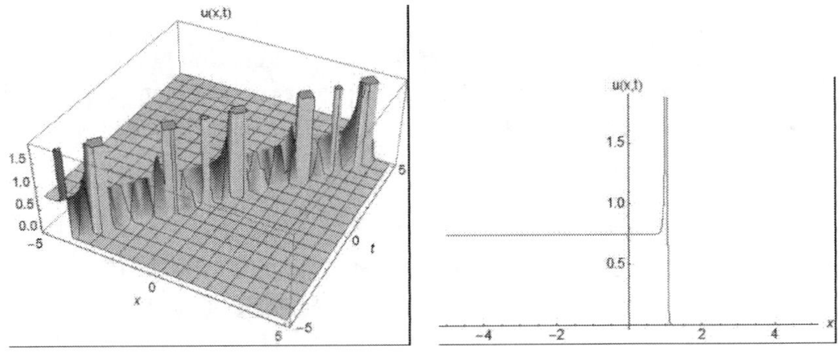

Figure 1. The 3D and 2D shapes of Eq. (16) with the values $k = 6$, $w = 2$, $b = 2$, $E = 8$, $c = 5$, $d = 4$, $a_4 = 3$, $-5 < x < 5$, $-5 < t < 5$ and $t = 1.2$ for the 2D graphic.

Case-2. when $b \neq d$
$a_0 = 0, a_1 = 0, a_2 = \frac{2a_4 b}{d}, a_3 = 0, w = -\frac{1}{10bc}, \delta = -\frac{d^2}{a_4 b^2}, \kappa = -\frac{12bk^2L^2}{5c}$,

substituting these values into Eq. (13), we get the following exponential function solution;

$$u_2(x,t) = \frac{\left(-1 + \frac{2bEe^{2bct-bkx}}{d}\right)a_4}{\left(\frac{d}{b} - Ee^{2bct-2bkx}\right)^2}. \tag{17}$$

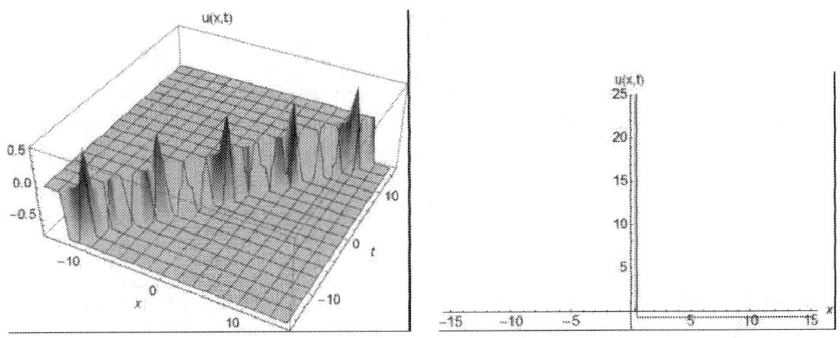

Figure 2. The 3D and 2D shapes of Eq. (17) with the values $k = 6$, $w = 2$, $b = 2$, $E = 8$, $c = 5$, $d = 4$, $a_4 = 3$, $-15 < x < 15$, $-15 < t < 15$ and $t = 0.5$ for the 2D graphic.

4 Conclusions

In this study, with the help of Wolfram Mathematica 9, the Bernoulli sub-equation function method is employed in investigating some new behavior to a nonlinear evolution equation that describe the dynamics of ionic currents along Microtubules. We obtained some new exponential function structure to the model by utilizing the above mentioned method. All the obtained analytical solutions have satisfied the model. When we compare our obtained results with the results in the literature [18, 19] we observed that our result are newly constructed with the different solution structure. Therefore, with our available results, we can say that the Bernoulli sub-equation function method is simple and efficient mathematical tool that can be employed to various mathematical models.

References

[1] A Bekir and F Uygun, Arab Journal of Mathematical Sciences **18**, 73-85 (2012)
[2] H Or-Roshid, MA Akbar, MdF Hoque and N Rahman, SpringerPlus **3(122)**, doi: 10.1186/2193-1801-3-122 (2014)
[3] H Zedan, SJ Monaquel, Applied Mathematics E-Notes **10**, 103-111 (2010)
[4] H Bulut, TA Sulaiman and HM Baskonus, Opt. Quant. Electron **48(564)**, 1-14 (2016)
[5] H Bulut, TA Sulaiman, HM Baskonus and AA Sandulyak, Optik **135**, 327-336 (2017)
[6] AH Khater, W Malfliet, DK Callebaut and ES Kamel, Chaos, Solitons and Fractals **14(3)**, 513-522 (2002)
[7] M Karimi, Mathematics Scientific Journal **9(1)**, 47-54 (2013)
[8] EME Zayed and HMA Rahman, Applied Mathematics E-Notes **10**, 235-245 (2010)
[9] Z Jin-Ming and Z Yao-Ming, Chin. Phys. B **20(1)**, 010205 (2011)
[10] M Mirzazadeh, Information Sciences Letters **3(1)**, 1-9 (2014)
[11] L Wazzan, Communications in Nonlinear Science and Numerical Simulation **14(6)**, 2642-2652 (2009)
[12] X. Yang, Z Deng and Y Wei, Advances Differential Equations **117**, DOI 10.1186/s13662-015-0452-4 (2015)
[13] HM Baskonus and M Askin, 6th International Youth Science Forum "LITTERIS ET ARTIBUS"' Computer Science and Engineering, Lviv, Ukraine, 24-26 November, (2016)
[14] MF El-Sabbagh, R Zait and RM Abdelazeem, Journal of Mathematics **10(4)**, 61-68 (2014)
[15] IE Inan and D Kaya, Physica A **381**, 104-115 (2007)
[16] M Akbari, Computational Methods for Differential Equations **2(1)**, 50-55 (2014)
[17] HM Baskonus, H Bulut, and M Kayhan, International Conference on Pure and Applied Mathematics (ICPAM-2015) Van/ Turkey,- 25-28 August (2015)
[18] MdN Alam and MdN Alam, Journal of Taibah University for Science (2017)
[19] E Tala-Tebue, ZI Djoufack, DC Tsobgni-Fozap, A Kenfack-Jiotsa, F Kapche-Tagne and TC Kofane, Chinese Journal of Physics doi: 10.1016/j.cjph.2017.03.004 (2017)
[20] B Zheng, WSEAS Transactions on Mathematics **7(11)**, 618-626 (2012)

Some Prototype Results of the Symmetric Regularized Long Wave Equation Arising in Nonlinear Ion Acoustic Waves

Erdem IŞIK[1], *Hasan BULUT*[2], Sibel Şehriban ATAŞ[2,*]

[1] Tunceli Vocational School, Munzur University, Tunceli, Turkey
[2] Faculty of Science, Department of Mathematics, Firat University, Elazig, Turkey

Abstract. In this manuscript, we consider the Bernoulli sub-equation function method for obtaining new exponential prototype structures to th Symmetric Regularized Long Wave mathematical model. We obtain new results by using this technique. We plot two- and three-dimensional surfaces of the results by using Wolfram Mathematica 9. At the end of this manuscript, we submit a conclusion in the comprehensive manner.

1. Introduction

In the recent years, the investigation of exact travelling wave solutions to nonlinear partial differential equations plays an important role in the study of nonlinear modelling physical phenomena. Also the study of the travelling wave solutions plays an important role in nonlinear science. A variety of powerful methods has been presented, such as Hirota's bilinear method[1]. the inverse scattering transform [2], sine-cosine method [3], homotopy analysis method [4,5], variational iteration method [6-8]. We study different aspects of the Symmetric Regularized-Long-Wave equation (SRLW). The Symmetric Regularized-Long-Wave equation is a model for the weakly nonlinear ion acoustic and space-charge waves[9]. This equation was introduced by Seyler and Fenstermacher in[10]. Moreover, some important results have been submitted to the literature by various scientists [11,21]

The paper is organized as follows: In section 2, we describe the Bernoulli Sub-Equation function method (BSEFM). In section 3, we consider the following Symmetric

* Corresponding author: suzundag90@gmail.com

Regularized Long Wave equation [22] defined by

$$u_{tt} + u_{xx} + uu_{xt} + u_x u_t + u_{xxtt} = 0. \tag{1}$$

Moreover, in section 4 we give the physical interpretations and remarks of the solutions obtained by BSEFM. Also a comprehensive conclusion is given in section 5.

2. Fundamental Properties of the Method

An approach to the mathematical models including partial differential equations will be presented in this sub-section of paper. The steps of the this technique which is partially new modified can be taken as follows [10].

Step 1. We consider the following nonlinear partial differential equation (NLPDE) in two variables and a dependent variable u ;

$$P(u_x, u_t, u_{xt}, u_{xx}, \cdots) = 0, \tag{2}$$

and take the travelling wave transformation

$$u(x,t) = U(\eta),\ \eta = x - ct, \tag{3}$$

where $c \neq 0$. Substituting Eq.(3) in Eq.(2) yields a nonlinear ordinary differential equation (NLODE) as following;

$$N(U, U', U'', \cdots) = 0, \tag{4}$$

where $U = U(\eta), U' = \dfrac{dU}{d\eta}, U'' = \dfrac{d^2U}{d\eta^2}, \cdots.$

Step 2. Take trial equation of solution for Eq.(4) as following:

$$U(\eta) = \sum_{i=0}^{n} a_i F^i = a_0 + a_1 F + a_2 F^2 + \cdots + a_n F^n, \tag{5}$$

and

$$F' = bF + dF^M,\ b \neq 0,\ d \neq 0,\ M \in \mathbb{R} - \{0,1,2\}, \tag{6}$$

where $F(\eta)$ is Bernoulli differential polynomial. Substituting Eq.(5) along with Eq.(6) in Eq.(4), it yields an equation of polynomial $\Omega(F(\eta))$ of $F(\eta)$ as following;

$$\Omega(F(\eta)) = \rho_s F(\eta)^s + \cdots + \rho_1 F(\eta) + \rho_0 = 0. \tag{7}$$

According to the balance principle, we can obtain a relationship between n and M .

Step 3. Let the coefficients of $\Omega(F(\eta))$ all be zero will yield an algebraic equations system;

$$\rho_i = 0,\ i = 0, \cdots, s. \tag{8}$$

Solving this system, we will specify the values of $a_0, \cdots, a_n$.

Step 4. When we solve Eq.(6), we obtain the following two situation according to b and d ;

$$F(\eta)=\left[\frac{-d}{b}+\frac{E}{e^{b(M-1)\eta}}\right]^{\frac{1}{1-M}},\ b\neq d, \tag{9a}$$

$$F(\eta)=\left[\frac{(E-1)+(E+1)\tanh\left(\frac{b(1-M)\eta}{2}\right)}{1-\tanh\left(\frac{b(1-M)\eta}{2}\right)}\right]^{\frac{1}{1-M}},\ b=d,\quad E\in\mathbb{R}. \tag{9b}$$

Using a complete discrimination system for polynomial, we obtain the solutions to Eq.(4) with the help of Wolfram Mathematica 9 programming and classify the exact solutions to Eq.(4). For a better understanding of results obtained in this way, we can plot two- and three-dimensional surfaces of solutions by taking into consideration suitable values of parameters.

3. Application

When we consider the travelling wave transformation and perform the transformation $u(x,t)=U(\eta),\eta=kx-ct$ which c is constant non-zero in Eq.(1), we obtain NLODE as following;

$$2\left(k^2+c^2\right)U-ckU^2+c^2k^2U''=0. \tag{10}$$

When we consider to Eq.(5) and Eq.(6) for balance principle between U'' and U^2, we obtain the following relationship between n and M;

$$2M=n+2. \tag{11}$$

This relationship gives us some new different complex travelling wave solutions for Eq.(1).

Case 1: If we take as $n=4$ and $M=3$ in Eq. (11), we can write following equations;

$$U=a_0+a_1F+a_2F^2+a_3F^3+a_4F^4, \tag{12}$$

$$\begin{aligned}U'&=a_1bF+a_1dF^3+2a_2bF^2+2a_2dF^4+3a_2bF^3+\\&+3a_3dF^5+4a_4bF^4+4a_4dF^6,\end{aligned} \tag{13}$$

and

$$\begin{aligned}U''&=b^2Fa_1+4bdF^3a_1+3d^2F^5a_1+4b^2F^2a_2+12bdF^4a_2+\\&+8d^2F^6a_2+9b^2F^3a_3+24bdF^5a_3+15d^2F^7a_3+16b^2F^4a_4+\\&+40bdF^6a_4+24d^2F^8a_4,\\&\vdots\end{aligned} \tag{14}$$

where $a_2\neq 0,\ b\neq 0,\ d\neq 0$. When we put Eq.(12-14) in Eq.(10), we obtain a system of algebraic equations for Eq.(13). Therefore, we obtain a system of algebraic equations from these coefficients of polynomial of F . Solving this system with the help of wolfram Mathematica 9, we find the following coefficients and solutions;

Case 1a. For $b \neq d$, it can be considered the following coefficients;

$$a_1 = 0, a_2 = \frac{24\sqrt{2}dka_0}{\sqrt{a_0(a_0 - \sqrt{-16 + a_0^2})}},\ a_3 = 0,$$
$$a_4 = 12d^2k^2\left(a_0 + \sqrt{-16 + a_0^2}\right), \tag{15}$$
$$c = \frac{1}{4}k(a_0 + \sqrt{-16 + a_0^2}), b = \frac{\sqrt{a_0\left(a_0 - \sqrt{-16 + a_0^2}\right)}}{4\sqrt{2}k}.$$

Substituting these coefficients in Eq.(12) along with Eq.(9a), we obtain the following new exponential function solution for Eq.(1);

$$u_1(x,t) = a_0 + \frac{24a_0d\sqrt{2}e^{\frac{\rho\left(-4x+t\left(a_0+\sqrt{-16+a_0^2}\right)\right)}{8\sqrt{2}}}Ek\rho}{\left(-4\sqrt{2}dk + e^{\varpi\left(-4x+t\left(a_0+\sqrt{-16+a_0^2}\right)\right)}E\rho\right)^2}, \tag{16}$$

where a_0, d, k are real constants and $\rho = \sqrt{a_0\left(a_0 - \sqrt{-16 + a_0^2}\right)}$, $\varpi = \frac{\rho}{8\sqrt{2}}$.

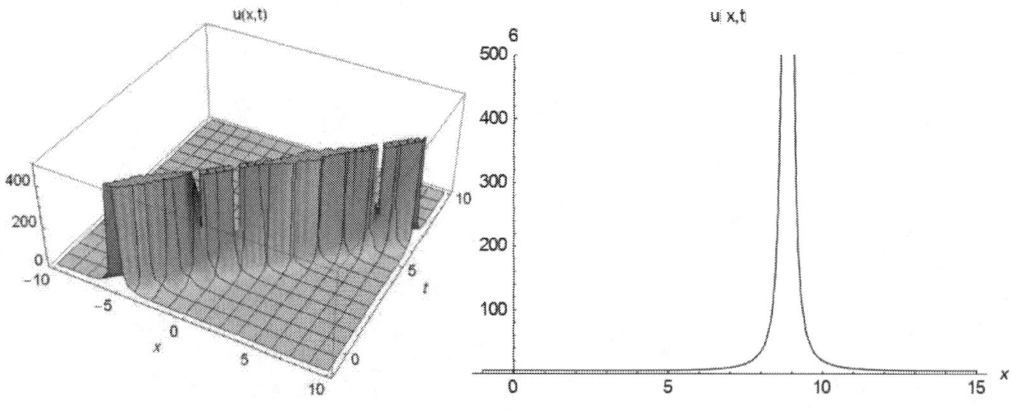

Figure 1. The 2D and 3D surfaces of Eq.(16) under the values of $a_0 = 5, k = 0.3, d = 50, E = 0.8,\ -10 < x < 10, -1 < t < 10$ and $t = 6$ for 2D surfaces.

Case 1b. Another coefficients for Eq.(1) and for $b \neq d$, it can be considered follows;

$$a_0 = \frac{2c}{k} + \frac{2k}{c}, a_1 = 0,\ a_2 = -24d\sqrt{c^2 + k^2}, a_3 = 0,$$
$$a_4 = 48cd^2k, b = -\frac{\sqrt{c^2 + k^2}}{2ck}. \tag{17}$$

When we regulate Eq.(12) under the terms of Eq.(9a,17), we find the other new exponential function solution for Eq.(1);

$$u_2(x,t) = \frac{2c}{k} + \frac{2k}{c} + \frac{48cd^2k}{\left(e^{\upsilon(-ct+kx)}E + \frac{2d}{\upsilon}\right)^2} - \frac{24d\upsilon ck}{2cdk + \upsilon cke^{\upsilon(-ct+kx)}E}, \tag{18}$$

where c, d, k are real constants and $\upsilon = \frac{\sqrt{c^2 + k^2}}{ck}$.

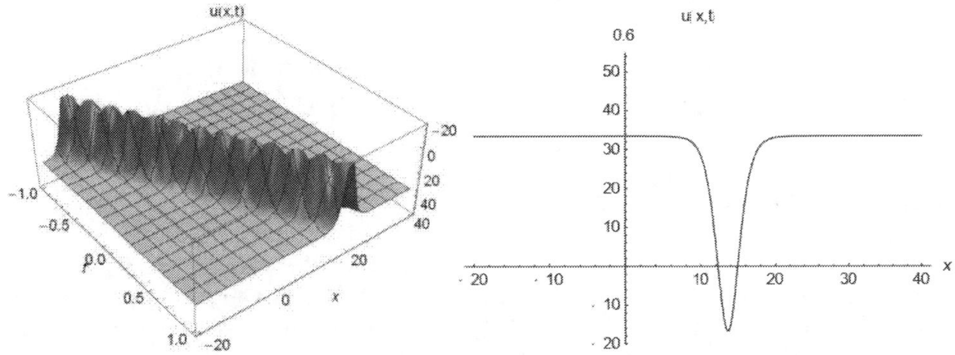

Figure 2. The 2D and 3D surfaces of Eq.(18) under the values of $c = 5, k = 0.3, d = 50, E = 0.8,\ -20 < x < 40, -1 < t < 1$, $t = 0.6$ for 2D surfaces.

Case 1c. Another coefficients to the Eq.(1) for $b \neq d$, it can be considered as follows;

$$a_0 = \frac{8b^2c^2}{\sqrt{-1 + 4b^2c^2}}, a_1 = 0,\ a_2 = \frac{48bc^2d}{\sqrt{-1 + 4b^2c^2}}, a_3 = 0,$$
$$a_4 = \frac{48c^2d^2}{\sqrt{-1 + 4b^2c^2}}, k = \frac{c}{\sqrt{-1 + 4b^2c^2}}. \tag{19}$$

Considering these coefficients in Eq.(12) along with Eq.(9a,19), we gain other new complex exponential function solution to Eq.(1);

$$u_3(x,t)=\frac{8b^2c^2\left(d^2e^{4bc(-t+\kappa x)}+4bde^{2bc(-t+\kappa x)}E+b^2E\right)}{\sqrt{-1+4b^2c^2}\left(de^{2bc(-t+\kappa x)}-bE\right)^2}, \tag{20}$$

where b,c,d are real constants and $\kappa=\dfrac{1}{\sqrt{-1+4b^2c^2}}$.

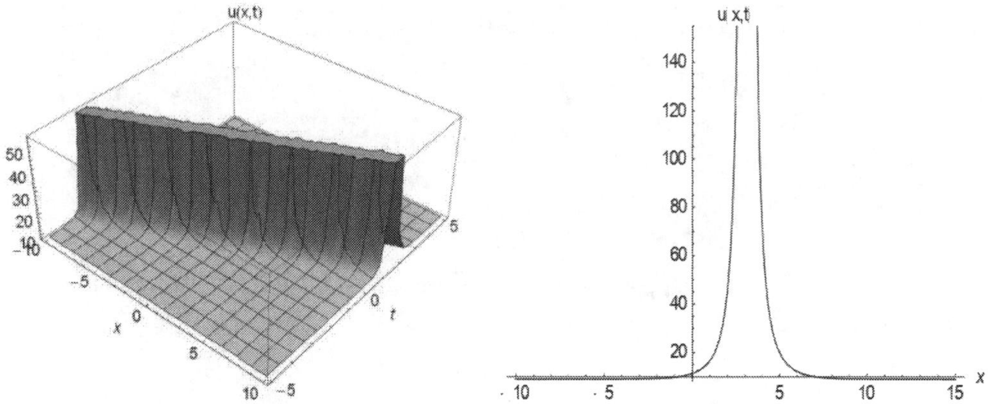

Figure 3. The 2D and 3D surfaces of real part of Eq.(20) under the values of $b=2, d=0.1, c=1.1, E=0.8, -10<x<10, -5<t<5$, $t=0.1$ for 2D surfaces.

4. DISCUSSION AND CONCLUSIONS

The solutions such as Eq.(16,18,20) obtained by using BSEFM are the new exponential function structures solutions for Eq.(1) when we compare with the paper submitted to the literature by [22]. It has been observed that all solutions have verified the Eq. (1) by using Wolfram Mathematica 9. To the best of our knowledge, the application of BSEFM to the Eq.(1) has not been previously submitted to the literature. The method proposed in this paper can be used to seek more travelling wave solutions of NLEEs because the method has some advantages such as easily calculations, writing programme for obtaining coefficients and so many. Eq.(18) can be rewritten as hyperbolic functions by using the fundamental properties of hyperbolic functions as follows;

$$u_1(x,t)=\frac{2c}{k}+\frac{2k}{c}+\frac{48cd^2k}{\left(Ee^{\upsilon(-ct+kx)}+\dfrac{2d}{\upsilon}\right)^2}-\frac{24d\left(c^2+k^2\right)}{2cdk+E\upsilon kce^{\upsilon(-ct+kx)}},$$

where c,d,k are real constants and $\upsilon=\dfrac{\sqrt{c^2+k^2}}{ck}$. When we consider the general properties of hyperbolic and exponential functions, we can rewrite $u_1(x,t)$ solution in the form of hyperbolic function as following.

$$u_1(x,t)=\frac{2c}{k}+\frac{2k}{c}+\frac{48cd^2k}{\left(\dfrac{E\pm E\sqrt{1-\sec h^2(T)}}{\sec h(T)}+\delta\right)^2}-\frac{24d\left(c^2+k^2\right)}{2cdk+\dfrac{\rho\pm\rho\sqrt{1-\sec h^2(T)}}{\sec h(T)}},$$

$$T=\frac{(-ct+kx)}{ck}\sqrt{c^2+k^2},\delta=\frac{2cdk}{\sqrt{c^2+k^2}},\rho=E\sqrt{c^2+k^2}.$$

References

1. R. Hirota, The Direct Method in Soliton Theory. *Cambridge Univ. Press*, (2004).
2. M.J. Ablowitz, P.A. Clarkson, Solitons, nonlinear evolution equations and inverse scattering, *Cambridge: Cambridge University Press*, (1991).
3. A. M. Wazwaz, Travelling wave solutions for combined and double combined sine-cosineGordon equations by the variable separated ODE method. *Appl. Math. Comput*, **177**, 755-760 (2006).
4. M. Dehghan, J. Manafian, A. Saadatmandi, The solution of the linear fractional partial differential equations using the homotopy analysis method. *Z. Naturforsch*, **65a**, 935949 (2010).
5. M. Dehghan, J. Manafian, A. Saadatmandi, Solving nonlinear fractional partial differential equations using the homotopy analysis method. *Num. Meth. Partial Differential Eq.*, **26**, 448-479 (2010).
6. J. H. He, Variational iteration method a kind of non-linear analytical technique: some examples. *Int. J. Nonlinear Mech*, **34**, 699-708 (1999).
7. M. Dehghan, M. Tatari, Identifying an unknown function in a parabolic equation with overspecified data via He's variational iteration method. *Chaos Solitons Fractals*, **36,** 157-166 (2008).
8. M. Dehghan, J. Manafian, A. Saadatmandi, Application of semi-analytic methods for the Fitzhugh-Nagumo equation, which models the transmission of nerve impulses, *Math. Meth. Appl. Sci*, **33**, 1384-1398 (2010).
9. Carlos Banquet Brango, The Symmetric Regularized-Long-Wave equation: Well-posedness and nonlinear stability. *Physica*, **D(241)**, 125-133 (2012).
10. C. Seyler, D. Fenstermacher, A symmetric regularized-long-wave equation. *Phys. Fluids* **27**, 4–7 (1984).

11. G. K. Watugala, Sumudu Transform: A New Integral Transform to Solve Differantial Equations and Control Engineering Problems. *International Journal of Mathematical Education in Science and Technology,* **24**, 35-43 (1993).
12. Y. Pandir, New exact solutions of the generalized Zakharov–Kuznetsov modified equal-width equation. *Pramana journal of physics*, **82**(6), 949–964 (2014).
13. H. Bulut, H. M. Baskonus and F. B. M. Belgacem, The Analytical Solutions of Some Fractional Ordinary Differential Equations by Sumudu Transform Method. *Abstract and Applied Analysis,*(2013).
14. A. M. Wazwaz, The tanh method: solitons and periodic solutions for Dodd-Bullough-Mikhailov and Tzitzeica- Dodd-Bullough equations. *Chaos, Solitons and Fractals*, **25**, 55-56 (2005).
15. C.S. Liu, A new trial equation method and its applications. *Communications in Theoretical Physics* , **45**(3), 395-397 (2006).
16. C.S. Liu, Trial Equation Method to Nonlinear Evolution Equations with Rank Inhomogeneous: Mathematical Discussions and Its Applications. *Communications in Theoretical Physics*, **45**(2), 219–223.
17. H. Bulut, Y. Pandir, H. M. Baskonus, Symmetrical Hyperbolic Fibonacci Function Solutions of Generalized Fisher Equation with Fractional Order. *AIP Conf.*, **1558**, 1914 (2013).
18. Y. Pandir, Y. Gurefe, U, Kadak, and E. Misirli, Classifications of exact solutions for some nonlinear partial differential equations with generalized evolution. *Abstract and Applied Analysis*, **2012**, 16 pages, (2012).
19. Ryabov, P. N., Sinelshchikov, D. I., and Kochanov, M. B., Application of the Kudryashov method for finding exact solutions of the high order nonlinear evolution equations. *Applied Mathematics and Computation*, **218**(7), 3965–3972 (2011).
20. Kudryashov, N. A., One method for finding exact solutions of nonlinear differential equations. *Communications in Nonlinear Science and Numerical Simulation,* **17**(6), 2248–2253 (2012).
21. Lee J., and Sakthivel, R., Exact travelling wave solutions for some important nonlinear physical models. *Pramana—Journal of Physics*, **80**(5), 757–769 (2013).
22. H. Bulut, H.M.Baskonus, E. Cuvelek, On The Prototype Solutions of Symmetric Regularized Long Wave Equation by Generalized Kudryashov Method. *Mathematics letters*, **1**(2), 10-16 (2015).

ITM Web of Conferences **13**, 01017 (2017) DOI: 10.1051/itmconf/20171301017
CMES2017

Some Novel Exponential and Complex Structural Properties of the Fisher Equation Arising in Mathematical Bioscience

Hasan Bulut[1,⋆], *Sibel Sehriban* Atas[1,⋆⋆], and *Haci Mehmet* Baskonus[2,⋆⋆⋆]

[1] *Faculty of Science, Firat University, Elazig, Turkey*
[2] *Faculty of Engineering, Munzur University, Tunceli, Turkey*

Abstract. In this manuscript, we consider the Bernoulli sub-equation function method for obtaining new exponential and complex prototype structures to the Fisher Equation arising in Mathematical biosciences. We obtain new results by using the technique for new properties of model and for more understanding of properties of model. We plot two- and three-dimensional surfaces of the results by using Wolfram Mathematica 9. At the end of this manuscript, we submit a conclusion in the comprehensive manner.

1 Introduction

The studies of the solutions to the nonlinear partial differential equation have become of signifiance important in the fields of science, engineering, mathematical physics and biosciences, such as chemistry, plasma physics, fluid mechanics, etc.

Many studies have been presented to the literature in terms of Mathematical modal of humankind problem. For example, E.C. Alvord et al have investigated the density or concentration of cancer cells [1, 2]. M. Beren et al one characteristic of glioblastoma cells which has gained considerable attention is the 'go or grow'- hypotesis, which states that proliferation and migration are mutually exclusive phenotypes of glioblastoma cells [3]. Many different models of glibastoma growth have been proposed by D.Basanta et al have searched for game theoretical models [4], and K.R.Swanson et al sytems of partial differantial equaitons [5], E. Khain et al to individual-based models [6]. The effects of the density driven swithcing have observed by Pham et al [7].

In recent years, some researchers have focused on individual-based model. For example, Nelander et al have concerned with the analysis of an individual-based model [8]. Moreover, some important results have been submitted to the literature by various scientists [9, 10].

The paper is organized as follows: In section 2, we describe the Bernoulli Sub-Equation function method (BSEFM). In section 3, we consider the following Fisher equaiton defined by [11]

$$\omega u_{xx} - u_t + \rho u - \rho u^2 = 0, \tag{1}$$

Moreover, in section 4 we give the physical interpretations and remarks of the solutions obtained by BSEFM. Also a comprehensive conclusion is given in section 5.

⋆e-mail: hbulut@firat.edu.tr
⋆⋆e-mail: suzundag90@gmail.com
⋆⋆⋆e-mail: hmbaskonus@gmail.com

2 Fundamental Properties of Bernoulli Sub-Equation Function Method

An approach to the mathematical models including partial differential equations will be presented in this sub-section of paper. The steps of the this technique which is partially new modified can be taken as follows [12].

Step-1. We consider the partial differential equation in two variables such as x, t and a dependent variable u;

$$P(u_x, u_t, u_{xx}, u_{xt}, \ldots), \tag{2}$$

and take the wave transformation

$$u(x,t) = U(\eta), \ \eta = x - ct, \tag{3}$$

where $c \neq 0$. Substituting Eq. (3) into Eq. (2), it gives us the following nonlinear ordinary differential equation;

$$N(U, U', U'', U''', \ldots). \tag{4}$$

where $U = U(\eta), U' = \frac{dU}{d\eta}, U'' = \frac{d^2U}{d\eta^2}, \ldots$

Step-2. Taking the trial solution for Eq. (4) as:

$$U(\eta) = \sum_{i=0}^{n} a_i F^i(\eta) = a_0 + a_1 F(\eta) + a_2 F^2(\eta) + \ldots + a_n F^n(\eta), \tag{5}$$

where

$$F' = bF(\eta) + dF^M(\eta), \tag{6}$$

where $b \neq 0, \ d \neq 0, \ M \in \mathbb{R} \setminus \{0, 1, 2\}$, and $F(\eta)$ is Bernoulli differential polynomial. Substituting above relations into Eq. (4), we obtain an equation of polynomial $\Omega(F(\eta))$ of $F(\eta)$;

$$\Omega(F(\eta)) = \rho_s F^s(\eta) + \ldots + \rho_1 F(\eta) + \rho_0 = 0. \tag{7}$$

According to the balance principle, we can get values of and M and n.

Step-3. Let us consider the coefficients of $\Omega(F(\eta))$ all be zero, we will obtain an algebraic equations system:

$$\rho_i = 0, \ \ i = 0, \ldots, s. \tag{8}$$

Solving this system, we will determine the values of $a_0, \ldots a_n$.

Step-4. When we solve nonlinear Bernoulli differential equation that is Eq. (6), we obtain the following two situations according to b and d;

$$F(\eta) = \left[\frac{-d}{b} + \frac{E}{e^{b(M-1)\eta}}\right]^{\frac{1}{1-M}}, \quad b \neq d \tag{9}$$

$$F(\eta) = \left[\frac{(E-1) + (E+1)tanh\left(\frac{b(1-M)\eta}{2}\right)}{1 - tanh\left(\frac{b(1-M)\eta}{2}\right)}\right]^{\frac{1}{1-M}}, \quad b = d, E \in \mathbb{R}. \tag{10}$$

Using a complete discrimination system for polynomial, we obtain the solutions to Eq.(4) with the help of Wolfram Mathematica 9 programming and classify the exact solutions to Eq.(4). For a better understanding of results obtained in this way, we can plot two- and three-imensional surfaces of solutions by taking into consideration suitable values of parameters.

3 Application

In this section, we have successfully applied the BSEFM to the Fisher equation for getting some new complex travelling wave solutions.

When we consider the travelling wave transformation and perform the transformation $u = U(\eta), \eta = x - ct$, where is constant non-zero in Eq. Eq. (1), we obtain NLODE as following;

$$\omega U' + cU' + \rho U - \rho U^2 = 0. \tag{11}$$

Applying the homogeneous balancing technique on Eq. (11) by using the highest derivative U'' and the highest power nonlinear term U^2, yields the following relation between n and M;

$$2M = n + 2 \quad n, M \in \mathbb{Z}^+. \tag{12}$$

If we take as $n = 4$ and $M = 3$ in Eq. (5), we can write following equation;

$$U(\eta) = a_0 + a_1 F + a_2 F^2 + a_3 F^3 + a_4 F^4, \tag{13}$$

differentiating Eq. (13) twice, gives the following equations;

$$\begin{aligned} U' = {} & a_1 bF + a_1 dF^3 + 2a_2 bF^2 + 2a_2 dF^4 + 3a_3 F^3 \\ & +3a_3 dF^5 + 4a_4 bF^4 + 4a_4 dF^6, \end{aligned} \tag{14}$$

$$\begin{aligned} U'' = {} & a_1 bF' + a_1 dF^2 F' + 4a_2 bFF' + 16a_4 dF^3 F' + 8a_2 dF^3 F' \\ & +9a_3 bF^2 F' + 15a_3 F^4 F' + 24a_4 dF^5 F', \end{aligned} \tag{15}$$

where $a_2 \neq 0$, $b \neq 0$, $d \neq 0$. When we put Eqs. (12-14) in Eq.(10), we obtain a system of algebraic equations for Eq. (13). Therefore, we obtain a system of algebraic equations from these coefficients of polynomial of . Solving this system with the help of wolfram Mathematica 9, we find the following coefficients and solutions;

Case 1a: For $b \neq d$, it can be considered the following coefficients;

$$a_0 = 1, a_1 = 0, a_2 = 0, a_3 = 0, a_4 = \frac{24d^2\omega}{\rho}, b = -\frac{i\sqrt{\rho}}{2\sqrt{6\omega}}, c = \frac{5i\sqrt{\rho\omega}}{\sqrt{6}}.$$

Substituting these coefficients in Eq. (13) along with Eq. (9), we obtain the following new complex exponential function solution for Eq. (1);

$$u_1(x,t) = 1 + 24d^2\omega\left(E\sqrt{\rho}e^{\frac{5\rho t}{6} + \frac{\rho}{\sqrt{6\omega}}ix} - 2\sqrt{6\omega}\, di\right)^{-2}, \tag{16}$$

where d, ω, ρ are real constants.

DOI: 10.1051/itmconf/20171301017

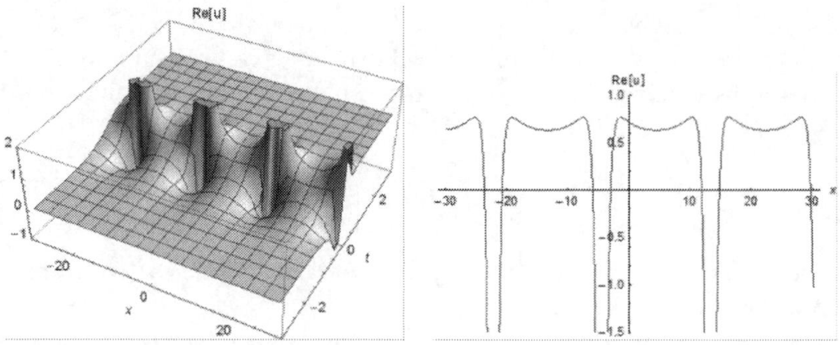

Figure 1. The The 2D and 3D surfaces of real part of Eq. (16) under the values of $\omega = 4$, $\rho = 3$, $E = -0.7$, $d = 0.2$, $-30 < x < 30$, $-3 < t < 3$ and $t = 0.02$ for the 2D surface.

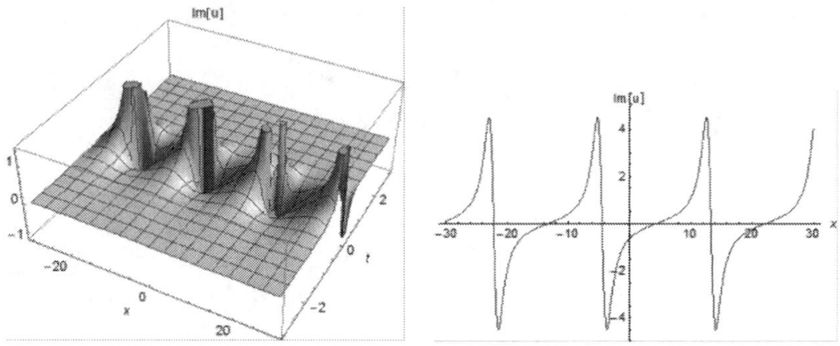

Figure 2. The The 2D and 3D surfaces of imaginary part of Eq. (16) under the values of $\omega = 4$, $\rho = 3$, $E = -0.7$, $d = 0.2$, $-30 < x < 30$, $-3 < t < 3$ and $t = 0.02$ for the 2D surface.

Case 1b: Another coefficients for Eq. (1) and for $b \neq d$, it can be considered follows;

$$a_0 = 1, a_1 = 0, a_2 = \frac{4d\sqrt{6\omega}}{\sqrt{\rho}}, a_3 = 0, a_4 = \frac{24d^2\omega}{\rho}, b = \frac{\sqrt{\rho}}{2\sqrt{6\omega}}, c = \frac{5\sqrt{\rho\omega}}{\sqrt{6}}.$$

Substituting these coefficients in Eq. (13) along with Eq. (9), we obtain the new exponential function solution for Eq. (1);

$$u_2(x,t) = E^2\sqrt{\rho}\left(E\sqrt{\rho} - 2\sqrt{6\omega}\, de^{-\frac{5\rho}{6}t + \frac{\sqrt{\rho}}{\sqrt{6\omega}}x}\right)^{-2}, \tag{17}$$

where d, ω, ρ are real constants.

Case 1c: Another coefficients for Eq. (1) and for $b \neq d$, it can be considered follows;

$$a_0 = 0, a_1 = 0, a_2 = \frac{4di\sqrt{6\omega}}{\sqrt{\rho}}, a_3 = 0, a_4 = \frac{24d^2\omega}{\rho}, b = \frac{i\sqrt{\rho}}{2\sqrt{6\omega}}, c = \frac{5i\sqrt{\rho\omega}}{\sqrt{6}}.$$

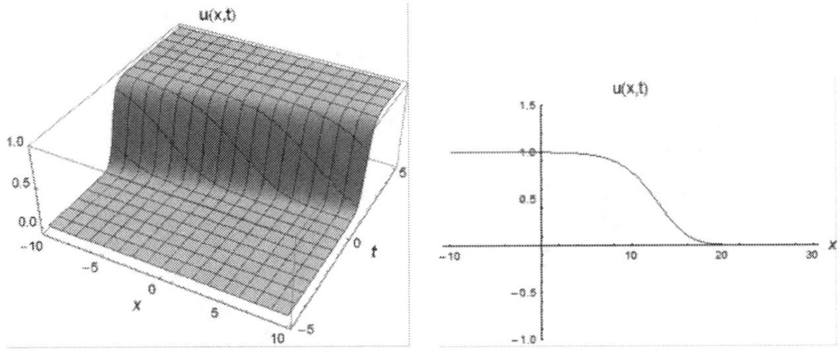

Figure 3. The The 2D and 3D surfaces of Eq. (17) under the values of $\omega = 4, \rho = 5, E = -0.7, d = 0.9$, $-10 < x < 10, -5 < t < 5$ and $t = 2$ for the 2D surface.

Substituting these coefficients in Eq. (13) along with Eq. (9), we obtain the following new complex exponential function solution for Eq. (1);

$$u_3(x,t) = 1 - E^2\sqrt{\rho}\left(E\sqrt{\rho} - 2i\sqrt{6\omega}\, de^{\frac{5\rho}{6}t + \frac{i\sqrt{\rho}}{\sqrt{6\omega}}x}\right)^{-2}, \tag{18}$$

where d, ω, ρ are real constants.

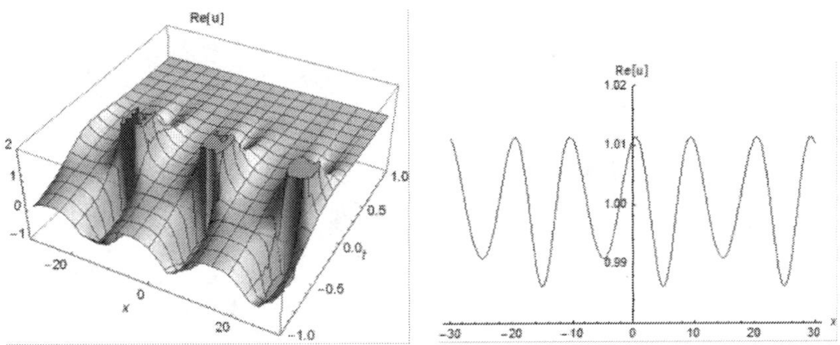

Figure 4. The The 2D and 3D surfaces of real part of Eq. (18) under the values of $\omega = 5, \rho = 3, E = 27, d = 10$, $-30 < x < 30, -1 < t < 1$ and $t = 0.02$ for the 2D surface.

Case 1d: Another coefficients for Eq. (1) and for $b \neq d$, it can be considered follows;

$$a_0 = 0, a_1 = 0, a_2 = -\frac{4di\sqrt{6\omega}}{\sqrt{\rho}}, a_3 = 0, a_4 = \frac{24d^2\omega}{\rho}, b = -\frac{i\sqrt{\rho}}{2\sqrt{6\omega}}, c = -\frac{5i\sqrt{\rho\omega}}{\sqrt{6}}.$$

Substituting these coefficients in Eq. (13) along with Eq. (9), we obtain the following new complex exponential function solution for Eq. (1);

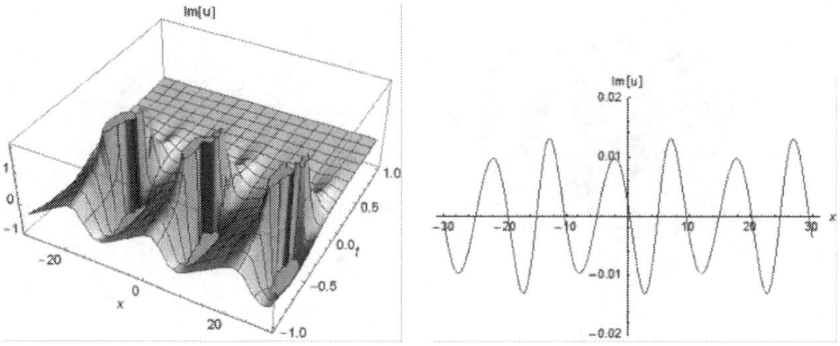

Figure 5. The The 2D and 3D surfaces of imaginary part of Eq. (18) under the values of $\omega = 5, \rho = 3, E = 27$, $d = 10, -30 < x < 30, -1 < t < 1$ and $t = 0.02$ for the 2D surface.

$$u_4(x,t) = \frac{4E\sqrt{6\rho\omega}die^{-\frac{5\rho}{6}t+\frac{i\sqrt{\rho}}{\sqrt{6\omega}}x} + 24d^2\omega}{\left(E\sqrt{\rho}ie^{-\frac{5\rho}{6}t+\frac{i\sqrt{\rho}}{\sqrt{6\omega}}} - 2\sqrt{6\omega}d\right)^2}, \tag{19}$$

where d, ω, ρ are real constants.

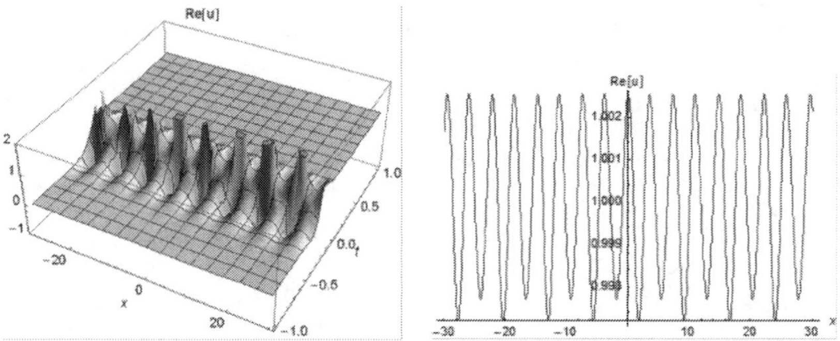

Figure 6. The The 2D and 3D surfaces of real part of Eq. (19) under the values of $\omega = 3, \rho = 13, E = 7, d = 20$, $-30 < x < 30, -1 < t < 1$ and $t = 0.1$ for the 2D surface.

DOI: 10.1051/itmconf/20171301017

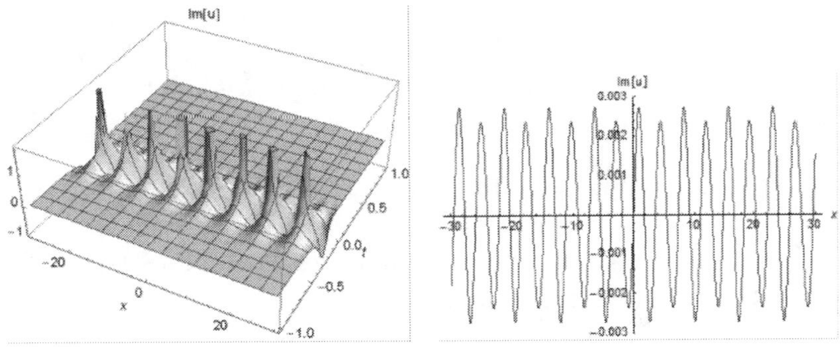

Figure 7. The The 2D and 3D surfaces of imaginary part of Eq. (19) under the values of $\omega = 3, \rho = 13, E = 7, d = 20, -30 < x < 30, -1 < t < 1$ and $t = 0.1$ for the 2D surface.

4 Conclusions

The solutions such as Eq. (16, 17, 18, 19) obtained by using BSEFM are the new complex and exponential function structures solutions for Eq. (1) when we compare with the paper submitted to the literature by Nelander et al [11]. It has been observed that all solutions have verified the Eq. (1) by using Wolfram Mathematica 9. To the best of our knowledge, the application of BSEFM to the Eq. (1) has not been previously submitted to the literature. The method proposed in this paper can be used to seek more travelling wave solutions of NLEEs because the method has some advantages such as easily calculations, writing programme for obtaining coefficients and so many. Eq (17) can be re-rewritten as hyperbolic functions by using the fundamental properties of hyperbolic functions as follows;

$$u_2(x,t) = E^2 \sqrt{\rho}\left(E\sqrt{\rho} - 2\sqrt{6\omega}\, d e^{-\frac{5\rho}{6}t + \frac{\sqrt{\rho}}{\sqrt{6\omega}}x}\right)^{-2}, \tag{20}$$

where d, ω, ρ are real constants. When we consider the general properties of hyperbolic and exponential functions, we can re-write $u_2(x,t)$ solution as follows;

$$u_2(x,t) = E^2 \sqrt{\rho}\left(E\sqrt{\rho} - 2\sqrt{6\omega}\, d\left(\frac{1 \pm \sqrt{1 - sech^2(T)}}{sech(T)}\right)\right)^{-2}, \tag{21}$$

where $T = -\frac{5\rho}{6}t + \frac{\sqrt{\rho}}{\sqrt{6\omega}}x$.

These travelling wave solutions have been introduced to the literature some important physical information about Eq. (1). As mentioned in section 1, it is estimated that this solution is related to physical features of hyperbolic function [13, 14].

When we compare these structures found in this paper with the paper which has been published by P. Gerlee and S. Nelander [11], we can say that the complex solutions is new results for the considered model. We think that they will give more properties of model considered. Therefore, the method can be applied to the another models in terms of giving new features of models.

References

[1] A Giese, R Bjerkving and M Westphal, J. Clin. Oncol **21(8)**, 1624 (2003)
[2] A Farin, SO Suzuki, M Weiker, JE Goldman, JN Bruce and P Canoll, A dynamic analysis, Glia **53(8)**, 799-808 (2006)
[3] H Hatzikirou, D Basanta, M Basanta, M Simon, K Schaller and A Deutsch, Math, Med, Biol **29**, 49-65 (2012)
[4] D Basanta, S JG, R Rockne, K Swanson and A Alexander, Phys. Biol **8**, 015016 (2011)
[5] KR Swanson, RC Rockne, J Claridge, MA Chaplain, EC Alvord and ARA Anderson, Cancer Res. **71(24)**, 7366-7375 (2011)
[6] E Khain, M Katakowski, N Charteries, F Jiang and M. Chopp, Phys. Rev. E **86(1)**, 011904 (2012)
[7] K Pham, A Chauviere, H Hatzikirou, X Li, HM Byrne, V Cristini and J Lowengrub, J. Biol, Dynam **6**, 1-18 (2011)
[8] P Gerlee and S Nelander, Plos Comput, Biol **8**, e1002556 (2012)
[9] HM Baskonus, H Bulut, FM Belgacem, J. Comput. Appl. Math **312(1)**, 257-266 (2017)
[10] FM Belgacem, H Bulut, HM Baskonus, Adv. Math. Stat. Sci, 445-453 (2015)
[11] P Gerlee and S Nelander, Mathematical Biosciences **276**, 75-81 (2016)
[12] B Zheng, WSEAS Transactions on Mathematics **7(11)**, 618-626 (2012)
[13] EW Weisstein, Concise Encyclopedia of Mathematics $\mathbf{2}^{nd}$ edition, Newyork: CRC (2002)
[14] WH Beyer, CRC Standard Mathematical Tables $\mathbf{28}^{th}$ edition, Boca Raton, FL: CRC (1987)

Regarding on The Novel Forms of the (3+1) - Dimensional Kadomstev-Petviashvili Equation

Hasan Bulut[1,*], *Betul* Demirdag[2], *Haci Mehmet* Baskonus[3]

[1,2] Department of Mathematics, University of Firat, Elazig, Turkey

[3] Department of Computer Engineering, University of Munzur, Tunceli, Turkey

hbulut@firat.edu.tr, betuldenizd@gmail.com, hmbaskonus@gmail.com

Abstract This article, we have applied the Bernoulli Sub-Equation method to the (3+1)-Dimensional Kademstev-Petviashvili equation. We have obtained some new analytical solutions such as exponential function and rational solutions by using this technique. We have observed that two analytical solutions have been verified the (3+1)-Dimentional Kadomstev-Petviashvili equations by using Wolfram Mathematica 9. At the end of this manuscript, we submitted a conclusion in a comprehensive manner.

Keywords: Bernoulli Sub-Equation function method; (3+1)-Dimensional Kadomstev-Petviashvili equation; Exponential function solution; rational function solution

1 Introduction

With the aid of powerful devices including computer technology and computational in nonlinear sciences, linear and nonlinear equations have been become an important area among scientists and engineers for obtaining various solutions. For example, a solitary wave has been firstly investigated by the Scottish engineer John Russel [1]. He has followed a water wave travelling through a canal. The nonlinear partial differential equation systems give us the better physical explanations of significant mathematical models [2]. Many powerful methods have been developed to find the exact solutions to the NLEEs. Some of these methods such as (G'/G)-expansion method [3,4], the improved (G'/G)-expansion method, the Sumudu transform method [5-6] have been submitted to the literature.

In the rest of this work, we apply the Bernoulli-Sub equation function method (BSFM) to the (3+1)-dimensional Kadomstev–Petviashvili equation [7];

$$u_{xt}+6u_x^2+6uu_{xx}-u_{xxxx}-u_{yy}-u_{zz}=0. \tag{1}$$

* Corresponding author: hbulut@firat.edu.tr

Xie et al. [8] have investigated the some wave solutions of Eq. (1) by the improved tanh function method. We introduce the general properties of BSEFM in section 2. We apply BSEFM in section 3 for obtaining some exponential and rational function solutions. At the end of this study, we submit a conclusion in the comprehensive manner.

2 Fundamental Properties of Bernoulli Sub-Equation Function Method

Step 1.

We consider the following nonlinear partial differential equation (NLPDE) in two variables and a dependent variable u;

$$P(u_x, u_t, u_{xt}, u_{xx}, \cdots) = 0, \tag{2}$$

and take the travelling wave transformation

$$u(x, y, z, t) = U(\eta),\ \eta = x + y + z - ct, \tag{3}$$

where $c \neq 0$. Substituting Eq.(3) in Eq.(2) yields a nonlinear ordinary differential equation (NLODE) as following;

$$N(U, U', U'', \cdots) = 0, \tag{4}$$

where

$$U = U(\eta), U' = \frac{dU}{d\eta}, U'' = \frac{d^2U}{d\eta^2}, \cdots.$$

Step 2.

Take trial equation of solution for Eq.(4) as following:

$$U(\eta) = \sum_{i=0}^{n} a_i F^i = a_0 + a_1 F + a_2 F^2 + \cdots + a_n F^n, \tag{5}$$

$$F' = bF + dF^M,\ b \neq 0,\ d \neq 0,\ M \in R - \{0, 1, 2\}, \tag{6}$$

where $F(\eta)$ is Bernoulli differential polynomial. Substituting Eq.(5) along with Eq.(6) in Eq.(4), it yields an equation of polynomial $\Omega(F(\eta))$ of $F(\eta)$ as following;

$$\Omega(F(\eta)) = \rho_s F(\eta)^s + \cdots + \rho_1 F(\eta) + \rho_0 = 0. \tag{7}$$

according to the balance principle, we can obtain a relationship between n and M.

Step 3.

Let the coefficients of $\Omega(F(\eta))$ all be zero will yield an algebraic equations system

$$\rho_i = 0,\ i = 0, \cdots, s. \tag{8}$$

solving this system, we will specify the values of $a_0, \cdots, a_n$.

Step 4.

When we solve Eq.(6), we obtain the following two situation according to b and d;

$$F(\eta)=\left[\frac{-d}{b}+\frac{E}{e^{b(M-1)\eta}}\right]^{\frac{1}{1-M}},\ b\neq d,$$
$$F(\eta)=\left[\frac{(E-1)+(E+1)\tanh\left(\frac{b(1-M)\eta}{2}\right)}{1-\tanh\left(\frac{b(1-M)\eta}{2}\right)}\right]^{\frac{1}{1-M}},\ b=d,\ E\in R. \tag{9}$$

Using a complete discrimination system for polynomial, we obtain the solutions to Eq.(4) with the help of Wolfram Mathematica 9 programming and classify the exact solutions to Eq.(4). For a better understanding of results obtained in this way, we can plot two- and three-dimensional surfaces of solutions by taking into consideration suitable values of parameters.

3 Implementation of the BSEFM

In this section, we have successfully considered the BSEFM to the (3+1)-Dimensional Kademstev-Petviashvili equation for getting some new travelling wave solutions.

Application When we consider the travelling wave transformation and perform the transformation $u(x,y,z,t)=U(\eta),\eta=x+y+z-ct$ which c is constant non-zero, we obtain NLODE as following;

integrating twice and setting the constants of integration to zero, we obtain

$$-(c+2)U+3U^2-U''=0. \tag{10}$$

when we consider to Eq.(5) and Eq.(6) for balance principle between U'' and U^2, we obtain the following relationship between n and M;

$$2M=n+2. \tag{11}$$

this relationship gives us some new different analytical solutions for Eq. (1).

Case 1:

If we take as $n=4$ and $M=3$ in Eq. (11), we can write following equations;

$$U=a_0+a_1F+a_2F^2+a_3F^3+a_4F^4, \tag{12}$$

$$U'=a_1bF+a_1dF^3+2a_2bF^2+2a_2dF^4+3a_3bF^3$$
$$+3a_3dF^5+4a_4bF^4+4a_4dF^6, \tag{13}$$

$$U''=a_1b^2F+4a_1bdF^3+3a_1d^2F^5+4a_2b^2F^2+12a_2bdF^4+8a_2d^2F^6+9a_3b^2F^3+24a_3bdF^5$$
$$+15a_3d^2F^7+16a_4b^2F^4+40a_4bdF^6+24a_4d^2F^8,$$
$$\vdots \tag{14}$$

where $a_4 \neq 0, b \neq 0, c \neq 0, d \neq 0$. When we put Eq. (12-14) in Eq. (10), we obtain a system of algebraic equations. Therefore, we obtain a system of algebraic equations from these coefficients of polynomial of F. Solving this system with the help of wolfram Mathematica 9, we find the following coefficients and solutions;

Case 1a.

For b ≠ d, it can be considered the following coefficients;

$$a_0 = \frac{4b^2}{3}, a_1 = 0, a_2 = -2\sqrt{2}b\sqrt{a_4}, a_3 = 0, c = -2 + 4b^2, d = -\frac{\sqrt{a_4}}{2\sqrt{2}}. \tag{15}$$

substituting these coefficients in Eq.(12) along with Eq.(9), we obtain the following new exponential function solution for Eq.(1)

$$u_1(x,y,z,t) = \left(\frac{4b^2}{3} - \frac{2\sqrt{2}b\sqrt{a_4}}{e^{-2b\left(-\left(-2+4b^2\right)t+x+y+z\right)}EE + \frac{\sqrt{a_4}}{2\sqrt{2}b}} + \frac{a_4}{\left(e^{-2b\left(-\left(-2+4b^2\right)t+x+y+z\right)}EE + \frac{\sqrt{a_4}}{2\sqrt{2}b}\right)^2} \right). \tag{16}$$

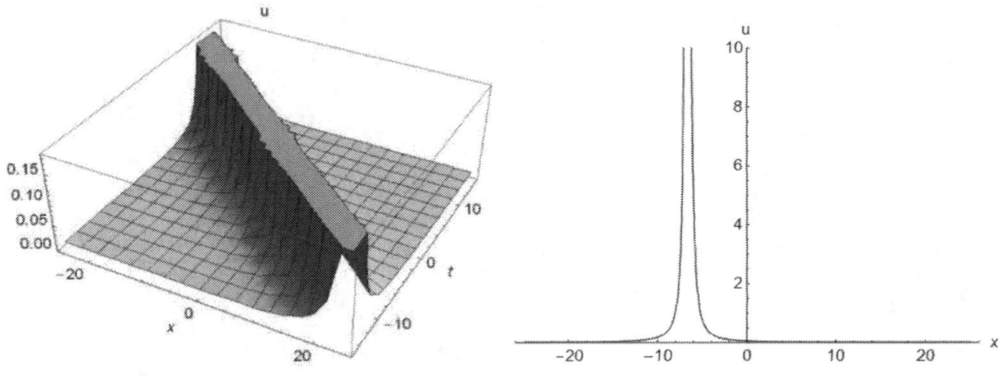

Figure 1. The 2D and 3D surfaces of Eq.(16) under the values of $a_4 = 0.2,\ b = 0.1, EE = -1, y = 0.03, z = 0.4, -10 < t < 10, -20 < x < 20.$ and $t = 2$ for 2D surfaces.

Case 1b.

Another coefficients for Eq.(1) and for $b \neq d$, it can be considered follows

$$a_0 = \frac{2+c}{3}, a_1 = 0,\ a_2 = -4d\sqrt{2+c}, a_3 = 0, a_4 = 8d^2, b = -\frac{\sqrt{2+c}}{2}. \tag{17}$$

when we regulate Eq.(12) under the terms of Eq.(9,17), we find the other new exponential function solution for Eq.(1);

$$u_2(x,y,z,t)=\left[\frac{4b^2}{3}-\frac{2b\sqrt{2}\sqrt{a_4}}{e^{-2b\left(-\left(-2+4b^2\right)t+x+y+z\right)}EE+\frac{\sqrt{a_4}}{2\sqrt{2b}}}+\frac{a_4}{\left(e^{-2b\left(-\left(-2+4b^2\right)t+x+y+z\right)}EE+\frac{\sqrt{a_4}}{2\sqrt{2b}}\right)^2}\right]. \quad (18)$$

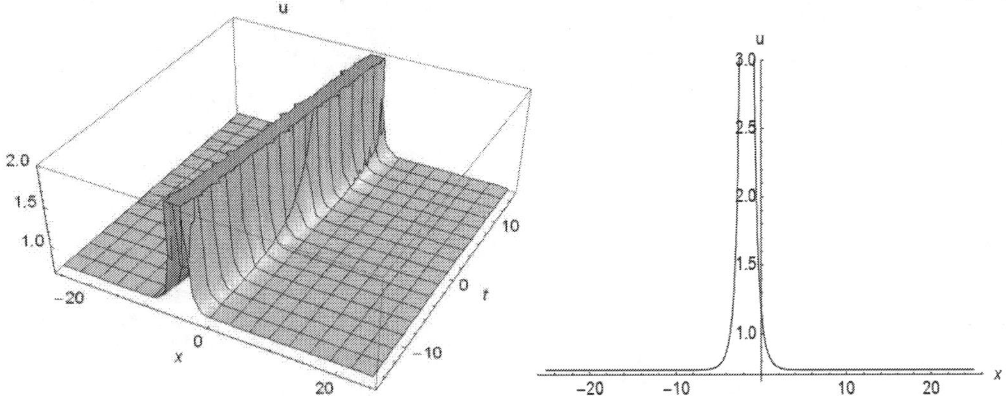

Figure 2. The 2D and 3D surfaces of Eq.(18) under the values of $a_4=0.2, EE=-1, b=0.1, y=0.03, z=0.4, -20<x<20, -10<t<10.$ $t=2$ for 2D surfaces.

4 CONCLUSION

In this paper, we have successfully applied the BSEFM to the Eq. (1). We have obtained some exponential and rational solutions. This method has given a lot of coefficients which gives many results. Some of them have been considered in this paper for obtaining new solutions. If other coefficients is considered, of course, ones can obtain many various solutions for Eq. (1). Therefore, it can be said that this method is a powerful tool for obtaining the solutions of such Eq. (1).

REFERENCES

1. J.S. Russell, Report on waves, 14th Mtg of the British Association for the Advancement of Science, (1844)
2. C. X. Xue, E. Pan and S. Y. Zhang, Solitary waves in a magneto-electro-elastic circular rod, Smart Materials and Structures, 20(105010), **1-7**, (2011)
3. Y.J. Yang, New application of the $(G'/G,1/G)$-expansion method to KP equation, Applied Mathematical Sciences, **Vol. 7**, no. 20, 959-967, (2013)
4. L.X. Li, E.Q. Li and M.L. Wang, the $(G'/G,1/G)$-expansion method and its application to traveling wave solutions of the Zakharov equations, Applied Mathematics, **Vol. 25**, no. 4, 454-462, (2010)
5. F.B.M., Belgacem, 2010. Sumudu transform applications to Bessel functions and equations. Applied Mathematical Sciences **4 (74)**, 3665–3686.
6. F.B.M. Belgacem, R. Silambarasan, Sumudu transform and trigonometric functions, Journal of Computational and Applied Mathematics, http://dx.doi.org/10.1016/j.cam.2015.12.036, (2016)

7. Ahmet Bekir, Ferhat Uygun, Exact travelling wave solutions of nonlinear evolution equations by using the $(\frac{G'}{G})$- expansion method, Arab Journal of Mathematical Sciences, **Vol:18**, 73–85, (2012)
8. F.D. Xie, Y. Zhang, Z.S. Lu, Symbolic computation in non-linear evolution equation: application to (3+1)-dimensional Kadomtsev-Petviashvili equation, Chaos, Solitons & Fractals 24 (2005) 257.

On Some Complex Aspects of the (2+1)-dimensional Broer-Kaup-Kupershmidt System

Hasan Bulut[1,*], *Hilal Arslanoglu* Isik[2,**], and *Tukur Abdulkadir* Sulaiman[3,4,***]

[1] *Firat University, Department of Mathematics, Elazig, Turkey*
[2] *Faculty of Engineering, Munzur University, Tunceli, Turkey*
[3] *Firat University, Department of Mathematics, Elazig, Turkey*
[4] *Federal University Dutse, Department of Mathematics, Jigawa, Nigeria*

Abstract. The improved Bernoulli sub-equation function method is used in extracting some new exponential function solutions to the (2+1)-dimensional Broer-Kaup-Kupershmidt system. It is of vital effort to look for more solutions of the (2+1)-dimensional Broer-Kaup-Kupershmidt system, which are very helpful for coastal and civil engineers to apply the nonlinear water models in a harbor and coastal design. All the obtained solutions satisfied the (2+1)-dimensional Broer-Kaup-Kupershmidt system. The two- and three-dimensional shapes of all the obtained solutions in this paper are also presented. All the computations and the graphics plots in this study are carried out with the aid of the Wolfram Mathematica 9.

1 Introduction

Investigations of the solutions of various nonlinear evolution equations (NLEEs) have become of great important because of the roles the solutions play in our real life situations. Nonlinear evolution equations often used to describe complex aspects in the field of nonlinear sciences such as Biological sciences, chemistry, mathematical physics, engineering and physics. For the past decades, various scholars have displayed their different effort for seeking the solutions of such types of equations, many analytical methods have been developed for this task such as sine-Gordon expansion method [1–3], the Bell-polynomial method [4], the new generalized Jacobi elliptic function expansion method [5], the Exp-function method [6], the modified Exp-function method [7], the (G'/G)-expansion method [8–10], the sub equation method [11], the simplified Hirota's method [12], the simplest equation method [13], the modified simplest equation method [14–16], the improved (G'/G)-expansion method [17], the multiple exp-function algorithm [18], the Lie group analysis and symmetry reductions [19]. In general, various methods have been developed to explore the search of different types of solutions to different kind of NLEEs [20–26].

However, this study is devoted in using the improved Bernoulli sub-equation function method (IBSEFM) [27] to extract some new solutions to the Broer-Kaup-Kupershmidt system [28, 29]. The

*e-mail: hbulut@firat.edu.tr
**e-mail: harslanoglu@hotmail.com
***e-mail: sulaiman.tukur@fud.edu.ng

IBSEFM is designed by expanding the Bernoulli sub-equation function method [27]. The solutions to the Broer-Kaup-Kupershmidt system are assume to be very helpful to coastal and civil engineers in applying to the nonlinear water models in a harbor and coastal design [29]. Several efforts have been made to tackle the Broer-Kaup-Kupershmidt system this include the exp-function method [30], the improved mapping approach [31], the generalized modified CK direct method [32] and the the bifurcation theory of planar dynamical systems [33].

2 Analysis of the Method

In this section, we present the steps of the algorithm to be followed in exploring the search of the new solutions to the given NLEE.

Step 1. Let us consider the following nonlinear partial differential equation;

$$P(u,\ u_x u^2\ u_x,\ u_t,\ \ldots), \tag{1}$$

performing the wave transformation $u(x,\ t) = U(\eta),\ \eta = x - ct$, reduces Eq. (1) to the following nonlinear ordinary differential equation (NODE);

$$Q(U,\ UU'\ U'',\ U^3,\ \ldots), \tag{2}$$

Step 2. Consider the following as a trial solution to Eq. (1);

$$U(\eta) = \frac{\sum_{i=0}^{n} a_i G^i}{\sum_{j=0}^{m} b_j G^j} = \frac{a_0 + a_1 G + a_2 G^2 + \ldots + a_n G^n}{b_0 + b_1 G + b_2 G^2 + \ldots + b_m G^m}, \tag{3}$$

according to the Bernoulli theory,

$$G' = wG + dG^M,\ w \neq 0,\ d \neq 0,\ M \in \mathbb{R} \setminus \{0, 1, 2\}, \tag{4}$$

where $F = F(\eta)$ is a Bernoulli differential polynomial. Substituting Eq. (4) into (2) produces an equation of polynomial $\Psi(G)$ function of G:

$$\Psi(G) = \rho_s G^s + \ldots + \rho_1 G + \rho_0 = 0. \tag{5}$$

We obtain the values of n, m and M by using the balancing technique.

Step 3. We get a system of equations by letting the summations of every coefficients of $\Psi(G)$ with the same power to zero. We obtain the values of the coefficients $a_0, \ldots, a_n$ and $b_0, \ldots, b_n$ by simplifying the system, using the Wolfram Mathematica.

Step 4. Solving the nonlinear Bernoulli differential equation (Eq. (4)), gives the following two cases based on w and d:

$$G(\eta) = \left[\frac{-d}{w} + \frac{\epsilon}{e^{w(M-1)\eta}}\right]^{\frac{1}{1-M}},\quad w \neq d \tag{6}$$

$$G(\eta) = \left[\frac{(\epsilon - 1) + (\epsilon + 1) tanh\left(\frac{w(1-M)\eta}{2}\right)}{1 - tanh\left(\frac{w(1-M)\eta}{2}\right)}\right]^{\frac{1}{1-M}},\quad w = d, \epsilon \in \mathbb{R}. \tag{7}$$

Using the complete discrimination of the polynomial of G, we solve the system of equations by using the Wolfram Mathematica 9 and obtain the various cases for the results of the coefficients. We obtain the exact solutions to Eq. (1) by putting the obtained values of the coefficients into the trial solution that is Eq. (3).

3 Application

Consider the following Broer-Kaup-Kupershmidt system [28, 29]:

$$\begin{aligned} u_{ty} - u_{xxy} + 2(uu_x)_y + 2v_{xx} &= 0, \\ v_t + v_{xx} + 2(uv)_x &= 0. \end{aligned} \tag{8}$$

Applying the wave transformation $u = U(\eta)$, $v = V(\eta)$, $\eta = x + y - ct$ on Eq. (8), yields the following NODE:

$$U'' - 2U^3 + 3cU^2 - c^2U = 0, \tag{9}$$

balancing the terms U'' and U^3, yields the following relations between m and n; $n + 1 = M + m$. Choosing $M = 3$, $m = 1$ and inserting in the obtained relation, gives $n = 3$.

Using $m = 1, n = 3$ along with the Eq. (3), we get the following equation:

$$U(\eta) = \frac{a_0 + a_1G(\eta) + a_2G^2(\eta) + a_3G^3(\eta)}{b_0 + b_1G(\eta)} = \frac{\Phi(\eta)}{\Omega(\eta)}, \tag{10}$$

differentiating Eq. (10) twice, we get the following equations;

$$U'(\eta) = \frac{\Phi'(\eta)\Omega(\eta) - \Phi(\eta)\Omega'(\eta)}{\Omega^2(\eta)}, \tag{11}$$

$$U''(\eta) = \frac{\Phi''(\eta)\Omega(\eta) - \Phi(\eta)\Omega'(\eta)}{\Omega^2(\eta)} - \frac{[\Phi(\eta)\Omega'(\eta)]'\Omega^2(\eta) - 2\Phi(\eta)[\Omega'(\eta)]^2\Omega(\eta)}{\Omega^4(\eta)}, \tag{12}$$

where $G' = wG + dG^3$, $w \neq 0$, $d \neq 0$. Substituting Eq. (11) and (12) into Eq. (9) and simplifying, we get an equation that involves the polynomials of G. We therefore collect the system of equations from that polynomial there by equating the summation of each coefficients of G that are having the same power to zero. We simplify the system of equations with the help of Wolfram Mathematica 9 and get the following cases of solutions;

Case-1.1: when $w \neq d$

$$a_0 = -2wb_0, a_1 = -2wb_1, a_2 = -2db_0, a_3 = -2db_1, c = -2w,$$

inserting the values of these coefficients into Eq. (10), gives:

$$\begin{aligned} u_{1.1}(x,y,t) = &-\frac{1}{b_0 + \frac{b_1}{\sqrt{\epsilon e^{-2w(x+y+2wt)} - \frac{d}{w}}}}\left(\frac{2db_0}{\epsilon e^{-2w(x+y+2wt)} - \frac{d}{w}}\right. \\ &\left. +2wb_0 + \frac{2db_1}{\left(\epsilon e^{-2w(x+y+2wt)} - \frac{d}{w}\right)^{\frac{3}{2}}} + \frac{2wb_1}{\sqrt{\epsilon e^{-2w(x+y+2wt)} - \frac{d}{w}}}\right), \end{aligned} \tag{13}$$

$$v_{1.1}(x,y,t) = -u_{1.1}(x,y,t)\Big(w + \frac{u_{1.1}(x,y,t)}{2} + \frac{\chi_1(x,y,t)}{2}\Big) - \frac{\chi_2(x,y,t)}{2}, \tag{14}$$

where $\chi_1(x,y,t) = \dfrac{\epsilon w b_1 e^{-2w(x+y+2wt)}}{\left(\epsilon e^{-2w(x+y+2wt)} - \frac{d}{w}\right)^{\frac{3}{2}}\left(b_0 + \frac{b_1}{\sqrt{\epsilon e^{-2w(x+y+2wt)} - \frac{d}{w}}}\right)}$

and

$$\chi_2(x,y,t) = \frac{1}{b_0 + \frac{b_1}{\sqrt{\epsilon e^{-2w(x+y+2wt)} - \frac{d}{w}}}}\left(\frac{4d\epsilon w b_0 e^{-2w(x+y+2wt)}}{\left(\epsilon e^{-2w(x+y+2wt)} - \frac{d}{w}\right)^2} + \frac{6d\epsilon w b_1 e^{-2w(x+y+2wt)}}{\left(\epsilon e^{-2w(x+y+2wt)} - \frac{d}{w}\right)^{\frac{5}{2}}}\right.$$
$$\left. + \frac{2\epsilon w^2 b_1 e^{-2w(x+y+2wt)}}{\left(\epsilon e^{-2w(x+y+2wt)} - \frac{d}{w}\right)^{\frac{3}{2}}}\right)$$

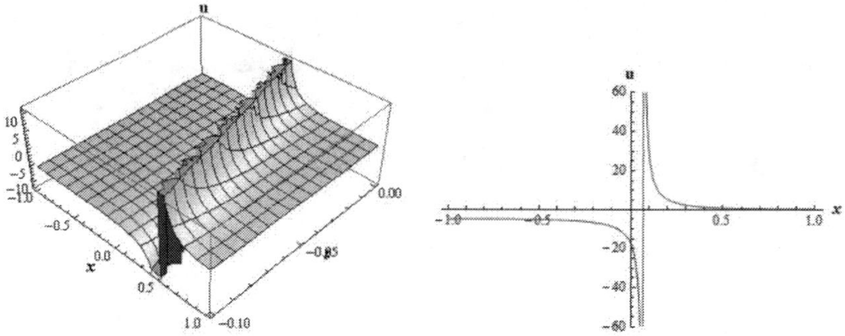

Figure 1. The 3-dimensional and 2-dimensional shapes of Eq. (13) by substituting these values $b_0 = 3$, $b_1 = 1.5$, $w = 2.5$, $d = 2$, $\epsilon = 3.3$, $y = 0.3$, $-1 < x < 1$, $-0.1 < t < 0$ and $t = 0.003$ for the 2-dimensional shape.

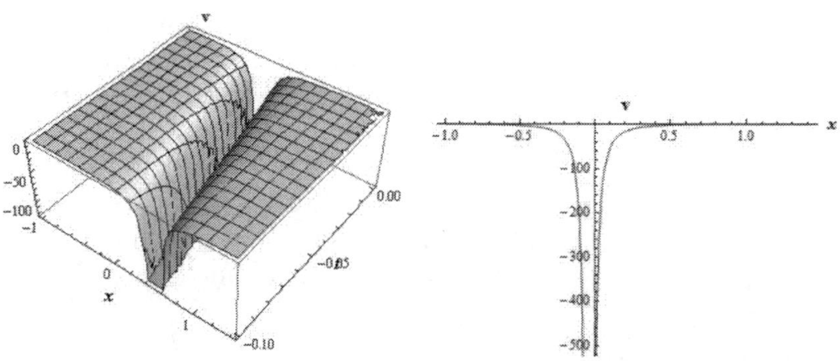

Figure 2. The 3-dimensional and 2-dimensional shapes of Eq. (14) by substituting these values $b_0 = 3$, $b_1 = 1.5$, $w = 2.5$, $d = 2$, $\epsilon = 3.3$, $y = 0.3$, $-1 < x < 1.42$, $-0.1 < t < 0$ and $t = 0.003$ for the 2-dimensional shape.

Case-2.1: when $w \neq d$

$$a_0 = 0, a_1 = 0, a_2 = -2db_0, a_3 = -2db_1, c = -2w,$$

inserting the values of these coefficients into Eq. (10), gives:

$$u_{2.1}(x,y,t) = -\frac{1}{b_0 + \frac{b_1}{\sqrt{\epsilon e^{-2w(x+y-2wt)} - \frac{d}{w}}}}\left(\frac{2db_0}{\epsilon e^{-2w(x+y-2wt)} - \frac{d}{w}} + \frac{2db_1}{\left(\epsilon e^{-2w(x+y-2wt)} - \frac{d}{w}\right)^{\frac{3}{2}}}\right), \tag{15}$$

$$v_{2.1}(x,y,t) = u_{2.1}(x,y,t)\left(w - \frac{u_{2.1}(x,y,t)}{2} - \frac{\chi_3(x,y,t)}{2}\right) - \chi_4(x,y,t), \tag{16}$$

where $\chi_3(x,y,t) = \frac{\epsilon w b_1 e^{-2w(x+y-2wt)}}{\left(\epsilon e^{-2w(x+y-2wt)} - \frac{d}{w}\right)^{\frac{3}{2}}\left(b_0 + \frac{b_1}{\sqrt{\epsilon e^{-2w(x+y-2wt)} - \frac{d}{w}}}\right)}$

and

$$\chi_4(x,y,t) = \frac{1}{b_0 + \frac{b_1}{\sqrt{\epsilon e^{-2w(x+y-2wt)} - \frac{d}{w}}}}\left(\frac{4d\epsilon w b_0 e^{-2w(x+y-2wt)}}{\left(\epsilon e^{-2w(x+y-2wt)} - \frac{d}{w}\right)^2} + \frac{6d\epsilon w b_1 e^{-2w(x+y-2wt)}}{\left(\epsilon e^{-2w(x+y-2wt)} - \frac{d}{w}\right)^{\frac{5}{2}}}\right)$$

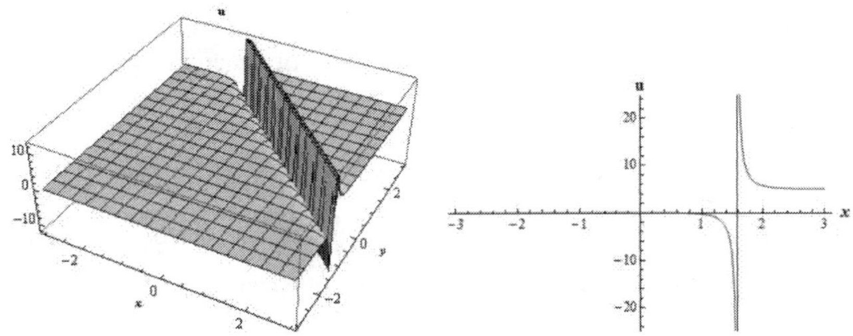

Figure 3. The 3-dimensional and 2-dimensional shapes of Eq. (15) by substituting these values $b_0 = 3$, $b_1 = 1.5$, $w = 2.5$, $d = 2$, $\epsilon = 3.3$, $t = 0.3$, $-3 < x < 3$, $-3 < y < 3$ and $t = 0.2$ for the 2-dimensional shape.

4 Conclusions

The improved Bernoulli sub-equation function method is used in this study to explore a search for the new solutions of the Broer-Kaup-Kupershmidt system and some new exponential function solutions are constructed. All the solutions obtained are verified to have satisfied the Broer-Kaup-Kupershmidt system. We also plot the three and two dimensional shapes of all the existing solutions in this paper. We performed all the computations and the graphics plot in this work with the aid of Wolfram Mathematica 9. We compare our results with the results obtained in [28, 29] and we observe that our results are indeed newly constructed solutions with some novel exponential function structures. From the results obtained, we observe that the improved Bernoulli sub-equation function method is a powerful mathematical tool that can be applied to the various nonlinear evolutions problems that arise in different fields of nonlinear sciences.

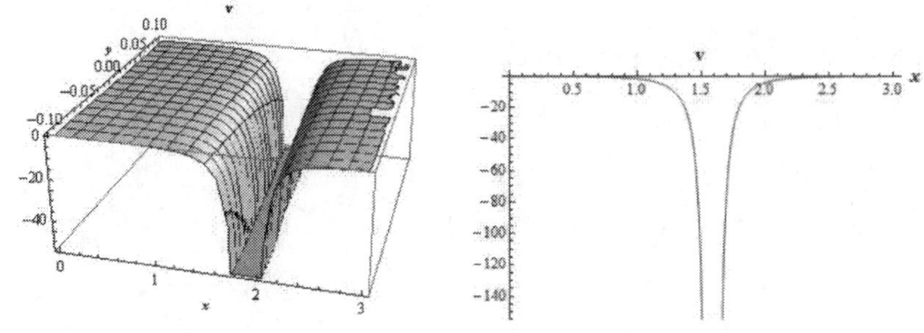

Figure 4. The 3-dimensional and 2-dimensional shapes of Eq. (16) by substituting these values $b_0 = 3$, $b_1 = 1.5$, $w = 2.5$, $d = 2$, $\epsilon = 3.3$, $t = 0.3$, $0 < x < 3$, $-0.1 < y < 0.1$ and $t = 0.2$ for the 2-dimensional shape.

References

[1] H Bulut, TA Sulaiman and HM Baskonus, Opt Quant Electron **48(564)**, 1-14 (2016)
[2] HM Baskonus, TA Sulaiman and H Bulut, Optik **131**, 1036-1043 (2017)
[3] H Bulut, TA Sulaiman, HM Baskonus and AA Sandulyak, Optik **135**, 327-336 (2017)
[4] Z Zhao, Y Zhang and W Rui, Applied Mathematics and Computation **248**, 456-462 (2014)
[5] D Lu and Q Shi, Journal of Information and Computational Science **7(8)**, 1733-1737 (2010)
[6] HA Ghany and M Zakarya, Proceedings of Basic and Applied Sciences **1(1)**, 73-87 (2013)
[7] F Ozpinar, HM Baskonus and H Bulut, Entropy **7(12)**, 8267-8277 (2015)
[8] H Jafari, N Kadkhoda and A Biswas, Journal of King Saud University-Science **25**, 57-62 (2015)
[9] MM Kabir and R Bagherzadeh, Middle-East Journal of Scientific Research **9(5)**, 602-610 (2011)
[10] R Arora and S Yadav, American Journal of Computational and Applied Mathematics **2(3)**, 119-123 (2012)
[11] M Mehdipoor and A Neirameh, American Journal of Computational and Applied Mathematics **126**, 4718-4722 (2015)
[12] AM Wazwaz, Applied Mathematics Letters **64**, 21-26 (2017)
[13] H Jafari and N Kadkhoda, New Trends in Mathematical Sciences 64-68 (2014)
[14] M Akbari, Quantum Physics Letters **3(3)**, 33-36 (2014)
[15] K Khan, MA Akbar and NHM Ali, Mathematical Physics **2013**, 146704 (2013)
[16] Y Zhao, Y He and Y Long, Journal of Applied Mathematics **2013**, 960798 (2013)
[17] Z Zeng and J Liu, Indian Journal of Pure and Applied Physics **53**, 713-717 (2015)
[18] W Ma and Z Zhu, Applied Mathematics and Computation **218**, 11871-11879 (2012)
[19] E Yasar, New Trends in Mathematical Sciences **4(4)**, 163-174 (2016)
[20] F Tchier, AI Aliyu, A Yusuf and M Inc, The European Physical Journal Plus **132(136)**, 163-174 (2017)
[21] H Bulut, Abstract and Applied Analysis **2013**, 742643 (2013)
[22] MF Aghdaei and J Manafianheris, Journal of Mathematical Extension **5(2)**, 91-104 (2011)
[23] AM Wazwaz, Physica Scripta **85(2)**, 025006 (2012)
[24] A Atangana, The European Physical Journal Plus **131(10)**, 373 (2016)
[25] KA Touchent and FBM Belgacem, Nonlinear Studies **22(4)**, 1-10 (2015)

[26] HM Baskonus and M Askin, 6th International Youth Science Forum "LITTERIS ET ARTIBUS"' Computer Science and Engineering, Lviv, Ukraine, 24-26 November, (2016).

[27] HM Baskonus and H. Bulut, Waves in Random and Complex Media **26(2)**, 189-196 (2016)

[28] H Bulut and TA Sulaiman, International Youth Science Forum "Litteris et Artibu", 24-26 November, Lviv, Ukraine (2016)

[29] M Song, S Li and J Cao, Abstract and Applied Analysis **2010**, 652649 (2010)

[30] AG Davodi, DD Ganji, AG Davodi and A Asgari, Applied Mathematics and Computation **217(15)**, 1415-1420 (2015)

[31] M Song-Hua, F Jian-Ping and Z Chun-Long, Zeitschrift fur Naturforschung A **62**, 8-12 (2007)

[32] J Yu, International Journal of Bifurcation and Chaos **23(9)**, doi: 10.1142/S0218127413501563 (2013)

[33] T Ya-ning and C Yan-na, Basic Sciences Journal of Textile Universities **26(3)**, 332-337 (2013)

ITM Web of Conferences **13**, 01020 (2017) DOI: 10.1051/itmconf/20171301020
CMES2017

A Three Factor Remote User Authentication Scheme Using Collision Resist Fuzzy Extractor in Single Server Environment

Debasis Giri[1] and *Tanmoy* Maitra[2,*]

[1] *Department of Computer Science and Engineering, Haldia Institute of Technology, Haldia-721657, India*
[2] *Department of Computer Science and Engineering, Jadavpur University, Kolkata-700032, India*

Abstract. Due to rapid growth of online applications, it is needed to provide such a facility by which communicators can get the services by applying the applications in a secure way. As communications are done through an insecure channel like Internet, any adversary can trap and modify the communication messages. Only authentication procedure can overcome the aforementioned problem. Many researchers have proposed so many authentication schemes in this literature. But, this paper has shown that many of them are not usable in real world application scenarios because, the existing schemes cannot resist all the possible attacks. Therefore, this paper has proposed a three factor authentication scheme using hash function and fuzzy extractor. This paper has further analyzed the security of the proposed scheme using random oracle model. The analysis shows that the proposed scheme can resist all the possible attacks. Furthermore, comparison between proposed scheme and related existing schemes shows that the proposed scheme has better trade-off among storage, computational and communication costs.

1 Introduction

Nowadays, online applications like bill payment, banking system, telecare medical system, social networking, e-voting and so on are rapidly used for their easy and efficient access. All the applications are going through a client/server environment and communications are done through public channel like Internet because of availability of public bandwidth. Therefore, all communication messages of the applications are public. As a result, any one can trap and modify the communication messages. For this purpose, in such communication system, authentication scheme is rapidly used by which after verifying the communicators and their messages, a secure communication can be done through public channel. In this regard, smart card and password based user authentication scheme is very much popular for online communication system. However, a suitable smart card based authentication scheme should satisfy the following property:

Low cost: Computational cost, communication cost and smart card storage cost are three basic network parameters to measure performances of an authentication scheme. Therefore, it has to considered that these three parameters are reduced as much possible when an authentication scheme is going to be designed.

Prevention of security attacks: During communication through public channel, it is needed to secure the message from outsider adversary. An authentication scheme needs to be designed such a way by which from the communication messages, any adversary will unable to extract useful information and the scheme can resist security attacks.

Session key agreement: After authentication in both ways, a common secret session key is needed to carry on the communications within the same session after encrypting the paintext messages by the session key.

Mutual authentication: Mutual authentication is a essential property for an authentication scheme by which all the communicators can authenticate or verify each other.

Efficient login procedure: For an authentication scheme, it is needed that smart card checks the wrong inputs before going to send a login message to server. By checking wrong inputs in login phase, extra communication overhead can be avoided. Therefore, it is crucial property of a good authentication scheme.

*e-mail: tanmoy.maitra@live.com

Efficient password change procedure: A valid user can change the password freely and securely without taking any help from server. For this purpose, the smart card should verify the old password in the password change phase, so that an unauthorized user cannot change the authorized user's password even if it gets the valid users' smart card. The afore mentioned property should present in an authentication scheme.

Traceability: In an authentication scheme, it is also needed to trace the sender of a message for corresponding receiver [1]. Otherwise, any one can mount denial of service (DoS) attack.

Furthermore, to enhanced the security of password based authentication scheme, biometric feature (i.e, finger print, ires, retina etc.) [2] is added with the password. Therefore, this research focus to design a biometric plus password based efficient authentication scheme by considering all the afore mentioned properties.

1.1 Literature Survey

A brief survey of existing authentication schemes is described in this section. First Lamport [3] proposed a password-based authentication scheme based on one way hash function. However, Shimizu et al. [4] showed that the Lamport's scheme [3] suffers from different attacks. After that so many remote user authentication schemes [5–16] have been proposed in this regard which are based on only password. But, the researchers have considered biometric feature [2] with the password to enhance the security label. Therefore, many researchers have proposed biometric and password based authentication schemes in [17–28]. Li and Hwang [17] proposed a biometrics-based remote user authentication scheme in 2010. However, in 2011, Das [18] showed that Li and Hwang's scheme [17] had flaws in the login phase, authentication phase and password change phase and therefore, Das also proposed an authentication scheme. An [19] showed that Das's scheme [18] cannot resist the server masquerading attack, user impersonation attack, password guessing attack and insider attack, and so proposed an improved scheme. Li et al. [20] found that An's scheme [19] suffered from the denial-of-service (DoS) attack, the forgery attack and also did not provide forward secrecy. Furthermore, in 2013, Lee and Hsu [21] pointed out that Das's scheme [18] is also suffering from privileged insider attack and the off-line password guessing attack. Therefore, Lee and Hsu [21] proposed a biometric based authentication scheme to overcome the weaknesses of Das's scheme [18]. In 2013, Tan [22] proposed a three-factor authentication scheme. But, Yan et al. [23] pointed out that Tan's scheme [22] is vulnerable to the Denial-of-Service (DoS) attack. However, recently, Mishra et al. [24] showed that Yan et al.'s scheme [23] suffers from off-line password guessing attack and has inefficient login and password change phases. Huang et al. [29] proposed an authentication scheme based on RSA. But Amin et al. [30] proved that Huang et al.'s scheme [29] unable to protect forgery attack and also introduced an authentication protocol in [30].

1.2 Contribution

This paper proposes a three factor authentication scheme using hash function and fuzzy extractor, where three factor means (1) users' password, (2) users' biometric and (3) smart card. This paper further analyzes the security of the proposed scheme using random oracle model. The analysis shows that an adversary cannot mount any attacks on the proposed scheme due to hardness of inversion of one-way hash function as well as it has to solve hardness of fuzzy factor. Furthermore, comparison between proposed scheme and related existing schemes shows that the proposed scheme has better trade-off among storage, computational and communication costs. It is a great contribution that the proposed scheme resists all the possible attacks with better trade-off among different costs.

1.3 Road Map

This section describes a road map which has been followed throughout this paper. Section 2 briefly introduces some preliminary mathematical concepts for introducing the proposed scheme. Section 3 describes a network model and an adversary model to analyze the proposed scheme. A proposed scheme is described in section 4. Section 5 describes cryptanalysis of the proposed scheme and Section 6 compares the performances of the proposed scheme with previously published schemes. Conclusion of this paper appears in section 7.

2 Preliminaries

In this section, a briefly review the basic concepts of cryptographic one-way hash function and collision resist fuzzy extractor are introduced.
Definition 1. A collision resistant cryptographic one-way hash function [25, 27] maps a string of arbitrary length to a string of fixed length called the hashed value. It can be symbolized as: $H : A \rightarrow B$, where A is a binary string of

arbitrary length and B is a binary string of fixed length n. If $Adv_{\mathcal{A}}^{H}(t_1)$ is the advantage to an adversary $\mathcal{A}$ to choose a random pair $(a, b) \in A \times A$ such that $H(a) = H(b)$, where $a \neq b$ for the time duration t_1, it can be considered that $Adv_{\mathcal{A}}^{H}(t_1)$ is the probability in the advantage which is computed over the random choices made by the adversary $\mathcal{A}$ for the time duration t_1. Then the cryptographic one-way hash function $H(\cdot)$ is called collision-resistant, if $Adv_{\mathcal{A}}^{H}(t_1) \leq \xi_1$, for any small $\xi_1 > 0$. $Adv_{\mathcal{A}}^{H}(t_1)$ is represented as:

$$Adv_{\mathcal{A}}^{H}(t_1) = Pr\Big[(a, b) \in_R A \times A \mid (a \neq b) \ \wedge \ H(a) \ = \ H(b)\Big], \tag{1}$$

where $Pr[\mathcal{E}]$ denotes the random event $\mathcal{E}$.

Definition 2. A collision resistant fuzzy extractor [2, 27] can be model as a procedure, known as Gen, which takes a binary string say, B of some metric space M as an input, where $M \in \{0, 1\}^k$, for some k bits and produces a random string say, $\phi \in_R \{0, 1\}^n$, for some n bits and an auxiliary string say, $\theta \in_R \{0, 1\}^r$, for some r bits, where $r = k$ or n bits. It can be mathematically represented by $Gen : M \longrightarrow \phi \times \theta$. Another procedure, known as Rep, takes a binary string say, B' of the metric space $M \in \{0, 1\}^k$, where $B \neq B'$ and a uniform distribution binary string say, $\theta' \in_R \{0, 1\}^r$ to produce the random string $\phi' \in_R \{0, 1\}^n$, symbolized as $Rep : M \times \theta' \longrightarrow \phi'$. If $Adv_{\mathcal{A}}^{FE}(t_2)$ is the advantage to an adversary $\mathcal{A}$ to choose a pair $(B, B') \in_R M \times M$ randomly such that $des(B, B') \leq \delta d$, $Gen(B) = Gen(B')$ and $Rep(B, \theta) = Rep(B', \theta')$, where δd is the difference tolerance level and $B \neq B'$ for the time duration t_2, it can be considered that $Adv_{\mathcal{A}}^{FE}(t_2)$ is the probability that the advantage is computed over the random choices made by $\mathcal{A}$ for the time duration t_2. Then the fuzzy extractor FE is called collision-resistant, if $Adv_{\mathcal{A}}^{FE}(t_2) \leq \xi_2$, for any small $\xi_2 > 0$. $Adv_{\mathcal{A}}^{FE}(t_2)$ is represented as:

$$\begin{aligned} Adv_{\mathcal{A}}^{EF}(t_2) = Pr\Big[&(B, B') \in_R M \times M \mid (B \neq B') \ \wedge \ des(B, B') \leq \delta d \\ &\wedge Gen(B) = Gen(B') \ \wedge \ Rep(B, \theta) = Rep(B', \theta')\Big], \end{aligned} \tag{2}$$

for all probabilistic polynomial-time algorithms Gen and Rep.

3 Model

This section will introduce two following models:

Network Model: Architecture of the proposed scheme is shown in Figure 1 where, users have to register to a remote server to get their smart card which is known as registration procedure (see Figure 1(a)). Whenever the registered users want to get service from the remote server by accessing their smart card through public channel like Internet, the smart card sends a login request message to the remote server. After verifying the login request message, the remote server sends corresponding reply message to the sender. After receiving the reply message, the corresponding smart card checks the validity of the reply message. Upon receiving correct reply message, both user and the remote server agree for a shared secret session key (See Figure 1(b)).

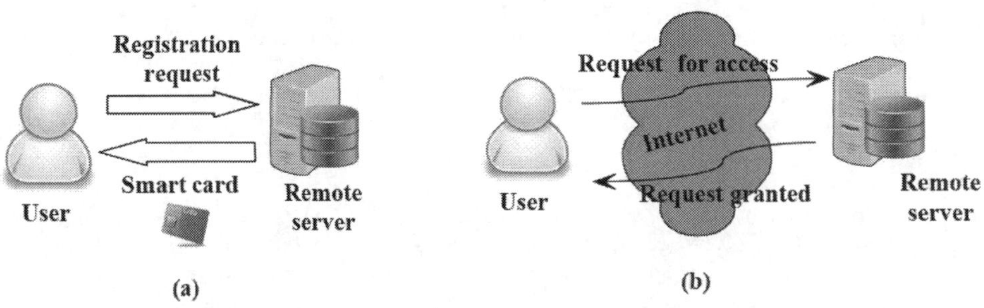

Figure 1. Network architecture of proposed scheme (a) registration procedure and (b) login and authentication procedure

Adversary Model: To analyze the security of the proposed scheme, Dolev-Yao threat model [31] has been considered in which the communicating parties communicate through an insecure channel. Therefore, an adversary $\mathcal{A}$ can trap the transmitted messages over the public or insecure channel, and furthermore he/she can modify, delete or change the contents of the transmitted messages. The adversary $\mathcal{A}$ also obtains the information which are stored in the user's smart card by monitoring the power consumption [32, 33]. Generally, identity and password of the user are low entropy in cryptography, that means the adversary can guess the identity and password individually using dictionary attack in polynomial time. But, the adversary cannot guess identity and password simultaneously in on-line/off-line within a polynomial time as pointed out in [34]. According to our adversary model, we consider two following cases:

- *Case 1.* A third party from outside of the system tries to mount various attacks on authentication system as an adversary $\mathcal{A}$.
- *Case 2.* A registered user from inside of the system tries to extract secret information of the server by which he/she can mount various attacks on authentication system as an other user or adversary $\widehat{\mathcal{A}}$.

4 Proposed Scheme

This section proposes an authentication scheme. A nomenclature is given in Table 1 to introduce the proposed scheme. The proposed scheme consists of five phases namely, 1) initialization phase, 2) registration phase, 3) login phase, 4) authentication and key agreement phase and 5) password update phase. The phases are as follows:

Table 1. Nomenclature

Term	*Usage*
U_i	*i-th* Patient
RS	Remote server
pw_i	Password of user U_i
ID_i	Identity of user U_i
B_i	Biometric parameter of U_i
r_i	Random number chosen by smart card
y_i	Random number chosen by RS
T	Current timestamp
$des(\cdot)$	Distance measurement function
δd	Estimated difference
X'	Parameter X computed or extracted by smart card
X^*	Parameter X computed or extracted by RS
δT	Estimated timestamp
SK_i	Shared session key between U_i and RS
$H(\cdot)$	Cryptographic one-way hash function
s	Secret key of remote server
$\|\|$	Concatenation operation
$\oplus$	Bit wise XOR operation

4.1 Initialization Phase

A remote server RS runs algorithm $\mathcal{G}$ to compute a large prime number q. Then it selects a random number s such that $s \in_R Z_q^*$. It further chooses a collision resist cryptographic one-way hash function $H : \{0,1\}^* \rightarrow \{0,1\}^n$, where n is a fixed length integer number. Finally, RS publishes $H(\cdot)$ as public and keeps s as secret.

4.2 Registration Phase

Whenever a new user U_i wants to register to the remote server RS, this phase is invoked. This phase is as follows:

1. The user inputs their biometric feature (i.e., finger print) to a sensor. The sensor generates a corresponding biometric information B_i and provides it to U_i.

2. The user U_i chooses an identity ID_i, password pw_i and generates an unique pair (θ_i, ϕ_i) from B_i by computing $(\phi_i, \theta_i) \leftarrow Gen(B_i)$. U_i then computes $pwr_i = H(pw_i \| \phi_i)$ and sends $\langle ID_i, pwr_i \rangle$ to RS through a secure channel.

DOI: 10.1051/itmconf/20171301020

U_i	RS
Input biometrics to a sensor to get biometric information B_i	
Choose ID_i and pw_i	
Compute $(\phi_i, \theta_i) \leftarrow Gen(B_i)$	
Calculate $pwr_i = H(pw_i \Vert \phi_i)$	
$\xrightarrow{\langle ID_i, pwr_i \rangle}$	Compute $A_i = H(s \Vert ID_i)$, $D_i = H(A_i)$
	Check D_i =? database
	Calculate $C_i = A_i \oplus pwr_i$, store D_i into its database
Calculate $\widehat{B_i} = B_i \oplus H(ID_i \Vert pw_i)$	$\xleftarrow{\langle C_i, D_i, des(\cdot) \rangle}$
$\widehat{\theta_i} = \theta_i \oplus H(ID_i \Vert pw_i)$	
Reserve $\langle C_i, D_i, \widehat{B_i}, \widehat{\theta_i}, des(\cdot) \rangle$ into smart card	$\rightarrow$ Secure channel

Figure 2. Registration phase of the proposed scheme

3. After getting a registration request $\langle ID_i, pwr_i \rangle$ from U_i, RS computes $A_i = H(s \| ID_i)$ and $D_i = H(A_i)$. RS then checks D_i is present in its database or not. If it is present in its database, RS sends a decline message to U_i because, ID_i is already used by another user. Therefore, the user U_i has to choose another identity until fresh identity is not obtained. If D_i is not present in its database, RS computes $C_i = A_i \oplus pwr_i$ and stores parameters $\langle C_i, D_i, des(\cdot) \rangle$ into the memory of smart card, where $des(\cdot)$ is a distance measurement function. Then RS issues the smart card for U_i and sends it through a secure channel or by person. RS then updates its database by adding D_i into the list.

4. After getting the smart card, U_i inserts it into a terminal or card reader and submits their identity ID_i and password pw_i.

5. The smart card computes $\widehat{B_i} = B_i \oplus H(ID_i \| pw_i)$, $\widehat{\theta_i} = \theta_i \oplus H(ID_i \| pw_i)$. Finally, the smart card stores $\langle \widehat{B_i}, \widehat{\theta_i} \rangle$ into its memory. Note that, smart card stored parameters are $\langle C_i, D_i, \widehat{B_i}, \widehat{\theta_i}, des(\cdot) \rangle$.

Figure 2 shows the registration phase of the proposed scheme.

4.3 Login Phase

Whenever a registered user U_i wants to access the remote server, this phase is invoked. U_i inserts their smart card into a card reader or terminal and provides their biometric information B_i^* through sensor, identity ID_i and password pw_i to the smart card. The smart card then executes following steps:

1. The smart card computes $B_i' = \widehat{B_i} \oplus H(ID_i \| pw_i)$ and checks $des(B_i^*, B_i') \leq \delta d$. If it does not hold, the smart card rejects U_i; otherwise, it follows next step.

2. The smart card computes $\theta_i' = \widehat{\theta_i} \oplus H(ID_i \| pw_i)$, $\phi_i' \leftarrow Rep(B_i^*, \theta_i')$, $pwr_i' = H(pw_i \| \phi_i')$, $A_i' = C_i \oplus pwr_i'$, $D_i' = H(A_i')$ and checks computed D_i' and stored D_i are equal or not. If equality does not hold, the smart card rejects U_i; otherwise, it follows next step.

3. The smart card chooses a random number $r_i \in_R Z_q^*$ and further computes $E_i = r_i \oplus A_i'$, $G_i = T_i^1 \oplus A_i' \oplus H(r_i)$ and $F_i = H(r_i \| A_i' \| T_i^1)$, where T_i^1 is the current login timestamp of U_i. The user U_i then sends a login request message $\langle ID_i, G_i, F_i, E_i \rangle$ to the registration server RS through a public channel.

Figure 3 shows the login phase of the proposed scheme.

4.4 Authentication and Key Agreement Phase

After receiving the login request message $\langle ID_i, G_i, F_i, E_i \rangle$ from the user U_i at timestamp T_s, the remote server RS computes following steps:

1. RS computes $A_i^* = H(s \| ID_i)$, $r_i^* = A_i^* \oplus E_i$, $T_i^{1^*} = A_i^* \oplus G_i \oplus H(r_i^*)$ and checks $(T_s - T_i^{1^*}) \leq \delta T$. It it does not hold, RS rejects U_i; otherwise, executes the next step.

U_i	Terminal	RS
Input ID_i, pw_i and biometric B_i^*		
	Compute $B_i' = \widehat{B_i} \oplus H(ID_i\|\|pw_i)$	
	Check $des(B_i^*, B_i') \leq \delta d$	
	Compute $\theta_i' = \widehat{\theta_i} \oplus H(ID_i\|\|pw_i)$, $\phi_i' \leftarrow Rep(B_i^*, \theta_i')$	
	$pwr_i' = H(pw_i\|\|\phi_i')$, $A_i' = C_i \oplus pwr_i'$	
	$D_i' = H(A_i')$	
	Check computed D_i' =? stored D_i	
	Choose random number $r_i \in_R Z_q^*$	
	Compute $E_i = r_i \oplus A_i'$, $G_i = T_i^1 \oplus A_i' \oplus H(r_i)$	
	$F_i = H(r_i\|\|A_i'\|\|T_i^1)$	
	$\langle ID_i, G_i, F_i, E_i\rangle \longrightarrow$	$\rightarrow$ Insecure channel

Figure 3. Login phase of the proposed scheme

Terminal	RS
$\langle ID_i, G_i, F_i, E_i\rangle \longrightarrow$	Compute $A_i^* = H(s\|\|ID_i)$, $T_i^{1^*} = A_i^* \oplus G_i$
	Check $(T_s - T_i^{1^*}) \leq \delta T$
	Compute $r_i^* = A_i^* \oplus E_i$, $F_i^* = H(r_i^*\|\|A_i^*\|\|T_i^{1^*})$
	Check computed F_i^* =? received F_i
	Chooses a random number $y_i \in_R Z_q^*$
	Compute $Q_i = A_i^* \oplus y_i$, $K_i = T_s \oplus A_i^* \oplus H(y_i)$
Compute $y_i' = A_i' \oplus Q_i$, $T_s' = A_i' \oplus K_i \oplus y_i'$	$SK_i = H(y_i\|\|r_i^*)$, $L_i = H(T_s\|\|SK_i\|\|A_i^*)$
Check $(T_i^2 - T_s') \leq \delta T$	$\longleftarrow \langle Q_i, L_i, K_i\rangle$
Compute $SK_i' = H(y_i'\|\|r_i)$	
$L_i' = H(T_s'\|\|SK_i'\|\|A_i')$	
Check computed L_i' =? received L_i	
U_i agrees upon the shared secret key SK_i	$\rightarrow$ Public channel

Figure 4. Authentication and key agreement phase of the proposed scheme

2. RS computes $F_i^* = H(r_i^*\|A_i^*\|T_i^{1^*})$ and checks computed F_i^* and received F_i are equal or not. If it does not hold, RS rejects login request message of U_i; otherwise, follows the next step.

3. RS chooses a random number $y_i \in_R Z_q^*$ and further computes $Q_i = A_i^* \oplus y_i$, $K_i = T_s \oplus A_i \oplus H(y_i)$, $SK_i = H(y_i\|r_i^*)$, $L_i = H(T_s\|SK_i\|A_i^*)$ and sends a reply message $\langle Q_i, L_i, K_i\rangle$ to U_i through a public channel. RS accepts SK_i as a shared secret session key.

After receiving the reply message $\langle Q_i, L_i, K_i\rangle$ from RS at timestamp T_i^2, the smart card of the user U_i further executes the following steps to verify the reply message of RS:

1. The smart card computes $y_i' = A_i' \oplus Q_i$, $T_s' = A_i' \oplus K_i \oplus H(y_i')$ and checks $(T_i^2 - T_s') \leq \delta T$. It it does not hold, the smart card rejects the reply message; otherwise, executes the next step.

2. The smart card computes $SK_i' = H(y_i'\|r_i)$, $L_i' = H(T_s'\|SK_i'\|A_i')$ and checks computed L_i' and received L_i are equal or not. If they are equal, the user U_i agrees upon the shared secret key SK_i; otherwise, rejects the reply message.

Figure 4 shows the authentication and key agreement phase of the proposed scheme.

4.5 Password Update Phase

Whenever a user U_i wants to change their password, this phase is invoked. U_i inserts their smart card into a card reader or terminal and provides their biometric information B_i^* through sensor, identity ID_i and password pw_i to the smart card. The smart card then executes following steps:

1. The smart card computes $B_i' = \widehat{B_i} \oplus H(ID_i\|pw_i)$ and checks $des(B_i^*, B_i') \leq \delta d$. If it does not hold, the smart card rejects U_i; otherwise, it follows next step.

U_i	**Terminal**
Input ID_i, pw_i and biometric B_i	
	Compute $B_i' = \widehat{B_i} \oplus H(ID_i \| pw_i)$
	Check $des(B_i^*, B_I') \leq \delta d$
	$\theta_i' = \widehat{\theta_i} \oplus H(ID_i \| pw_i)$, $\phi_i' \leftarrow Rep(B_i^*, \theta_i')$
	$pwr_i' = H(pw_i \| \phi_i')$, $A_i' = C_i \oplus pwr_i'$
	$D_i' = H(A_i')$
	Check computed D_i' =? stored D_i
Input a new $pw_i^{[new]}$	
	Compute $pwr_i^{[new]} = H(pw_i^{[new]} \| \phi_i')$
	$C_i^{[new]} = A_i' \oplus pwr_i^{[new]}$, $\widehat{B_i^{[new]}} = B_i' \oplus H(ID_i \| pw_i^{[new]})$
	$\widehat{\theta_i^{[new]}} = \theta_i' \oplus H(ID_i \| pw_i^{[new]})$
	store $C_i^{[new]}$, $\widehat{B_i^{[new]}}$, $\widehat{\theta_i^{[new]}}$
	by replacing C_i, $\widehat{B_i}$, $\widehat{\theta_i}$

Figure 5. Password update phase of the proposed scheme

2. The smart card computes $\theta_i' = \widehat{\theta_i} \oplus H(ID_i \| pw_i)$, $\phi_i' \leftarrow Rep(B_i^*, \theta_i')$, $pwr_i' = H(pw_i \| \phi_i')$, $A_i' = C_i \oplus pwr_i'$, $D_i' = H(A_i')$ and checks computed D_i' and stored D_i are equal or not. If equality does not hold, the smart card rejects U_i; otherwise, gives permission to enter their new password.

3. The user U_i selects their new password $pw_i^{[new]}$ and proves it to the smart card. The smart card then further proceeds to next step.

4. The smart card computes $pwr_i^{[new]} = H(pw_i^{[new]} \| \phi_i')$, $C_i^{[new]} = A_i' \oplus pwr_i^{[new]}$, $\widehat{B_i^{[new]}} = B_i' \oplus H(ID_i \| pw_i^{[new]})$ and $\widehat{\theta_i^{[new]}} = \theta_i' \oplus H(ID_i \| pw_i^{[new]})$. The smart card then stores $C_i^{[new]}$, $\widehat{B_i^{[new]}}$ and $\widehat{\theta_i^{[new]}}$ in the place of C_i, $\widehat{B_i}$ and $\widehat{\theta_i}$ respectively into the memory of smart card.

Figure 5 shows the password update phase of the proposed scheme.

5 Security Analysis of Proposed Scheme

The formal security analysis of the proposed scheme under the random oracle model is presented in this section. This security analysis uses the formal security analysis under the generic group model of cryptography. In the following, this work defines random oracles for the formal security analysis of the proposed scheme:

- *Oracle$\mathcal{H}$* is a random oracle which maintains a tuple $\langle x, y \rangle$ such that $y = H(x)$. It returns x from y upon receiving a query (qH, y) if $\langle x, y \rangle$ is present in the tuple; otherwise returns a random number r_1. Then it stores a new entry $\langle r_1, y \rangle$ into its tuple.
- *Oracle$\mathcal{FE}$* is a random oracle which contains two parts:

  1.*Oracle$\mathcal{FE}_{Gen}$* unconditionally outputs the pair (ϕ, θ) from the corresponding tuple $\langle B, \phi, \theta \rangle$ upon receiving a query $(qGen, B)$ such that $(\phi, \theta) \leftarrow Gen(B)$ if $\langle B, \phi, \theta \rangle$ is present in its tuple; otherwise returns two random numbers r_2 and r_3. Then it stores new entry $\langle B, r_2, r_3 \rangle$ into its tuple.

  2.*Oracle$\mathcal{FE}_{Rep}$* unconditionally outputs ϕ from the corresponding tuple $\langle B', \phi, \theta \rangle$ upon receiving a query $(qRep, B', \theta)$ such that $\phi \leftarrow Rep(B', \theta)$ if $\langle B', \phi, \theta \rangle$ is present in its tuple; otherwise returns random number r_4. Then it stores new entry $\langle B', r_4, \theta \rangle$ into its tuple.

Theorem 1. *Under the assumption that a cryptographic one-way hash function $H(\cdot)$ and fuzzy extractor FE act as random oracles, the proposed scheme is provably secure against an adversary $\mathcal{A}$ for deriving the password pw_i and biometric parameter B_i of a user U_i even if the adversary $\mathcal{A}$ gets parameters that are stored into the memory of U_i's smart card and traps the communication messages between U_i and the remote server RS.*

Proof 1. This research construct an adversary $\mathcal{A}$ who has the ability to derive the password pw_i and biometric parameter B_i of a user U_i. For this purpose, this research assumes that the smart card of a user U_i is lost or stolen. Thus, the adversary $\mathcal{A}$ can extract the stored parameters $\langle C_i, D_i, \widehat{B_i}, \widehat{\theta_i} \rangle$ from the memory of the smart card of the user U_i by power monitoring [32][33]. The adversary $\mathcal{A}$ also traps login request message $\langle ID_i, G_i, F_i, E_i \rangle$ and a reply message $\langle Q_i, L_i, K_i \rangle$. The adversary $\mathcal{A}$ runs the experiment, $EXP1^{oracle}_{\mathcal{A},\,TFUAS}$ for our three factor user authentication scheme (TFUAS) to derive the password pw_i and biometric parameter B_i of the user U_i as given in the Algorithm 1.

Algorithm 1 $EXP1^{oracle}_{\mathcal{A},\,TFUAS}$

Input: $C_i, D_i, \widehat{B_i}, \widehat{\theta_i}, ID_i, G_i, F_i, E_i, Q_i, L_i, K_i$
Output: 0 or 1

Calls $Oracle\mathcal{H}$ on the input D_i to retrieve the information $A_i = H(s\|ID_i)$ as $(A_i^*) \leftarrow Oracle\mathcal{H}(D_i)$
Calls $Oracle\mathcal{H}$ on the input F_i to retrieve the information A_i, r_i and T_i^1 as $(r_i^* \parallel A_i^{**} \| T_i^{1*}) \leftarrow Oracle\mathcal{H}(F_i)$
Calls $Oracle\mathcal{H}$ on the input L_i to retrieve the information SK_i, T_s and A_i as $(T_s^* \| SK_i^* \| A_i^{***}) \leftarrow Oracle\mathcal{H}(L_i)$
if $(A_i^{***} == A_i^{**} == A_i^*)$ **then**
  Computes $r_i^{**} = A_i^* \oplus E_i$
  if $(r_i^{**} == r_i^*)$ **then**
    Computes $y_i^* = A_i^* \oplus Q_i$, $T_i^{1**} = A_i^* \oplus G_i \oplus H(r_i^*)$ and $T_s^{**} = A_i^* \oplus K_i \oplus H(Y_i^*)$
    if $(T_i^{1**} == T_i^{1*})$ && $(T_s^{**} == T_s^*)$ **then**
      Computes $pwr_i^* = C_i \oplus A_i^*$
    else
      Return 0 (**Failure**)
    end if
  else
    Return 0 (**Failure**)
  end if
else
  Return 0 (**Failure**)
end if
repeat
  Chooses a password $pw_i^{[guess]}$
  Computes $B_i^{[guess]} = \widehat{B_i} \oplus H(ID_i \| pw_i^{[guess]})$ and $\theta_i^{[guess]} = \widehat{\theta_i} \oplus H(ID_i \| pw_i^{[guess]})$
  Calls $Oracle\mathcal{FE}_{Rep}$ on the input $B_i^{[guess]}$ and $\theta_i^{[guess]}$ to retrieve the information ϕ_i, as $(\phi_i^*) \leftarrow Oracle\mathcal{FE}_{Rep}(B_i^{[guess]}, \theta_i^{[guess]})$
  Computes $pwr_i^{[guess]} = H(pw_i^{[guess]} \| \phi_i^*)$
until $(pwr_i^{[guess]} == pwr_i^*)$
if $(pwr_i^{[guess]} == pwr_i^*)$ **then**
  Return 1 (**Success**)
else
  Return 0 (**Failure**)
end if

We define the success probability for $EXP1^{oracle}_{\mathcal{A},\,TFUAS}$ as $Succ1^{oracle}_{\mathcal{A},\,TFUAS} = Pr[EXP1^{oracle}_{\mathcal{A},\,TFUAS} = 1]$. Then the advantage of $EXP1^{oracle}_{\mathcal{A},\,TFUAS}$ is given by $Adv1^{oracle}_{\mathcal{A},\,TFUAS}(t, qH, qFE) = max_{\mathcal{A}}\{Succ1^{oracle}_{\mathcal{A},\,TFUAS}\}$, where the maximum is taken over all $\mathcal{A}$ with the execution time t, the number of queries qH made to the $Oracle\mathcal{H}$ oracle and the number of queries qFE made to the $Oracle\mathcal{FE}$. Our proposed scheme is said to be provably secure against the adversary $\mathcal{A}$ for deriving the password pw_i and biometric parameter B_i of a user U_i, if $Adv1^{oracle}_{\mathcal{A},\,TFUAS}(t, qH, qFE) \leq \xi$, for any small $\xi > 0$. According to algorithm $EXP1^{oracle}_{\mathcal{A},\,TFUAS}$ (see Algorithm 1), if the adversary $\mathcal{A}$ gets success to compute inversion of the cryptographic one-way hash function $H(\cdot)$ and also gets success to solve hardness of fuzzy extractor, he/she can successfully derive the password pw_i and biometric parameter B_i of the user U_i by using of the $Oracle\mathcal{H}$ random oracle and $Oracle\mathcal{FE}$ random oracle, and wins the game. But, according to Definition 1 and Definition 2, we know that $Adv^{Oracle\mathcal{H}}_{\mathcal{A}}(t) \leq \xi_1$, for any small $\xi_1 > 0$ and $Adv^{Oracle\mathcal{FE}}_{\mathcal{A}}(t) \leq \xi_2$, for any small $\xi_2 > 0$. Since, we get $Adv1^{oracle}_{\mathcal{A},\,TFUAS}(t, qH, qFE) \leq \xi$, for any small $\xi > 0$ because, the proposed scheme depends on both $Adv^{Oracle\mathcal{H}}_{\mathcal{A}}(t)$ and $Adv^{Oracle\mathcal{FE}}_{\mathcal{A}}(t)$. Thus, our proposed scheme is secure against the adversary $\mathcal{A}$ for deriving the password pw_i and biometric parameter B_i of the user U_i.

Theorem 2. *Under the assumption that a cryptographic one-way hash function $H(\cdot)$ acts as a random oracle, the proposed scheme is provably secure against an adversary $\mathcal{A}$ for deriving the secret key s of the remote server RS even if the adversary $\mathcal{A}$ gets parameters that are stored into the memory of U_i's smart card and traps the communication messages between a user U_i and the remote server RS.*

Proof 2. This research construct an adversary $\mathcal{A}$ who has the ability to derive the secret key s of the remote server RS. For this purpose, this research considers same assumptions as discussed in Theorem 1. The adversary $\mathcal{A}$ runs the experiment, $EXP2^{oracle}_{\mathcal{A},\,TFUAS}$ for our three factor user authentication scheme (TFUAS) to derive the secret key s of the remote server RS as given in the Algorithm 2.

Algorithm 2 $EXP2^{oracle}_{\mathcal{A},\,TFUAS}$

Input: $D_i, ID_i, G_i, F_i, E_i, Q_i, L_i, K_i$
Output: 0 or 1

Calls $Oracle\mathcal{H}$ on the input D_i to retrieve the information $A_i = H(s\|ID_i)$ as $(A_i^*) \leftarrow Oracle\mathcal{H}(D_i)$
Calls $Oracle\mathcal{H}$ on the input F_i to retrieve the information A_i, r_i and T_i^1 as $(r_i^* \parallel A_i^{**}\|T_i^{1^*}) \leftarrow Oracle\mathcal{H}(F_i)$
Calls $Oracle\mathcal{H}$ on the input L_i to retrieve the information SK_i, T_s and A_i as $(T_s^*\|SK_i^*\|A_i^{***}) \leftarrow Oracle\mathcal{H}(L_i)$
if $(A_i^{***} == A_i^{**} == A_i^*)$ **then**
  Computes $r_i^{**} = A_i^* \oplus E_i$
  if $(r_i^{**} == r_i^*)$ **then**
    Computes $y_i^* = Q_i \oplus A_i^*$, $T_i^{1^{**}} = A_i^* \oplus G_i \oplus H(r_i^*)$ and $T_s^{**} = A_i^* \oplus K_i \oplus H(y_i^*)$
    if $(T_i^{1^{**}} == T_i^{1^*})$ && $(T_s^{**} == T_s^*)$ **then**
      Calls $Oracle\mathcal{H}$ on the input A_i^* to retrieve the information s and ID_i as $(s^*\|ID_i^*) \leftarrow Oracle\mathcal{H}(A_i^*)$
      if $(ID_i == ID_i^*)$ **then**
        Accepts s^* as secret key of RS
        Return 1 (**Success**)
      else
        Return 0 (**Failure**)
      end if
    else
      Return 0 (**Failure**)
    end if
  else
    Return 0 (**Failure**)
  end if
else
  Return 0 (**Failure**)
end if

We define the success probability for $EXP2^{oracle}_{\mathcal{A},\,TFUAS}$ as $Succ2^{oracle}_{\mathcal{A},\,TFUAS} = Pr[EXP2^{oracle}_{\mathcal{A},\,TFUAS} = 1]$. Then the advantage of $EXP2^{oracle}_{\mathcal{A},\,TFUAS}$ is given by $Adv2^{oracle}_{\mathcal{A},\,TFUAS}(t, qH) = max_{\mathcal{A}}\{Succ2^{oracle}_{\mathcal{A},\,TFUAS}\}$, where the maximum is taken over all $\mathcal{A}$ with the execution time t, the number of queries qH made to the $Oracle\mathcal{H}$ oracle. The proposed scheme is said to be provably secure against the adversary $\mathcal{A}$ for deriving the secret key s of the remote server RS, if $Adv2^{oracle}_{\mathcal{A},\,TFUAS}(t, qH) \leq \xi$, for any small $\xi > 0$. According to algorithm $EXP2^{oracle}_{\mathcal{A},\,TFUAS}$ (see Algorithm 2), if the adversary $\mathcal{A}$ gets success to compute inversion of the cryptographic one-way hash function $H(\cdot)$, he/she can successfully derive the secret key s of the remote server RS by using of the $Oracle\mathcal{H}$ random oracle and wins the game. But, according to Definition 1, we know that $Adv^{Oracle\mathcal{H}}_{\mathcal{A}}(t) \leq \xi_1$, for any small $\xi_1 > 0$. Since, we get $Adv2^{oracle}_{\mathcal{A},\,TFUAS}(t, qH) \leq \xi$, for any small $\xi > 0$ because, the proposed scheme depends on $Adv^{Oracle\mathcal{H}}_{\mathcal{A}}(t)$. Thus, our proposed scheme is secure against the adversary $\mathcal{A}$ for deriving the secret key s of the remote server RS.

Theorem 3. *Under the assumption that a cryptographic one-way hash function $H(\cdot)$ acts as a random oracle, the proposed scheme is provably secure against an adversary $\mathcal{A}$ for deriving a shared secret session key SK_i between a user U_i and the remote server RS even if the adversary $\mathcal{A}$ gets parameters that are stored into the memory of U_i's smart card and traps the communication messages between U_i and the remote server RS.*

Proof 3. This research construct an adversary $\mathcal{A}$ who has the ability to derive the session key SK_i between a user U_i and the remote server RS. For this purpose, this research considers same assumptions as discussed in Theorem 1. The adversary $\mathcal{A}$ runs the experiment, $EXP3^{oracle}_{\mathcal{A},\,TFUAS}$ for our three factor user authentication scheme (TFUAS) to derive the session key SK_i between the user U_i and the remote server RS as given in the Algorithm 3.

Algorithm 3 $EXP3^{oracle}_{\mathcal{A},\,TFUAS}$

Input: $D_i, ID_i, G_i, F_i, E_i, Q_i, L_i, K_i$
Output: 0 or 1

Calls $Oracle\mathcal{H}$ on the input D_i to retrieve the information $A_i = H(s\|ID_i)$ as $(A_i^*) \leftarrow Oracle\mathcal{H}(D_i)$
Calls $Oracle\mathcal{H}$ on the input F_i to retrieve the information A_i, r_i and T_i^1 as $(r_i^* \| A_i^{**}\|T_i^{1*}) \leftarrow Oracle\mathcal{H}(F_i)$
Calls $Oracle\mathcal{H}$ on the input L_i to retrieve the information SK_i, T_s and A_i as $(T_s^*\|SK_i^*\|A_i^{***}) \leftarrow Oracle\mathcal{H}(L_i)$
if $(A_i^{***} == A_i^{**} == A_i^*)$ **then**
  Computes $r_i^{**} = A_i^* \oplus E_i$
  if $(r_i^{**} == r_i^*)$ **then**
    Computes $T_i^{1**} = A_i^* \oplus G_i \oplus H(r_i^*)$, $y_i^* = Q_i \oplus A_i^*$ and $T_s^{**} = A_i^* \oplus K_i \oplus y_i^*$
    if $(T_i^{1**} == T_i^{1*})$ && $(T_s^{**} == T_s^*)$ **then**
      Calls $Oracle\mathcal{H}$ on the input SK_i^* to retrieve the information y_i and r_i as $(y_i^{**}\|r_i^{***}) \leftarrow Oracle\mathcal{H}(SK_i^*)$
      if $(r_i^{***} == r_i^*)$&& $(y_i^* == y_i^{**})$ **then**
        Accepts SK_i^* as the shared secret session key
        Return 1 (**Success**)
      else
        Return 0 (**Failure**)
      end if
    else
      Return 0 (**Failure**)
    end if
  else
    Return 0 (**Failure**)
  end if
else
  Return 0 (**Failure**)
end if

We define the success probability for $EXP3^{oracle}_{\mathcal{A},\,TFUAS}$ as $Succ3^{oracle}_{\mathcal{A},\,TFUAS} = Pr[EXP3^{oracle}_{\mathcal{A},\,TFUAS} = 1]$. Then the advantage of $EXP3^{oracle}_{\mathcal{A},\,TFUAS}$ is given by $Adv3^{oracle}_{\mathcal{A},\,TFUAS}(t, qH) = max_{\mathcal{A}}\{Succ3^{oracle}_{\mathcal{A},\,TFUAS}\}$, where the maximum is taken over all $\mathcal{A}$ with the execution time t, the number of queries qH made to the $Oracle\mathcal{H}$ oracle. The proposed scheme is said to be provably secure against the adversary $\mathcal{A}$ for deriving the session key SK_i between the user U_i and the remote server RS, if $Adv3^{oracle}_{\mathcal{A},\,TFUAS}(t, qH) \leq \xi$, for any small $\xi > 0$. According to algorithm $EXP3^{oracle}_{\mathcal{A},\,TFUAS}$ (see Algorithm 3), if the adversary $\mathcal{A}$ gets success to compute inversion of the cryptographic one-way hash function $H(\cdot)$, he/she can successfully derive the session key SK_i between the user U_i and the remote server RS by using of the $Oracle\mathcal{H}$ random oracle and wins the game. But, according to Definition 1, we know that $Adv^{Oracle\mathcal{H}}_{\mathcal{A}}(t) \leq \xi_1$, for any small $\xi_1 > 0$. Since, we get $Adv3^{oracle}_{\mathcal{A},\,TFUAS}(t, qH) \leq \xi$, for any small $\xi > 0$ because, the proposed scheme depends on $Adv^{Oracle\mathcal{H}}_{\mathcal{A}}(t)$. Thus, our proposed scheme is secure against the adversary $\mathcal{A}$ for deriving the session key SK_i between the user U_i and the remote server RS.

Theorem 1 demonstrated that the proposed scheme is secure against the *off-line password guessing attack*. Theorem 3 demonstrates that the proposed scheme is secure against the *session key recovery attack* because, without knowing random numbers $\{r_i, y_i\}$ then $\mathcal{A}$ cannot compute the session key SK_i. In the proposed scheme, the communicating messages depend on random numbers and the timestamp. Therefore, the communication messages are guaranteed to be different for every session. Thus, $\mathcal{A}$ cannot mount a *replay attack* on this proposed scheme. In this proposed scheme, $\mathcal{A}$ cannot mount a *forgery attack* without knowing secret password pw_i and biometric parameter B_i of a user U_i and the secret key s of the remote server RS. Theorems 1 and 2 show that the secret information of the remote server and the user are secure from $\mathcal{A}$. Thus, it is infeasible to mount a *forgery attack* on this proposed scheme.

A valid user say, $U_{\widehat{A}}$ as an adversary $\widehat{A}$ cannot login into the proposed authentication scheme as an another user say, U_i because, to login into the system, $\widehat{A}$ has to know the secret key s of the remote server RS. As, $\widehat{A}$ is a valid user, it knows their identity $ID_{\widehat{A}}$, password $pw_{\widehat{A}}$ and biometric information $\phi_{\widehat{A}}$. Therefore, $\widehat{A}$ can compute

Table 2. Security Functionality comparison of the proposed scheme with related schemes

Schemes	Ref. [17]	Ref. [18]	Ref. [19]	Ref. [22]	Ref. [23]	Ref. [24]	Ref. [26]	Our scheme
A1	√	√	×	-	√	√	×	×
A2	√	√	×	×	×	√	×	×
A3	√	√	√	-	-	×	×	×
A4	√	√	×	-	√	×	×	×
A5	√	√	√	-	-	√	√	×
A6	-	-	×	√	-	×	×	×
EPC	×	√	×	×	×	√	√	√
ELP	×	√	×	×	×	√	√	√

A1: Password guessing attack, A2: Insider attack, A3: Forgery attack, A4: Smart card stolen attack, A5: Replay attack, A6: Denial of Service (DoS) attack, ELP: Efficient login phase, EPC: Efficient password change phase, ×: no, √: yes

Table 3. Computation, communication and storage cost comparison of the proposed scheme with related schemes

Schemes	***Storage cost*** (in bits)	***Communication Cost*** (in bits) Login + Authentication	***Computation cost***	
			Login	Authentication
Ref. [17]	448	576	$2T_H$	$5T_H$
Ref. [18]	576	832	$3T_H$	$8T_H$
Ref. [19]	576	704	$3T_H$	$6T_H$
Ref. [22]	384	576	$4T_H+1T_{enc}$	$7T_H+1T_{dec}$
Ref. [23]	640	960	$3T_H$	$8T_H$
Ref. [24]	800	1120	$4T_H$	$10T_H+1T_{enc}+1T_{dec}$
Ref. [26]	384	1216	$4T_H$	$7T_H+1T_{enc}+1T_{dec}$
Proposed scheme	512	832	$4T_H$	$6T_H$

$H(s\|ID_{\widehat{A}})$ by computing $C_{\widehat{A}} \oplus H(pw_{\widehat{A}}\|\phi_{\widehat{A}})$, where $C_{\widehat{A}}$ is the smart card store parameter of $\widehat{A}$. From $H(s\|ID_{\widehat{A}})$, $\widehat{A}$ cannot extract s due to hardness of inversion of one-way hash function. Furthermore, Theorem 2 shows that s cannot be extracted or computed from known parameters of $\widehat{A}$. Therefore, $\widehat{A}$ unable to mount any attacks on the proposed scheme.

6 Comparison

In this section, the performances of the proposed scheme with the existing authentication schemes namely, Li and Hwang's scheme [17], Das's scheme [18], An's scheme [19], Tan's scheme [22], Yan et al.'s scheme [23], Mishra et al.'s schemw [24] and He et al.'s scheme [26] are compared. However, the compared schemes in [17–19] and [22–24] and [26] are not suitable for practical use because, the schemes cannot resist the possible attacks as shown in Table 2. In the introduction part of this paper, it has been described that is insecure against security attacks. Moreover, security analysis of the proposed scheme (see Section 5) shows that the proposed scheme can resist all the possible attacks. Thus, the proposed scheme is more secure than other schemes.

Table 3 shows the computational cost, storage cost and communication overhead comparison of schemes in [17–19], [22–24], and [26] with our proposed scheme. For this purpose, only login and authentications phases have been consider due to maximum use. T_H, T_{enc} and T_{dec} are the time required for hash operation, symmetric key encryption and decryption respectively. The proposed scheme takes time for 10 hashing operations in two phases which is the lower among related and compared schemes in [18, 22–24, 26]. It can be reasonably assumed that the length of ID_i and pw_i are 64 bits each. Large prime number q, cryptographic one-way hash function $H(\cdot)$ like SHA-1[1], symmetric key encryption/decryption like AES with key size 128 bits[2], random numbers, symmetric key encryption/decryption and timestamp returns 128 bits each. The communication cost of proposed scheme for login message is (128+128+128+64) = 448 bits and message generated in authentication phase is (128+128+128)

[1]http://csrc.nist.gov/publications/fips/fips180-4/fips-180-4.pdf
[2]http://nvlpubs.nist.gov/nistpubs/FIPS/NIST.FIPS.197.pdf

= 384 bits. Therefore, total communication cost is (448+384) = 832 bits which is lower than existing and compared schemes in [23, 24, 26]. The storage cost of our proposed scheme is (128 + 128 + 128 + 128) = 512 bits which is also lower than related schemes in [18, 19, 23, 24].

After resisting all possible attacks as shown in Section 5, the proposed scheme provides better trade-off among storage, computational and communication costs than other related existing schemes. Hence, it can be claimed that the proposed scheme is more efficient and secure than other related existing schemes and also it is applicable for practical applications.

- **Advantages of proposed scheme**

In the following, the advantages of the proposed scheme have been discussed.

Efficient login phase: If a user U_i enters faulty password and faulty identity by some means in login phase of the proposed scheme, the smart easily can detect the wrong inputs before going to generate a login request message. For this purpose, smart card computes $B_i' = \widehat{B_i} \oplus H(ID_i \| pw_i)$ and checks $des(B_i^*, B_i') \leq \delta d$. If it does not hold, the smart card rejects U_i; otherwise, it computes $\theta_i' = \widehat{\theta_i} \oplus H(ID_i \| pw_i)$, $\phi_i' \leftarrow Rep(B_i^*, \theta_i')$, $pwr_i' = H(pw_i \| \phi_i')$, $A_i' = C_i \oplus pwr_i'$, $D_i' = H(A_i')$ and checks computed D_i' and stored D_i are equal or not. If equality does not hold, the smart card rejects U_i; otherwise, accepts the password pw_i and identity ID_i as correct inputs. Therefore, extra communication overhead due to wrong inputs can be avoided in the proposed scheme.

Efficient password change phase: If a user U_i enters faulty password and faulty identity by some means in password phase of the proposed scheme, the smart easily can detect the wrong inputs before going to give permission to the user to submit their new password. For this purpose, smart card computes executes the same steps as mentioned above to check the submitted inputs. If provided inputs are correct, then only the smart card give permission to enter new password to U_i. Furthermore, to change password of a user, smart card does not need to communicate with the remote server. Therefore, communication overhead is also reduced in proposed scheme with efficient wrong input detection.

Mutual authentication: In the proposed scheme, the remote server RS computes and accepts a secret session key SK_i after verifying legitimacy of a user U_i through login request message and then, RS sends a reply message to the user U_i. The user U_i agrees upon the same secret session key SK_i with RS after verifying legitimacy of RS through reply message. Therefore, both way authentication has been done in the proposed scheme. Furthermore, the proposed scheme can resist all the possible attacks (see, Section 5). Hence, the proposed scheme achieves mutual authentication.

7 Conclusion

This paper have proposed an authentication scheme. After analyzing the proposed scheme it can be stated that the proposed scheme can overcome the all possible attacks and has better trade-off among computational, storage and communication costs. Therefore, the proposed scheme is suitable for real world online applications.

References

[1] T. Maitra, Cryptanalysis of A Secure Remote User Authentication Scheme Using Smart Cards, CoRR **abs/1502.04820** (2015)

[2] Y. Dodis, L. Reyzin, A. Smith, in *Advances in Cryptology - EUROCRYPT 2004* (Springer Berlin Heidelberg, 2004), Vol. 3027 of *Lecture Notes in Computer Science*, pp. 523–540

[3] L. Lamport, Password Authentication with Insecure Communication, Commun. ACM **24**, 770 (1981)

[4] A. Shimizu, T. Horioka, H. Inagaki, A Password Authentication Method for Contents Communications on the Internet, IEICE Tran. Communications **E81-B**, 1666 (1998)

[5] J. Xu, W.T. Zhu, D.G. Feng, An improved smart card based password authentication scheme with provable security, Computer Standards & Interfaces **31**, 723 (2009)

[6] X. Li, J. Niu, M.K. Khan, J. Liao, An enhanced smart card based remote user password authentication scheme, Journal of Network and Computer Applications **36**, 1365 (2013)

[7] T. Maitra, R. Amin, D. Giri, P.D. Srivastava, An Efficient and Robust User Authentication Scheme for Hierarchical Wireless Sensor Networks without Tamper-Proof Smart Card, I. J. Network Security **18**, 553 (2016)

[8] O. Mir, T. van der Weide, C.C. Lee, A Secure User Anonymity and Authentication Scheme Using AVISPA for Telecare Medical Information Systems, Journal of Medical Systems **39**, 89 (2015)

[9] D. Giri, T. Maitra, R. Amin, P. Srivastava, An Efficient and Robust RSA-Based Remote User Authentication for Telecare Medical Information Systems, Journal of Medical Systems **39**, 145 (2014)

[10] K.H. Yeh, A lightweight authentication scheme with user untraceability, Frontiers of Information Technology & Electronic Engineering **16**, 259 (2015)

[11] C. Guo, C.C. Chang, Chaotic maps-based password-authenticated key agreement using smart cards, Communications in Nonlinear Science and Numerical Simulation **18**, 1433 (2013)

[12] R. Amin, T. Maitra, S.P. Rana, An Improvement of Wang. et. al.'s Remote User Authentication Scheme against Smart Card Security Breach, International Journal of Computer Applications **75**, 37 (2013)

[13] C.C. Chang, C.Y. Sun, A Secure and Efficient Authentication Scheme for E-coupon Systems, Wireless Personal Communications **77**, 2981 (2014)

[14] R. Amin, T. Maitra, D. Giri, An Improved Efficient Remote User Authentication Scheme in Multi-server Environment using Smart Card, International Journal of Computer Applications **69**, 1 (2013)

[15] T. Maitra, M.S. Obaidat, S.H. Islam, D. Giri, R. Amin, Security analysis and design of an efficient ECC-based two-factor password authentication scheme, Security and Communication Networks **9**, 4166 (2016)

[16] T. Maitra, M.S. Obaidat, R. Amin, S.H. Islam, S.A. Chaudhry, D. Giri, A robust ElGamal-based password-authentication protocol using smart card for client-server communication, International Journal of Communication Systems **30** (2017)

[17] C.T. Li, M.S. Hwang, An efficient biometrics-based remote user authentication scheme using smart cards, Journal of Network and Computer Applications **33**, 1 (2010)

[18] A. Das, Analysis and improvement on an efficient biometric-based remote user authentication scheme using smart cards, Information Security, IET **5**, 145 (2011)

[19] Y. An, Security Analysis and Enhancements of an Effective Biometric-Based Remote User Authentication Scheme Using Smart Cards, Journal of Biomedicine and Biotechnology **2012, Article ID 519723**, 1 (2012)

[20] X. Li, J. Niu, M.K. Khan, J. Liao, X. Zhao, Robust three-factor remote user authentication scheme with key agreement for multimedia systems, Security and Communication Networks (2014)

[21] C.C. Lee, C.W. Hsu, A secure biometric-based remote user authentication with key agreement scheme using extended chaotic maps, Nonlinear Dynamics **71**, 201 (2013)

[22] Z. Tan, An efficient biometrics-based authentication scheme for telecare medicine information systems, Przeglad Elektrotechniczny pp. 200–204 (2013)

[23] X. Yan, W. Li, P. Li, J. Wang, X. Hao, P. Gong, A Secure Biometrics-based Authentication Scheme for Telecare Medicine Information Systems, Journal of Medical Systems **37**, 9972 (2013)

[24] D. Mishra, S. Mukhopadhyay, A. Chaturvedi, S. Kumari, M. Khan, Cryptanalysis and Improvement of Yan et al.'s Biometric-Based Authentication Scheme for Telecare Medicine Information Systems, Journal of Medical Systems **38**, 24 (2014)

[25] T. Maitra, D. Giri, An Efficient Biometric and Password-Based Remote User Authentication using Smart Card for Telecare Medical Information Systems in Multi-Server Environment, Journal of Medical Systems **38**, 142 (2014)

[26] D. He, N. Kumar, J.H. Lee, R. Sherratt, Enhanced three-factor security protocol for consumer USB mass storage devices, Consumer Electronics, IEEE Transactions on **60**, 30 (2014)

[27] D. Giri, R.S. Sherratt, T. Maitra, R. Amin, Efficient biometric and password based mutual authentication for consumer USB mass storage devices, IEEE Transactions on Consumer Electronics **61**, 491 (2015)

[28] D. Giri, R.S. Sherratt, T. Maitra, A novel and efficient session spanning biometric and password based three-factor authentication protocol for consumer USB Mass Storage Devices, IEEE Transactions on Consumer Electronics **62**, 283 (2016)

[29] H.F. Huang, H.W. Chang, P.K. Yu, Enhancement of Timestamp-based User Authentication Scheme with Smart Card, International Journal of Network Security **16**, 463 (2014)

[30] R. Amin, T. Maitra, D. Giri, P.D. Srivastava, Cryptanalysis and Improvement of an RSA Based Remote User Authentication Scheme Using Smart Card, Wireless Personal Communications (2017)

[31] D. Dolev, A.C. Yao, On the security of public key protocols, Information Theory, IEEE Transactions on **29**, 198 (1983)

[32] P. Kocher, J. Jaffe, B. Jun, *Differential Power Analysis*, in *Advances in Cryptology "CRYPTO'99* (1999), Vol. 1666 of *Lecture Notes in Computer Science*, pp. 388–397

[33] T.S. Messerges, E.A. Dabbish, R.H. Sloan, Examining Smart-Card Security Under the Threat of Power Analysis Attacks, IEEE Trans. Comput. **51**, 541 (2002)

DOI: 10.1051/itmconf/20171301020

[34] S.K. Sood, A.K. Sarje, K. Singh, A secure dynamic identity based authentication protocol for multi-server architecture, Journal of Network and Computer Applications **34**, 609 (2011)

ITM Web of Conferences **13**, 01021 (2017) DOI: 10.1051/itmconf/20171301021
CMES2017

Classifications on the travelling wave solutions to the (3+1)-dimensional generalized KP and Jimbo-Miwa equations

Ozlem Isik[1,*], *Ozge Irmak* Degirmenci[2] , and *Hasan* Bulut[2]

[1]Kirklareli University, Department of Mathematics, 39100 Kirklareli, Turkey
[2]Firat University, Department of Mathematics, 23119 Elazig, Turkey

Abstract. In this paper, we use the powerful $\tan(F(\xi)/2)$ -expansion method with the help of Wolfram Mathematica 9 to investigate the solution structures of two well-known nonlinear evolution equations, namely ; the (3+1)-dimensional generalized KP and the (3+1)-dimensional Jimbo-Miwa equations. We obtain new solutions such as hyperbolic function, trigonometric function, exponential function and rational function solutions. We plot two- and three-dimensional graphics of some obtained results using the same program, Wolfram Mathematica 9.

1 Introduction

Many phenomena in nonlinear science can be expressed in the form of nonlinear partial differential equations. In recent years, various methods have been used to find the solitary wave solutions to such class of problems such as the sine-cosine function method [1], the variational iteration method [2], the Homotopy perturbation method [3], the exp-function method [4], the generalized tanh-coth method [5], the extended hyperbolic function method [6], the modified simple equation method [7], the sine-Gordon expansion method [8] and so on. However, in this study, we employ the effective $\tan(F(\xi)/2)$ -expansion method to investigate the new solutions of the (3+1)-dimensional generalized Kadomtsev-Petviashvili (KP) equation [9] and the (3+1)-dimensional Jimbo-Miwa equation [10]. The KP equation is introduced by [11] to describe the model of nonlinear wave motion that arises in the field of mathematical physics. The Jimbo-Miwa equation is used to describe (3+1)-dimensional wave model in the field of physics [12]. Various methods have been implemented to obtain different types of solutions to the equations mentioned above such as Backlund transformation [13], the simplified Hirota's method [14], the simplified Hirota's direct method [15], the exp-function method [4], the generalized tanh method [16], the Xu's stable-range method and logarithmic generalization of stable-range method [17], the Kudryashov method [18], the homogeneous balance method [19] and many more methods. Moreover fractional concepts are investigated in [20, 21, 22].

* Corresponding author: ozlem.isik@klu.edu.tr

2 The $\tan(F(\xi)/2)$-expansion method

In this section, we explain the $\tan(F(\xi)/2)$-expansion method and how it can be applied to obtain the solutions of a given nonlinear PDE as in the following

$$P\left(u,u^2,u_x,u_{xx},u_t,u_{tt},u_{xt},\ldots\right)=0. \tag{2.1}$$

Performing the travelling wave transformation $u(x,t)=U(\xi),\xi=x-kt$, Eq. (2.1) reduces to the following ODE

$$Q\left(U',U'',U''',\ldots\right)=0. \tag{2.2}$$

The solution of Eq. (2.2) is given as follows

$$U(\xi)=\sum_{i=0}^{m}A_i\left(p+\tan\left(\frac{F(\xi)}{2}\right)\right)^i+\sum_{i=1}^{m}B_i\left(p+\tan\left(\frac{F(\xi)}{2}\right)\right)^{-i},A_m\neq 0,B_m\neq 0, \tag{2.3}$$

where $m\in\mathbb{Z}^+$ that can be determined by balancing the highest order derivative and the highest nonlinear term in Eq. (2.2). The coefficients A_i, $B_i,(0\leq i\leq m)$ are constants to be determined and $F=F(\xi)$ satisfies the nonlinear ODE [23]:

$$F'(\xi)=a\sin(F(\xi))+b\cos(F(\xi))+c. \tag{2.4}$$

Substituting Eq. (2.3) into Eq. (2.2) produces a system of algebraic equations for $\tan(F(\xi)/2)^i$, $\cot(F(\xi)/2)^i$. After collecting this separated system of equations, we calculate $k,p,A_0,A_1,B_1,\ldots,A_m,B_m$. For solutions of Eq. (2.4), see [23].

3 Applications

In this section, implementation of the $\tan(F(\xi)/2)$-expansion method is given.
1. Let us consider the (3+1)-dimensional generalized KP equation given by [9]:

$$u_{xxxy}+3\left(u_xu_y\right)_x+u_{tx}+u_{ty}-u_{zz}=0, \tag{3.1}$$

performing the transformation $u(x,y,z,t)=U(\xi)$, $\xi=x+y+\beta z-kt$, (3.1) becomes:

$$U'''+3\left(U'\right)^2-\left(2k+\beta^2\right)U'=0. \tag{3.2}$$

Using the balancing principle on Eq. (3.2) by considering the terms $(U')^2$ and U''', we obtain *m=1*. Together with Eq. (2.3), we consider the solution of Eq. (3.1) to be:

$$U(\xi)=A_0+A_1\left(p+\tan\left(\frac{F(\xi)}{2}\right)\right)+B_1\left(p+\tan\left(\frac{F(\xi)}{2}\right)\right)^{-1}. \tag{3.3}$$

Substituting Eq. (3.3) into Eq. (3.2) and simplifying together with Eq. (2.4), we obtain a system of algebraic equations in A_1,B_1,p,k,a,b,c and solving this system we obtain families of solutions in each case by considering the solutions of Eq. (2.4) that can be found in [23]. One of the solutions of Eq. (3.1) for Case-1 is given below.

Case1: $A_1=0,B_1=b+c-2ap-bp^2+cp^2,k=\left(a^2+b^2-c^2+\beta^2\right)/2$

Solution4: when $a^2+b^2-c^2<0,c\neq 0$ and $b=0$

$$u_{1,4}(x,y,z,t)=c\left(c-2ap+cp^2\right)\Big/\left[cp-a^2+\sqrt{c^2-a^2}\,\tan\left(\omega_{1,4}(x,y,z,t)\right)\right], \tag{3.4}$$

where $\omega_{1,4}(x,y,z,t)=\sqrt{c^2-a^2}\left(2(x+y+\beta z)+\left(\beta^2+c^2-a^2\right)t\right)/4.$

2. Consider the (3+1)-dimensional Jimbo-Miwa equation [10]:

$$u_{xxxy}+3u_y u_{xx}+3u_x u_{xy}+2u_{yt}-3u_{xz}=0, \tag{3.5}$$

performing the transformation $u(x,y,z,t)=U(\xi),\xi=x+y+\beta z-kt$, Eq. (3.5) becomes

$$U'''+3(U')^2-(2k+3\beta)U'=0. \tag{3.6}$$

Using the balancing principle on Eq. (3.6) by considering the terms $(U')^2$ and U''', we obtain *m=1*. Together with Eq. (2.3), we consider the solution of Eq. (3.5) to be:

$$U(\xi)=A_0+A_1\left(p+\tan\left(\frac{F(\xi)}{2}\right)\right)+B_1\left(p+\tan\left(\frac{F(\xi)}{2}\right)\right)^{-1}. \tag{3.7}$$

Substituting Eq. (3.7) into Eq. (3.6) and simplifying together with Eq. (2.4), we obtain a system of algebraic equations in A_1,B_1,p,k,a,b,c. We solve this system and consequently families of solutions to Eq. (3.5) is obtained by considering the solutions of Eq. (2.4) that can be found in [23]. A family of solution to (3.5) is given for the Case1 as follows

Case1

$$A_1=b-c, B_1=-a^2-b^2-3cA_1+3bc-2c^2+2k+3\beta/(3A_1), a=-\sqrt{-b^2+c^2+2k+3\beta}$$

Solution-9: when $b=0$ and $a=c$

$$u_9(x,y,z,t)=c(1-p)+4/[2(x+y+\beta z)+3\beta t]. \tag{3.8}$$

4 Graphics Plot

In this section we present two- and three-dimensional graphics of some obtained solutions.

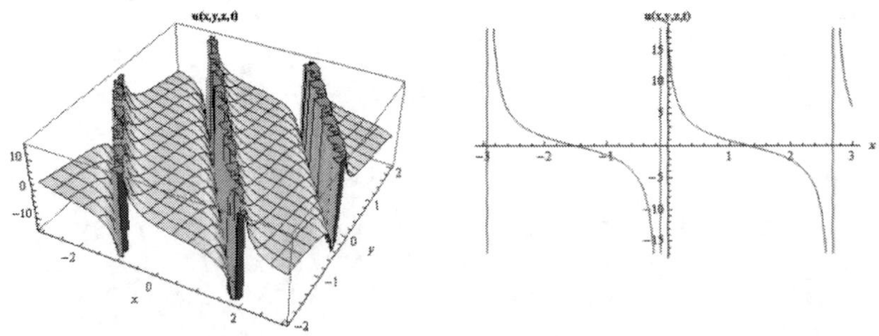

Fig. 1. 3D and 2D surfaces of Eq. (3.4) by considering the values *β=0.5, a=2, c=3, p=0.7, t=0.02, z=0.01, -3<x<3, -2<y<2* and *y=0.03* for the 2D graphic.

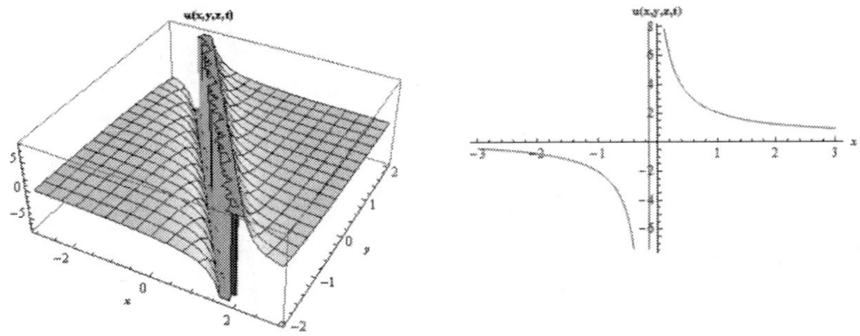

Fig. 2. 3D and 2D surfaces of Eq. (3.8) by considering the values *β=3, c=1, p=0.7, t=0.02, z=0.01, -3<x<3, -2<y<2* and *y=0.03* for the 2D graphic.

5 Conclusions

In this manuscript, we employed the $\tan(F(\xi)/2)$ -expansion method to (3+1)-dimensional generalized KP and Jimbo-Miwa equations. We obtained various types of solutions as, trigonometric, hyperbolic, exponential and rational functions. We plot two- and three-dimensional graphics of some obtained solutions by using Wolfram Mathematica9. These solutions are tested and it is verified that they satisfy their corresponding equations. We compared our results with the results obtained in [9] and [10] and we observe that our results are new with different solution structures.

References

1. A.J.M. Jawad, IJRRAS, **13**, 1 (2012)
2. A.M. Wazwaz, Computers and Mathematics with Applications, **54**, (2007)
3. T.A. Nofel, Applied Mathematitcs, **5**, (2014)
4. T. Ozis, I. Aslan, Phcysics Letters A, **372** (2008)
5. G.G. Hernan, Advances in Difference Equations, **9**, 138 (2015)
6. Y. Shang, Chaos, Solutions and Fractals, **36**, 3 (2008)
7. M. Mirzazadeh, Information Sciences Letters, **3**, 1 (2014)
8. H. Bulut, T.A. Sulaiman, H.M. Baskonus, Opt. Quant. Electron., **48**, 564 (2016)
9. W. Ma, Z. Zhu, Applied Mathematics and Computation, **218** (2012)
10. W. Ma, J. Lee, Chaos, Solitons and Fractals, **42** (2009)
11. B.B. Kadomtsev, V.I. Petviashvili, Sov. Phys. Dokl., **15** (1970)
12. M.T. Darvishi, M. Najafi, World Academy of Science, Engineering and Technology, **55**, (2011)
13. W. Ma, A. Abdeljabbar, Applied Mathematics Letters, **25**, 7 (2012)
14. A.M. Wazwaz, Commun. Nonlinear Sci. Numer. Simulat., **17** (2012)
15. A.M. Wazwaz, S.A. El-Tantawy, Nonlinear Dyn., **84** (2016)
16. W. Hong, K. Oh, Computers and Mathematics with Applications, **39** (2000)
17. B. Cao, Acta. Appl. Math., **112**, 2 (2010)
18. N. Kadkhoda, CJMS, **4,** 2 (2015)
19. M. Eslami, CJMS, **2**, 2 (2014)
20. E. Bas, F. Metin, Advances in Difference Equations, **2013** (2013)
21. E. Bas, ActaScientiarum. Technology, **37.2** (2015)
22. E. Bas, R.Yilmazer, E. Panakhov, The Scientific World Journal, **2013** (2013)
23. J. Manafian, Optik, **127** (2016)

Sequences with Random Indice in Classical Banach Space

Mehmet GÜRCAN[1,*], *Sinan ÇALIK*[1], *and Yunus GÜRAL*[1]

[1]Fırat University Faculty of Science, Department of Statistics, Elazığ, Turkey

Abstract. In this study we define sequences with random indice in classical Banach space and analyzing it's some properties. After then we define differences sequence space on sequences with random indice. The difference operation is based on indice, and the generated difference sequence has a free stepwise. And here an important definition is the embedding space. Consequently we showed that l_∞ is a embedding space to c.

1 Introduction

Throughout the study ω, l_∞, c, c_0 denote the spaces of all, bounded, convergent and null sequences, respectively. Kızmaz introduced the different sequence spaces on space A and later was generalized by Et and Çolak generalized different sequence spaces respectively $\Delta(A)$, $\Delta^r(A), r \geq 1$. These defined sequence spaces are often used in analysis, and the norm defined on them is usually supremum norm as follows Where convergence is defined according to norm and special sum of sequence. Let's say a few important following by Lorentz

$$\|a\| = sup_n |a_n| \, , a \in A \tag{1}$$

Where convergence is defined according to norm and special sum of sequence. Let's say a few important following by Lorentz

$$\hat{c} = \left\{a\colon lim_n \frac{1}{n} \sum_{k=1}^{n} a_{k+m} < \infty \; uniformly \; in \; m\right\} \tag{2}$$

By Maddox, strongly almost convergent to a number L,

* Corresponding author: mehmetgurcan2000@yahoo.com

$$[\hat{c}] = \left\{a: \frac{1}{n}\sum\nolimits_{k=1}^{n} |a_{k+m} - L| \to 0 \text{ uniformly } in\ m \text{ , as } n \to \infty\right\} \tag{3}$$

The following relationship exists between these sequence spaces

$$c \subset [\hat{c}] \subset \hat{c} \subset l_\infty \tag{4}$$

By Nanda

$$[\hat{c}, p] = \left\{a: \frac{1}{n}\sum\nolimits_{k=1}^{n} |a_{k+m} - L|^{p_k} \to 0 \text{ uniformly in } m \text{ , as } n \to \infty\right\} \tag{5}$$

$$[\hat{c}, p]_0 = \left\{a: \frac{1}{n}\sum\nolimits_{k=1}^{n} |a_{k+m}|^{p_k} \to 0 \text{ uniformly in } m \text{ , as } n \to \infty\right\} \tag{6}$$

$$[\hat{c}, p]_\infty = \left\{a: sup_{n,m} \frac{1}{n}\sum\nolimits_{k=1}^{n} |a_{k+m}|^{p_k} < \infty\right\} \tag{7}$$

Where $p = (p_k)$ positive reel sequence.

By M. Et and R. Çolak

$$A(\Delta^r) = \{a: \Delta^r a \in A\} \tag{8}$$

Were $\Delta^r x_k = \sum_{v=0}^{r} (-1)^r \binom{r}{v} a_{k+v}$. In addition the following sequence spaces have been defined by nodulus function f,

$$[\hat{c}, f, p](\Delta^r) = \left\{a: \frac{1}{n}\sum\nolimits_{k=1}^{n} [f(|\Delta^r a_{k+m} - L|)]^{p_k} \to 0\ uniformly\ in\ m\ , as\ n \to \infty\right\} \tag{9}$$

$$[\hat{c}, f, p]_0(\Delta^r) = \left\{a: \frac{1}{n}\sum\nolimits_{k=1}^{n} [f(|\Delta^r a_{k+m}|)]^{p_k} \to 0\ uniformly\ in\ m\ , as\ n \to \infty\right\} \tag{10}$$

$$[\hat{c}, f, p]_\infty(\Delta^r) = \left\{a: sup_{n,m} \frac{1}{n}\sum\nolimits_{k=1}^{n} [f(|\Delta^r a_{k+m}|)]^{p_k} < \infty\right\} \tag{11}$$

Where modulus f is a function from $[0, \infty)$ to $[0, \infty)$ such that

i) $f(x) = 0$ if and only if $x = 0$

ii) $f(x + y) \leq f(x) + f(y)$ for $x, y \geq 0$

iii) f is increasing

iv) f is a continuous the right at 0.

The related features, about above spaces, are shown in the following three theorems,

Theorem 1. The spaces $[\hat{c}, f, p](\Delta^r)$, $[\hat{c}, f, p]_0(\Delta^r)$ and $[\hat{c}, f, p]_\infty(\Delta^r)$ are linear spaces on the complex numbers.

Theorem 2. The spaces $[\hat{c}, f, p](\Delta^r)$, $[\hat{c}, f, p]_0(\Delta^r)$ are subspaces to $[\hat{c}, f, p]_\infty(\Delta^r)$ linear space.

Theorem 3. The set A' is a convex subset of linear space A then $\Delta^m(A') \subset \Delta^m(A)$ is a convex subset.

2 Sequences with Random Indice

$\xi = \{X_t : t \geq 0\}$ be a continuous time parameter stochastic process with state space $S = \{1, 2, \cdots\}$. Every X_t is a random variable on the probability space $(\Omega, \mathfrak{I}, P)$. Let be define sequence as following

$$(a, \xi) = (a_{X_t}), t \geq 0 \tag{12}$$

Where the realization sequence $(a, \xi(\omega)) = (a_{X_t(\omega)})$ is equal to (a) or subsequence of (a) with every realization $X_t(\omega)$ for arbitrary $\omega \in \Omega$. We defined special condition $(a) = (a_{X_t(\bar{\omega})})$. Every realization $(a_{X_t(\omega)}) \subset (a)$ for $\bar{\omega} \neq \omega$. Let we define following set and following spaces

$$\{a, \xi, \Omega\} = \{a_{X_t(\omega)} : \omega \in \Omega\} \tag{13}$$

$$\{A, \xi, \Omega\} = \{a_{X_t(\omega)} : \omega \in \Omega, a \in A\} \tag{14}$$

$$\{A, \xi\} = \{a_{X_t} : a \in A\} \tag{15}$$

Where it easily seen that $\{A, \xi, \Omega_1\} \subset \{A, \xi, \Omega_2\}$ for $\Omega_1 \subset \Omega_2$.

Definition: Let A and B sequence space with $A \subset B$. B is defined a embedding space to A if following equation is hold $\{A, \xi\} = \{B, \xi, \cap_{i=1}^{\infty} \Omega_i\}$, for every $\Omega = \cup_{i=1}^{\infty} \Omega_i$ and $\Omega \supset \Omega_1 \supset \Omega_2 \ldots$

The following theorem is a correctness the definition,

Theorem 1. The space l_∞ is a embedding space to c.

Proof: It's advisedly that $c \subset l_\infty$. For proof we must shown that $\{c, \xi\} = \{l_\infty, \xi, \cap_{i=1}^{\infty} \Omega_i\}$. For every $a \in l_\infty$, there is a $k_0 \geq 1$ and $\omega \in \Omega_{k_0}$ such that $a_{X_t(\omega)} \in \{l_\infty, \xi, \Omega_{k_0}\}$. For the sequence a is a bounded there is a at least one limit point. In this case the sequence $a_{X_t(\omega)}$ has a one limit point such that $a_{X_t(\omega)} \in \{c, \xi\}$. Otherwise the sequence $a_{X_t(\omega)}$ has a one limit point for $k > k_0$. So that $a_{X_t(\omega)} \in \{c, \xi\}$. As a result $\{c, \xi\} = \{l_\infty, \xi, \cap_{i=1}^{\infty} \Omega_i\}$. Then the space l_∞ is a embedding space to c. Proof is complete.

Let us some of important theorems for space $\{A, \xi, \Omega\}$

Theorem 2. If A is a Banach space with norm $\| \ \|$ then $\{A, \xi, \Omega\}$ is a Banach space.

Proof: Let be $(e_n) \subset A$ is a Cauchy sequence then $e_n \to e \in A, n \to \infty$ fort this $(A, \| \ \|)$ Banach space. Such that $e_{X_t}, e_{nX_t} \in \{A, \xi, \Omega\}$, $n \geq 1$, following inequality hold

$$\|e_{nX_t} - e_{X_t}\| \leq \|e_n - e\| \tag{16}$$

The proof is complete.

Theorem 3. If A is a linear space on $\mathbb{C}$ complex numbers then $\{A, \xi, \Omega\}$ is a linear space.

Proof: The proof is easily seen that $(\lambda a + \mu b)_{X_t(\omega)} = \lambda a_{X_t(\omega)} + \mu b_{X_t(\omega)}$, $\lambda, \mu \in \mathbb{C}, a, b \in A$.

Important Result: The following sequence spaces is a linear space on $\mathbb{C}$

$$\{\hat{c}, \xi, \Omega\} = \left\{a_{X_t(\omega)} : n^{-1} \sum_{k=1}^{n} a_{k+X_t(\omega)} < \infty, n \to \infty, uniformly\ \omega \in \Omega\right\} \quad (17)$$

$$\{[\hat{c}], \xi, \Omega\} = \left\{a_{X_t(\omega)} : n^{-1} \sum_{k=1}^{n} \left|a_{k+X_t(\omega)} - L\right| \to 0, n \to \infty, uniformly\ \omega \in \Omega\right\} \quad (18)$$

$$\{[\hat{c}, p], \xi, \Omega\} = \left\{a_{X_t(\omega)} : n^{-1} \sum_{k=1}^{n} \left|a_{k+X_t(\omega)} - L\right|^{p_k} \to 0, n \to \infty, uniformly\ \omega \in \Omega\right\} \quad (19)$$

$$\{[\hat{c}, p]_0, \xi, \Omega\} = \left\{a_{X_t(\omega)} : n^{-1} \sum_{k=1}^{n} \left|a_{k+X_t(\omega)}\right| \to 0, n \to \infty, uniformly\ \omega \in \Omega\right\} \quad (20)$$

$$\{[\hat{c}, p]_\infty, \xi, \Omega\} = \left\{a_{X_t(\omega)} : sup_{n,\omega} n^{-1} \sum_{k=1}^{n} \left|a_{k+X_t(\omega)}\right|^{p_k} < \infty\right\} \quad (21)$$

References

1. H. Kizmaz, On certain sequence space, Canad. Math. Bull, **24**(2), 169-76(1981).
2. M. Et and C.A. Bektaş, Generalized strongly (V, λ) summable sequences defined by Orlicz function, Mathematica Slovaca, **54**(4), 411-422(2004).
3. H. Altınok, M. Et and Y. Altın, Strongly almost summable difference sequences, Vietnam J. of Math., **34**(3), 331-339(2006).
4. B. Chandra Tripathy, M. Et and Y. Altın, Generalized difference sequence spaces by Orliczh function in a locally concex space, J. of An. And Ap. **1**(3), 175-192(2003).
5. B. Chandra Tripathy, Generalized difference paranormed statistically convergent sequences defined by Orliczh function in locally convex space, Soochow J. of Math. **30**(4), 431-446(2004).

Useful Inequalities for Upper and Lower Bounds for the Repairable Aging Cold Standby System

Gökhan GÖKDERE[*]

Fırat University Faculty of Science, Department of Statistics, Elazığ, Turkey

Abstract. In this paper, we study a cold standby repairable system consisting of two non-identical components and a single repair facility with one repairman. It is assumed that, one of the two components is operating while the other is in cold standby and also two components follow a geometric process. Under these assumptions, at first we present Laplace-Stieltjes transform of the system lifetime and then we give some useful inequalities about the mean lifetime of the system.

1 Introduction

Cold standby redundancy technique uses one or more redundant components that are unpowered, do not consume any energy and do not fail until being activated to replace a faulty online component. Whenever working component fails, then an available cold standby component, i.e. inactive standby component, is immediately powered up to take over the mission task. Some recent works on the research of the cold standby systems are in [1-3].

In practice after the repair, most repairable components are not "as good as new". To pay attention this problem, much work has been done by [4-7]. For an imperfect repair model, it is more acceptable to consider that the successive operating times of the component after repair will be even shorter while the consecutive repair times of the component after fail will be even longer [8-11].

Consider a redundant system consisting of two non-identical components called component 1 and component 2 and a single repair facility with one repairman. Initially, one component begins to operate, the other component is inactive standby. As soon as the operating component fails, the standby component is switched on immediately and failed component is taken for repair. When repair of one component is completed and the other component is still operating the repaired component returns to the standby pool and is once again available to be used as operating component. The system is breakdown when the working component fails while the other component is under repair.

[*] Corresponding author: g.g.gokdere@gmail.com

DOI: 10.1051/itmconf/20171301023

2 Lifetime of the system

In this section, at first we will investigate possible malfunction situations of the system and then we will define the system lifetime measured from the moment when component 1 fails.

We make the following assumptions concerning failures and repairs.

Assumption 1. Failure time of component k is the random variable $X_k^{(n)}$ with the distribution function

$$F_{X_k}^{(n)}(t) = F_k(m_k t), k = 1,2;\ t \geq 0,$$

where $m_k = \begin{cases} a^n, & k = 1 \\ a^{n-1}, & k = 2 \end{cases}$ and $a > 1$ is the ratio of the geometric process (GP) for component k.

Assumption 2. Repair time of component k is the random variable $Y_k^{(n)}$ with the distribution function

$$G_{Y_k}^{(n)}(t) = G_k(b^{n-1}t), k = 1,2;\ t \geq 0,$$

where $0 < b < 1$ is the ratio of the GP for component k.

Assumption 3. $X_k{}^{(1)}$, $X_k{}^{(2)}$, ... be successive working times and $Y_k{}^{(1)}$, $Y_k{}^{(2)}$, ... be successive repair times of component k. It is supposed that all these random variables are independent.

Let $\tau_{12}(a,b)$ denote the system lifetime under the assumption that at initial instant $t = 0$, component 1 begins to work and component 2 is inactive standby. Let $\tau(a,b)$ denote the system lifetime measured from the moment when component 1 fails. It is evident that $\tau_{12}(a,b) = X_1 + \tau(a,b)$, where X_1 and $\tau(a,b)$ are independent random variables.

The system described above breakdown when for some n either event

$$A_n = \left\{X_2^{(j)} > Y_1^{(j)}, X_1^{(j)} > Y_2^{(j)}, j = 1,2,\dots,n-1;\ X_2^{(n)} \leq Y_1^{(n)}\right\}$$

or event

$$B_n = \left\{X_2^{(j)} > Y_1^{(j)}, X_1^{(j)} > Y_2^{(j)}, j = 1,2,\dots,n-1;\ X_2^{(n)} > Y_1^{(n)}, X_1^{(n)} \leq Y_2^{(n)}\right\}$$

occurs. Note that $\alpha_{12}(n)$ is the probability that after the *n-th* repair working component 2 does not fail while component 1 is under $(n+1)$*-th* repair and $\alpha_{21}(n)$ is the probability that after the $(n+1)$*-th* repair, working component 1 does not fail while component 2 is under *n-th* repair. Then $\alpha_{12}(n)$ and $\alpha_{21}(n)$ can be represented as for $i = 1,2$ and $i \neq k$

$$\alpha_{ik}(n) = P\left\{X_k^{(n)} > Y_i^{(n)}\right\} = \int_0^\infty G_{Y_i}^{(n)}(t)d\,F_{X_k}^{(n)}(t) \tag{1}$$

Then, the probability that the system breakdown is obtained as

$$\alpha(n) = 1 - \alpha_{12}(n)\alpha_{21}(n). \tag{2}$$

Thus using (1) and (2), the probabilities of events A_n and B_n are found to be

$$P\{A_n\} = \prod_{j=1}^{n-1}(1-\alpha(j))\bar{\alpha}_{12}(n) \quad \text{and} \quad P\{B_n\} = \prod_{j=1}^{n-1}(1-\alpha(j))\alpha_{12}(n)\bar{\alpha}_{21}(n)$$

where $\bar{\alpha}_{12}(n) = 1 - \alpha_{12}(n)$, $\bar{\alpha}_{21}(n) = 1 - \alpha_{21}(n)$ and $n \geq 1$.

Then the lifetime of the system is represented as

$$P\left\{\tau(a,b)=\sum_{j=1}^{n-1}\zeta_j+\hat{X}_2^{(n)}\right\}=\prod_{j=1}^{n-1}(1-\alpha(j))\bar{\alpha}_{12}(n), \quad (3)$$

$$P\left\{\tau(a,b)=\sum_{j=1}^{n-1}\zeta_j+X_2^{(n)}+\hat{X}_1^{(n)}\right\}=\prod_{j=1}^{n-1}(1-\alpha(j))\alpha_{12}(n)\bar{\alpha}_{21}(n) \quad (4)$$

where $\zeta_j=X_2^{(j)}+X_1^{(j)}$ and for $n=1$ the sum and product vanishes.

3 Laplace-Stieltjes transform of the system lifetime

In this section, our objective is to find Laplace-Stieltjes (LS) transform of the system lifetime $\tau_{12}(a,b)$ with the help of the $\tau(a,b)$ which is defined previous section. We now introduce the following notations;

$$q_{ik(p)}(s)=\int_0^\infty e^{-st}G_{Y_i}^{(n)}(t)dF_{X_k}^{(n)}(t); \quad (5)$$

$$\bar{q}_{ik(p)}(s)=\int_0^\infty e^{-st}\bar{G}_{Y_i}^{(n)}(t)dF_{X_k}^{(n)}(t)=f_{X_k}^{(n)}(s)-q_{ik(p)}(s); \quad (6)$$

$$f_{X_k}^{(n)}(s)=\int_0^\infty e^{-st}dF_{X_k}^{(n)}(t), \quad (7)$$

where $i=1,2;\ k=1,2;\ i\neq k$ and $\bar{G}_{Y_i}^{(n)}(t)=1-G_{Y_i}^{(n)}(t)$. Using the formula of total expectation and (3)-(4) we have

$$Ee^{-s\tau(a,b)}=\sum_{n=1}^{\infty}\left(Ee^{-s\left(\zeta_1+\cdots+\zeta_{n-1}+\hat{X}_2^{(n)}\right)}\prod_{j=1}^{n-1}(1-\alpha(j))\bar{\alpha}_{12}(n)\right.$$

$$\left.+Ee^{-s\left(\zeta_1+\cdots+\zeta_{n-1}+X_2^{(n)}+\hat{X}_1^{(n)}\right)}\prod_{j=1}^{n-1}(1-\alpha(j))\alpha_{12}(n)\bar{\alpha}_{21}(n)\right). \quad (8)$$

Clearly, using (1) and (5)-(7), following equations can be obtained as

$$Ee^{-sX_k^{(j)}}=\int_0^\infty e^{-st}dP\left\{X_k^{(j)}<t\middle|X_k^{(j)}>Y_i^{(j)}\right\}=\frac{q_{ik(j)}(s)}{\alpha_{ik}(j)} \quad (9)$$

$$Ee^{-s\hat{X}_k^{(n)}}=\int_0^\infty e^{-st}dP\left\{X_k^{(n)}<t\middle|X_k^{(n)}<Y_i^{(n)}\right\}=\frac{\bar{q}_{ik(n)}(s)}{\bar{\alpha}_{ik}(n)} \quad (10)$$

The proof of (9) and (10) are presented in [12].

Taking into consideration Assumption 3 and substituting (9)-(10) in (8) we find that

$$Ee^{-s\tau(a,b)} = \sum_{n=1}^{\infty}\left(\bar{q}_{12(n)}(s) + q_{12(n)}(s)\bar{q}_{21(n)}(s)\right)\prod_{j=1}^{n-1}\left(q_{12(j)}(s)q_{21(j)}(s)\right).$$

In view of $\tau_{12}(a,b) = X_1 + \tau(a,b)$ and $f_{X_k}(t) = Ee^{-sX_k} = \int_0^{\infty} e^{-st}dF_{X_k}(t)$, $k = 1,2$ the LS transform of the system lifetime $\tau_{12}(a,b)$ is obtained as

$$Ee^{-s\tau_{12}(a,b)} = f_{X_1}(s)$$

$$\times\left(\sum_{n=1}^{\infty}\left(\bar{q}_{12(n)}(s) + q_{12(n)}(s)\bar{q}_{21(n)}(s)\right)\prod_{j=1}^{n-1}\left(q_{12(j)}(s)q_{21(j)}(s)\right)\right). \quad (11)$$

The LS transform of the system lifetime given by the equation (11) is required to calculate the expected value of the system's lifetime, i.e, the average life expectancy. However, calculation of this transform is not always possible. Hence, it is very important to determine the lower and upper bounds for the LS transform of the system lifetime and depending on this the average life expectancy of the system. In GP model, because of the machine operating time will be reduced gradually after each repair, we can write the following result.

Remark 1. For mean lifetime of the system, we have

$$E\{\tau_{12}(a,b)\} \le E\{\tau_{12}(1,1)\}.$$

Remark 2. From (11) and GP model, we have

$$Ee^{-s\tau_{12}(a,b)} \ge Ee^{-s\tau_{12}(1,1)}.$$

4 Analyzing the lifetime of the system when the failure time is fixed

Let failure time of component k is fixed and failure repair for component k follows GP repair. Then, the LS transform of the system lifetime is given by the following theorem.

Theorem 1. Let the failure time of component k is the random variable X_k with the distribution function

$$F_{X_k}(t) = F_k(m_k t) = \begin{cases}1 & m_k t \ge c\\ 0 & m_k t < c\end{cases}, \quad (12)$$

where $k = 1,2$; $t \ge 0$; $c > 0$; $m_k = \begin{cases}a^n, & k = 1\\ a^{n-1}, & k = 2\end{cases}$ and failure repair for component k holds Assumption 2.2. Then

$$Ee^{-s\tau_{12}(a,b)} = e^{-sc}\left(\sum_{n=1}^{\infty}\left(e^{-sc/a^{n-1}}\left(1 - G_1\left(\frac{b^{n-1}c}{a^{n-1}}\right)\right) + e^{-sc(a+1)/a^n}G_1\left(\frac{b^{n-1}c}{a^{n-1}}\right)\right.\right.$$

$$\left.\left.\times\left(1 - G_2\left(\frac{b^{n-1}c}{a^n}\right)\right)\right)\prod_{j=1}^{n-1}e^{-sc(a+1)/a^j}G_1\left(\frac{b^{j-1}c}{a^{j-1}}\right)G_2\left(\frac{b^{j-1}c}{a^j}\right)\right). \quad (13)$$

Proof. Substituting the expressions (12) in (5)-(7) and also considering Assumption 2.2, we obtain

$$q_{12(p)}(s) = \int_0^\infty e^{-st} G_{Y_1}^{(p)}(t) d\, F_{X_2}^{(p)}(t) = e^{-sc/a^{p-1}} G_1\left(\frac{b^{p-1}}{a^{p-1}} c\right),$$

$$q_{21(p)}(s) = \int_0^\infty e^{-st} G_{Y_2}^{(p)}(t) d\, F_{X_1}^{(p)}(t) = e^{-sc/a^{p}} G_2\left(\frac{b^{p-1}}{a^{p}} c\right),$$

$$\bar{q}_{12(p)}(s) = \int_0^\infty e^{-st} \bar{G}_{Y_1}^{(p)}(t) dF_{X_2}^{(p)}(t) = e^{-sc} - q_{12(p)}(s),$$

$$\bar{q}_{21(p)}(s) = \int_0^\infty e^{-st} \bar{G}_{Y_2}^{(p)}(t) dF_{X_1}^{(p)}(t) = e^{-sc} - q_{21(p)}(s),$$

$$f_{X_1}^{(p)}(s) = e^{-sc/a^p}, f_{X_2}^{(p)}(s) = e^{-sc/a^{p-1}}, f_{X_k}(s) = e^{-sc}.$$

Now substituting above expressions in (11) and then making suitable simplifications, we get the proof.

Theorem 2. Mean time to failure of the system for $a = 1$ and $b = 1$, we have

$$E\{\tau_{12}(1,1)\} = c + \frac{c(1 + G_1(c))}{1 - G_1(c)G_2(c)}. \tag{14}$$

Proof. Let $a = 1$ and $\mathrm{b} = 1$ in (13), then

$$Ee^{-s\tau_{12}(1,1)} = \frac{e^{-2sc}(1 - G_1(c)) + e^{-3sc}G_1(c)(1 - G_2(c))}{1 - e^{-2sc}G_1(c)G_2(c)}. \tag{15}$$

From (15) and $E\tau_{12}(1,1) = -\frac{d}{ds}Ee^{-s\tau_{12}(1,1)}(0)$ we can get the proof.

Remark 3. From Remark 2, we have

$$Ee^{-s\tau_{12}(a,b)} \geq \frac{e^{-2sc}(1 - G_1(c)) + e^{-3sc}G_1(c)(1 - G_2(c))}{1 - e^{-2sc}G_1(c)G_2(c)}.$$

Remark 4. For mean lifetime of the system, we have

$$2c \leq E\{\tau_{12}(a,b)\} \leq \frac{2c + cG_1(c) - cG_1(c)G_2(c)}{1 - G_1(c)G_2(c)}.$$

Remark 4 is very important in terms of giving the upper and lower bounds of the average operating time of the system under the restrictions of fixed operating time and repair time which is defined depending on GP. Now let us examine the following exception for the mean lifetime of the system. Let failure repair for component k be exponential, then the following theorem holds for the upper bound of the mean lifetime of the system.

Theorem 3. Let failure time of component k is fixed as described in (12) and failure repair for component k be exponential with cumulative distribution function

$$G_{Y_k}(t) = 1 - e^{-\mu_k t}, \tag{16}$$

where $t \geq 0$ and $\mu_k \geq 0;\ k = 1,2$. Then, under the restriction $\mu_2 \geq \mu_1$ and $\mu_1 + \mu_2 = u$ (u is constant),

$$E\{\tau_{12}(1,1)\} \leq c + \frac{c}{e^{-\frac{uc}{2}}}$$

Proof. According (16), (14) takes the form

$$E\{\tau_{12}(1,1)\} = c + \frac{c + c(1 - e^{-\mu_1 c})}{e^{-\mu_1 c} + e^{-\mu_2 c} - e^{-(\mu_1+\mu_2)c}}\ .$$

From $\mu_2 \geq \mu_1$ and $\mu_1 + \mu_2 = u$ it follows that

$$E\{\tau_{12}(1,1)\} \leq c + \frac{2c - ce^{-\mu_1 c}}{2e^{-\frac{uc}{2}} - e^{-\mu_1 c}e^{-\mu_2 c}}$$

$$\leq c + \frac{2c - ce^{-\mu_1 c}}{2e^{-\frac{uc}{2}} - e^{-\mu_1 c}e^{-\frac{uc}{2}c}}.$$

From this we get the proof.

Conclusions

In this article, we have studied a repairable aging cold standby system with a single standby unit when the components are independent. We have presented LS transform of the system lifetime and compute its expected value when the failure time is fixed. Finally, we aim to construct upper and lower bounds belongs to the system reliability by finding some inequalities corresponding to reliability of the technical system. The results of the study are important and can be applicable to a cold standby repairable system with preventive repair.

References

1. G. Levitin, L. Xing and Y. Dai, Cold-standby sequencing optimization considering mission cost, Reliability Engineering and System Safety, **118**, 28–34(2013).
2. S. Eryılmaz, A study on reliability of coherent systems equipped with a cold standby component, Metrika, **77**, 349-359(2014).
3. Q. Wu and S. Wu, Reliability analysis of two-unit cold standby repairable systems under Poisson shocks, Applied Mathematics and Computation, **218**, 171-182(2011).
4. M. Brown and F. Proschan, Imperfect repair, Journal of Applied Probability, **20**, 851-859(1983).
5. K.S. Park, Optimal number of minimal repairs before replacement, IEEE Transactions on Reliability, **28**, 137-140(1979).
6. M. Kijima, Some results for repairable system with general repair, Journal of Applied Probability, **26**, 89-102(1989).

7. V. Makis and A.K.S. Jardine, A note on optimal replacement policy under general repair, European Journal of Operational Research, **69**, 75-82(1993).
8. Y. Lam, Geometric processes and replacement problem, Acta Mathematicae Applicatae Sinica, **4**, 366-377(1988).
9. Y. Lam, A note on the optimal replacement problem, Advances in Applied Probability, **20**, 479-482(1988).
10. Y. L. Zhang, An optimal geometric process model for a cold standby repairable system, Reliability Engineering and System Safety, **63**, 107-110(1999).
11. Y.L. Zhang, A geometric process repair model for a repairable system with delayed repair, Computers and Mathematics with Applications, **55**, 1629-1643(2008).
12. G. Gökdere and M. Gürcan, Laplace-Stieltjes transform of the system mean lifetime via geometric process model, Open Mathematics, **14**, 384-392(2016).

ITM Web of Conferences **13**, 01024 (2017) DOI: 10.1051/itmconf/20171301024
CMES2017

On Asymptotically Lacunary Statistical Equivalent Sequences of Order α in Probability

Mahmut Işık[1,⋆] and *Kübra Elif* Akbaş[2,3,⋆⋆]

[1]*Faculty of Education, Harran University, Osmanbey Campus, 63190 Şanlıurfa, Turkey*
[2]*Faculty of Medicine, Fırat University, 23119 Elazığ, Turkey*
[3]*Faculty of Medicine, Inönü University, 44280, Malatya, Turkey*

Abstract. In this study, we introduce and examine the concepts of asymptotically lacunary statistical equivalent of order α in probability and strong asymptotically lacunary equivalent of order α in probability. We give some relations connected to these concepts.

1 Introduction and Background

The idea of statistical convergence was given by Zygmund [22] in the first edition of his monograph published in Warsaw in 1935. The concept of statistical convergence was introduced by Steinhaus [21] and Fast [9] and later reintroduced by Schoenberg [19] independently. Over the years and under different names statistical convergence has been discussed in the theory of Fourier analysis, Ergodic theory, Number theory, Measure theory, Trigonometric series, Turnpike theory and Banach spaces. Later on it was further investigated from the sequence spaces point of view and linked with summability theory by Altınok et al. [1], Connor [4], Çolak [5], Das et al. [6], Et et al. ([7],[20]) Fridy [11], Işık et al. ([13],[14],[15]) and many others. In recent years, generalizations of statistical convergence have appeared in the study of strong integral summability and the structure of ideals of bounded continuous functions on locally compact spaces. Statistical convergence and its generalizations are also connected with subsets of the Stone-Čech compactification of the natural numbers. Moreover, statistical convergence is closely related to the concept of convergence in probability.

Marouf [16] introduced definitions for asymptotically equivalent sequences and asymptotic regular matrices. Patterson [17] extend these concepts by presenting an asymptotically statistically equivalent analog of these definitions and natural regularity conditions for nonnegative summability matrices. Later on asymptotically equivalent sequences have been studied in ([2], [3],[18]).

By a lacunary sequence we mean an increasing integer sequence $\theta = (k_r)$ such that $h_r = (k_r - k_{r-1}) \to \infty$ as $r \to \infty$. Throught this paper the intervals determined by θ will be denoted by $I_r = (k_{r-1}, k_r]$ and the ratio $\frac{k_r}{k_{r-1}}$ will be abbreviated by q_r. Recently lacunary sequences have been studied in ([8],[10],[12],[20]).

Let X_n $(n \in \mathbb{N})$ be a random variable which is defined on a given event space S with respect to a given class of events Δ and a probability function $P : \Delta \to \mathbb{R}$, then we say that $X_1, X_2, X_3, ..., X_n$ ... is a sequence of random variables. A sequence of random variables is denoted by $\{X_n\}_{n\in\mathbb{N}}$.

⋆e-mail: misik63@yahoo.com
⋆⋆e-mail: elifet41@gmail.com

Definition 1 *Let (S, Δ, P) be a probability space, $\theta = (k_r)$ be a lacunary sequence, $\alpha \in (0, 1]$ be any real number. Two nonnegative sequences of random variables $\{X_n\}_{n\in\mathbb{N}}$ and $\{Y_n\}_{n\in\mathbb{N}}$ are said to be asymptotically lacunary statistical equivalent of order α in probability provided that for every $\varepsilon, \delta > 0$,*

$$\lim_r \frac{1}{h_r^\alpha} \left| \left\{ k \in I_r : P\left(\left| \frac{X_k}{Y_k} - X \right| \geq \varepsilon \right) \geq \delta \right\} \right| = 0.$$

In this case we write $X \overset{PS_\theta^\alpha}{\sim} Y$.

Definition 2 *Let (S, Δ, P) be a probability space, $\theta = (k_r)$ be a lacunary sequence and let α be any real number such that $\alpha > 0$. Two nonnegative sequences of random variables $\{X_n\}_{n\in\mathbb{N}}$ and $\{Y_n\}_{n\in\mathbb{N}}$ are said to be strong asymptotically lacunary equivalent of order α in probability provided that for every $\varepsilon > 0$,*

$$\lim_{r\to\infty} \frac{1}{h_r^\alpha} \sum_{k\in I_r} P\left(\left| \frac{X_k}{Y_k} - X \right| \geq \varepsilon \right) = 0.$$

In this case we write $X \overset{PN_\theta^\alpha}{\sim} Y$.

2 Main Results

We give the following results without proof.

Theorem 1 *Let $\theta = (k_r)$ be a lacunary sequence and let α and β be fixed real numbers such that $0 < \alpha \leq \beta \leq 1$, then $X \overset{PN_\theta^\alpha}{\sim} Y$ implies $X \overset{PS_\theta^\beta}{\sim} Y$.*

Theorem 2 *Let $\theta = (k_r)$ be a lacunary sequence and let α and β be fixed real numbers such that $0 < \alpha \leq \beta \leq 1$. If $\liminf_r q_r > 1$, then $X \overset{PS^\alpha}{\sim} Y$ implies $X \overset{PS_\theta^\beta}{\sim} Y$.*

Theorem 3 *Let $\theta = (k_r)$ be a lacunary sequence and $\alpha \in (0, 1]$. If $\limsup_r q_r < \infty$, then $X \overset{PS_\theta^\alpha}{\sim} Y$ implies $X \overset{PS}{\sim} Y$.*

Theorem 4 *Let $\theta = (k_r)$ be a lacunary sequence and let α be fixed real number such that $0 < \alpha \leq 1$. If $\liminf_r q_r > 1$, then $X \overset{PN^\alpha}{\sim} Y$ implies $X \overset{PN_\theta^\alpha}{\sim} Y$.*

Theorem 5 *Let $\theta = (k_r)$ be a lacunary sequence. If $\limsup_r \frac{k_r}{k_{r-1}^\alpha} < \infty$, then $X \overset{PN_\theta}{\sim} Y$ implies $X \overset{PN^\alpha}{\sim} Y$.*

Theorem 6 *Let $\theta = (k_r)$ be a lacunary sequence and $\alpha \in (0, 1]$. If*

$$\liminf_{r\to\infty} \frac{h_r^\alpha}{k_r} > 0$$

then $X \overset{PS}{\sim} Y$ implies $X \overset{PS_\theta^\alpha}{\sim} Y$.

Now let $\theta = (k_r)$ and $\theta' = (s_r)$ be two lacunary sequences such that $I_r \subseteq J_r$ for all $r \in \mathbb{N}$, α and β be fixed real numbers such that $0 < \alpha \leq \beta \leq 1$. Now we give some inclusion relations between the concepts of asymptotically lacunary statistical equivalent of order α in probability and strong asymptotically lacunary equivalent of order β in probability.

Theorem 7 *Let $\theta = (k_r)$ and $\theta' = (s_r)$ be two lacunary sequences such that $I_r \subset J_r$ for all $r \in \mathbb{N}$ and let α and β be fixed real numbers such that $0 < \alpha \le \beta \le 1$. Then each of the following assertions holds true*

(*i*) *If*

$$\liminf_{r\to\infty} \frac{h_r^\alpha}{\ell_r^\beta} > 0 \tag{1}$$

then $X \overset{PS^\beta_{\theta'}}{\sim} Y$ implies $X \overset{PS^\alpha_\theta}{\sim} Y$.

(*ii*) *If*

$$\lim_{r\to\infty} \frac{\ell_r}{h_r^\beta} = 1 \tag{2}$$

then $X \overset{PS^\alpha_\theta}{\sim} Y$ implies $X \overset{PS^\beta_{\theta'}}{\sim} Y$.
where $I_r = (k_{r-1}, k_r]$, $J_r = (s_{r-1}, s_r]$, $h_r = k_r - k_{r-1}$ and $\ell_r = s_r - s_{r-1}$.

Proof (*i*) Suppose that $I_r \subset J_r$ for all $r \in \mathbb{N}$ and let (1) be satisfied. For given $\varepsilon, \delta > 0$ we have

$$\left\{k \in J_r : P\left(\left|\frac{X_k}{Y_k} - X\right| \ge \varepsilon\right) \ge \delta\right\} \supseteq \left\{k \in I_r : P\left(\left|\frac{X_k}{Y_k} - X\right| \ge \varepsilon\right) \ge \delta\right\}$$

and so

$$\frac{1}{\ell_r^\beta}\left|\left\{k \in J_r : P\left(\left|\frac{X_k}{Y_k} - X\right| \ge \varepsilon\right) \ge \delta\right\}\right| \ge \frac{h_r^\alpha}{\ell_r^\beta}\frac{1}{h_r^\alpha}\left|\left\{k \in I_r : P\left(\left|\frac{X_k}{Y_k} - X\right| \ge \varepsilon\right) \ge \delta\right\}\right|$$

for all $r \in \mathbb{N}$. Now taking the limit as $r \to \infty$ in the last inequality and using (1) we get $X \overset{PS^\alpha_\theta}{\sim} Y$.

(*ii*) Let $X \overset{PS^\alpha_\theta}{\sim} Y$and (2) be satisfied. Since $I_r \subset J_r$, for $\varepsilon, \delta > 0$ we may write

$$\begin{aligned}
\frac{1}{\ell_r^\beta}\left|\left\{k \in J_r : P\left(\left|\frac{X_k}{Y_k} - X\right| \ge \varepsilon\right) \ge \delta\right\}\right| &= \frac{1}{\ell_r^\beta}\left|\left\{s_{r-1} < k \le k_{r-1} : P\left(\left|\frac{X_k}{Y_k} - X\right| \ge \varepsilon\right) \ge \delta\right\}\right| \\
&\quad + \frac{1}{\ell_r^\beta}\left|\left\{k_r < k \le s_r : P\left(\left|\frac{X_k}{Y_k} - X\right| \ge \varepsilon\right) \ge \delta\right\}\right| \\
&\quad + \frac{1}{\ell_r^\beta}\left|\left\{k_{r-1} < k \le k_r : P\left(\left|\frac{X_k}{Y_k} - X\right| \ge \varepsilon\right) \ge \delta\right\}\right| \\
&\le \frac{k_{r-1} - s_{r-1}}{\ell_r^\beta} + \frac{s_r - k_r}{\ell_r^\beta} \\
&\quad + \frac{1}{\ell_r^\beta}\left|\left\{k \in I_r : P\left(\left|\frac{X_k}{Y_k} - X\right| \ge \varepsilon\right) \ge \delta\right\}\right| \\
&\le \left(\frac{\ell_r}{h_r^\beta} - 1\right) + \frac{1}{h_r^\alpha}\left|\left\{k \in I_r : P\left(\left|\frac{X_k}{Y_k} - X\right| \ge \varepsilon\right) \ge \delta\right\}\right|
\end{aligned}$$

for all $r \in \mathbb{N}$. Hence $X \overset{PS^\beta_{\theta'}}{\sim} Y$.

Theorem 8 *Let $\theta = (k_r)$ and $\theta' = (s_r)$ be two lacunary sequences such that $I_r \subseteq J_r$ for all $r \in \mathbb{N}$, α and β be fixed real numbers such that $0 < \alpha \le \beta \le 1$. Then we have*

(*i*) *Let the condition* (1) *holds, then $X \overset{PN^\beta_{\theta'}}{\sim} Y$ implies $X \overset{PN^\alpha_\theta}{\sim} Y$,*

(ii) Let the condition (2) *holds,* $\{X_n\}_{n\in\mathbb{N}}$ *and* $\{Y_n\}_{n\in\mathbb{N}}$ *be two bounded sequences of random variables, then* $X \overset{PN_\theta^\alpha}{\sim} Y$ *implies* $X \overset{PN_{\theta'}^\beta}{\sim} Y$.

Proof Omitted.

Theorem 9 *Let* $\theta = (k_r)$ *and* $\theta' = (s_r)$ *be two lacunary sequences such that* $I_r \subseteq J_r$ *for all* $r \in \mathbb{N}$, α *and* β *be fixed real numbers such that* $0 < \alpha \le \beta \le 1$.

(i) Let the condition (1) *holds, then* $X \overset{PN_{\theta'}^\beta}{\sim} Y$ *implies* $X \overset{PS_\theta^\alpha}{\sim} Y$.

(ii) Let the condition (2) *holds and let* $\{X_n\}_{n\in\mathbb{N}}$ *and* $\{Y_n\}_{n\in\mathbb{N}}$ *be two bounded sequences of random variables then*$X \overset{PS_\theta^\alpha}{\sim} Y$ *implies* $X \overset{PN_{\theta'}^\beta}{\sim} Y$.

Proof *(i)* For any sequence $\{X_n\}_{n\in\mathbb{N}}$ of random variables and $\varepsilon, \delta > 0$, we have

$$\begin{aligned}\sum_{k\in J_r} P\left(\left|\frac{X_k}{Y_k} - X\right| \ge \varepsilon\right) &\ge \sum_{\substack{k\in I_r \\ P\left(\left|\frac{X_k}{Y_k} - X\right| \ge \varepsilon\right) \ge \delta}} P\left(\left|\frac{X_k}{Y_k} - X\right| \ge \varepsilon\right) \\ &\ge \left|\left\{k \in I_r : P\left(\left|\frac{X_k}{Y_k} - X\right| \ge \varepsilon\right) \ge \delta\right\}\right| \varepsilon\end{aligned}$$

and so that

$$\frac{1}{\ell_r^\beta} \sum_{k\in J_r} P\left(\left|\frac{X_k}{Y_k} - X\right| \ge \varepsilon\right) \ge \frac{h_r^\alpha}{\ell_r^\beta} \frac{1}{h_r^\alpha} \left|\left\{k \in I_r : P\left(\left|\frac{X_k}{Y_k} - X\right| \ge \varepsilon\right) \ge \delta\right\}\right| \varepsilon.$$

Hence $X \overset{PS_\theta^\alpha}{\sim} Y$.

ii) Omitted.

References

[1] H. Altınok, Y. Altın, M. Işık, Iran. J. Fuzzy Syst., **9**(2), 63-73 (2012)
[2] M. Başarır, S. Altundağ, Filomat, **22**(1), 161–172 (2008)
[3] N. L. Braha, Appl. Math. Comput., **219**(1), 280-288 (2012)
[4] J. S. Connor, Analysis, **8**, 47-63 (1988)
[5] R. Çolak, Modern Methods in Analysis and Its Applications, New Delhi, India: Anamaya Pub, 121-129 (2010)
[6] P. Das, S. Ghosal, S. Som, Arab J. Math. Sci., **21**(2), 253–265 (2015)
[7] M. Et, H. Altınok, Y. Altın, Appl. Math. Comput., **154**(1), 167–173 (2004)
[8] M. Et, H. Şengül, Filomat, **28**(8), 1593–1602 (2014)
[9] H. Fast, Colloq. Math., **2**, 241-244 (1951)
[10] A. R. Freedman, J. J. Sember, M. Raphael, Proc. Lond. Math. Soc., **37**, 508-520 (1978)
[11] J. A. Fridy, Analysis, **5** , 301-313 (1985)
[12] J. A. Fridy, C. Orhan, Pacific J. Math. **160**, 43-51 (1993)
[13] M. Işık, Math. Slovaca, **61**(5), 779–788 (2011)
[14] M: Işık, J. Inequal. Appl., **2010**, Art. ID 457892, 7 pp. (2010)

[15] M. Işık, K. E. Et, AIP Conference Proceedings, **1676**(1), 020045 (2015)
[16] M. S. Marouf, Internat. J. Math. Math. Sci., **16**(4), 755–762 (1993)
[17] R. F. Patterson, Demonstratio Math., **36**(1), 149–153 (2003)
[18] R. Savaş, M. Başarır, Filomat, **20**(1), 35-42 (2006)
[19] I. J. Schoenberg, Amer. Math. Monthly, **66**, 361-375 (1959)
[20] H. Şengül, M. Et, Acta Math. Sci. Ser. B Engl. Ed., **34**(2), 473-482 (2014)
[21] H. Steinhaus, Colloquium Mathematicum, **2**, 73-74 (1951)
[22] A. Zygmund, *Trigonometric Series*, (Cambridge University Press, Cambridge, UK, 1979)

Numerical Investigation on MHD Flow and Heat Transfer over an Exponentially Stretching Sheet with Viscous Dissipation and Radiation Effects

***İnci* Çilingir Süngü** [1]

[1] Department of Mathematics, University of Ondokuz Mayis, Samsun, Turkiye

Abstract. This study is to examine the steady two dimensional laminar flow of a viscous incompressible electrically conducting fluid over a continuous surface. In this study DTM-Padé method is used to solve which is a combination of differential transform method (DTM) and Padé approximant. Comparisons between the solutions obtained by DTM and DTM-Padé and are shown that DTM-Padé is the completely powerful method then DTM for solving the problems in which boundary conditions at infinity. Also in this study the effect of Magnetic and Radiation parameters, Prandtl number and Eckert number for velocity and temperature distributions are investigated.

1 Introduction

The nonlinear equations have found numerous applications of Applied Mathematics, Physics and technological processes. Nonlinear equations cover the cases of the following types: surface waves incompressible fluids, hydro magnetic waves in cold plasma, acoustic waves in inharmonic crystal, plastic sheets, paper production and many others. However, these equations are difficult to be solved analytically and sometimes it is impossible then application must be made with relevant numerical methods such as shooting method, finite difference etc. The flow of a viscous incompressible fluid over a stretching sheet has many interests over 10 years and was firstly considered by Sakiadis [1]. After many authors were focused this problem with extended conditions/situations like: two dimensional flow over a stretching sheet problem [2],heat transfer in the flow over stretching surface [3], boundary layer problem over an exponentially stretching sheet with thermal radiation [4], magneto hydrodynamic (MHD) boundary layer flow of a viscous fluid over an exponentially stretching sheet with radiation effect [5], MHD flow and heat transfer over an exponentially stretching sheet with viscous dissipation and radiation effects[6],etc.

The differential transform method was used first by Zhou [7] who solved linear and nonlinear initial value problems in electric circuit analysis. This method constructs an analytical solution in the form of a polynomial. The differential transform is well suited to

Corresponding Author : incicilingir@gmail.com

combine with other numerical techniques. The main advantages of these hybrid methods are that they can be used to solve both linear and nonlinear equations without linearization. But DTM solutions are not enough to wide region problem. In these domains one should use multi-step DTM or another combine. DTM and Padé approximant was used to solve MHD flow problems in [8-11]. In this study the same problem with [6] which was solved numerically with Runge-Kutta method is considered and solved analytically with Differential Transform (DTM)-Padé method. The validity of the proposed approach has been confirmed by comparing the results [4- 6].

2 Mathematical Formulation

2.1 Governing Equations and Boundary Conditions

Consider the two-dimensional steady flow of viscous incompressible electrically conducting fluid over a continuous stretching sheet at $y=0$. A uniform magnetic field of strength B_0 has been applied to perpendicular to the stretching sheet. The induced magnetic field is neglected. We have the momentum equation and energy equation as follows,

$$u\frac{\partial u}{\partial x}+v\frac{\partial u}{\partial y}=v\frac{\partial^2 u}{\partial y^2}-\frac{\sigma B_0}{\rho}u \tag{1}$$

$$\rho c_p\left(u\frac{\partial T}{\partial x}+v\frac{\partial T}{\partial y}\right)=\kappa\frac{\partial^2 T}{\partial y^2}-\frac{\partial q_r}{\partial y}+\mu\left(\frac{\partial u}{\partial y}\right)^2+\sigma B_0{}^2u^2 \tag{2}$$

along with the incompressibility condition

$$\frac{\partial u}{\partial x}+\frac{\partial v}{\partial y}=0. \tag{3}$$

Here u and v are the velocity components in x and y directions, respectively, ρ is the density, μ and $v\left(=\frac{\mu}{\rho}\right)$ are dynamic and kinematic viscosity, respectively, c_p is the specific heat, q_r is the radiative heat flux, κ is the thermal conductivity and T is temperature.

The boundary conditions are

DOI: 10.1051/itmconf/20171301025

$$u(0)=U_0 e^{x/L} \quad v(0)=0 \quad T(0)=T_\infty + T_0 e^{x/2L}$$
$$for\ y\to\infty \quad u\to 0 \quad T\to 0 \tag{4}$$

where U_0 is the reference velocity, T_0 is the temperature at the plate and T_∞ is the temperature far away from the plate while L is the reference length.

The radiative heat flux q_r is

$$q_r = -\frac{4\sigma^*}{3k^*}\frac{\partial T^4}{\partial y} \tag{5}$$

[12]. Here k^* is the mean absorption coefficient and σ^* is the Stefan_Boltzman constant. T^4 is expressed as a linear function of temperature, hence

$$T^4 \cong 3T_\infty^{\ 3}T - 3T_\infty^{\ 4} \tag{6}$$

is obtained by Taylor series at T_∞ neglecting higher order terms.

Using equations (5) and (6) in equation (2) heat transport equation is given as

$$\rho c_p\left(u\frac{\partial T}{\partial x}+v\frac{\partial T}{\partial y}\right)=\left(\kappa+\frac{16\sigma^* T_\infty^{\ 3}}{3k^*}\right)\frac{\partial^2 T}{\partial y^2}+\mu\frac{\partial^2 T}{\partial y^2}+\sigma B_0^{\ 2}u^2 \tag{7}$$

Using the similarity variables

$$u=U_0 e^{x/L} f'(\eta) \quad ,v=-\sqrt{\frac{\nu U_0}{2L}}e^{x/2L}\{f(\eta)+\eta f'(\eta)\}$$
$$T=T_0 e^{x/2L}\theta(\eta)+T_\infty \quad ,\eta=\sqrt{\frac{U_0}{2\nu L}}e^{x/2L}y \tag{8}$$

equations (1), (2) and (7) are reduced to

$$f'''-2(f')^2+ff''-Mf'=0 \tag{9}$$

$$\left(1+\frac{4}{3}R\right)\theta''+\Pr\left(f\theta'-f'\theta+Ec\left(f''\right)^2+M.Ec\left(f'\right)^2\right)=0 \tag{10}$$

with the boundary conditions

$$\begin{array}{llll} f(0)=0 & f'(0)=1 & \theta(0)=1 \\ for\ \eta\rightarrow\infty & f'\rightarrow 0 & \theta\rightarrow 0 \end{array}. \tag{11}$$

Here the primes denote differentiation with respect to η and dimensionless parameters are

$M=\dfrac{2\sigma B_0^{\ 2}L}{\rho U_0 e^{x/L}}$: Magnetic Parameter

$Ec=\dfrac{U_0^{\ 2}}{T_0 c_p}$: Eckert Number

$\Pr=\dfrac{\mu c_p}{\kappa}$: Prandtl Number

$R=\dfrac{4\sigma^* T_\infty^{\ 3}}{k^*\kappa}$: Radiation Parameter.

2.2 DTM with the Padé Approximant

In the following, we introduce the main features of the differential transformation method according to the differential transform of the *n*th derivative of a function $y(x)$ in one variable is defined as follows:

$$Y(n)=\frac{1}{n!}\left[\frac{d^n y(x)}{dx^n}\right]_{x=x_0}. \tag{12}$$

In Equation (12), $y(x)$ is the original function and $Y(n)$ is the transformed function and the differential inverse transform of $Y(n)$ is defined as follows:

$$y(x)=\sum_{n=0}^{\infty}Y(n)(x-x_0)^n. \tag{13}$$

In real applications, function $y(x)$ is expressed by a finite series and Equation (13) can be written as

$$y(x)=\sum_{n=0}^{i}Y(n)(x-x_0)^n. \tag{14}$$

From Equation (14), we obtain

$$y(x)=\sum_{n=0}^{\infty}\frac{(x-x_0)^n}{n!}\left[\frac{d^n y(x)}{dx^n}\right]_{x=x_0}. \tag{15}$$

Actually Equations (15) implies that the concept of differential transform is derived from Taylor series expansion. Although DTM is not able to evaluate the derivatives symbolically, relative derivatives can be calculated by an iterative way which is described by the transformed equations of the original function. DTM is a semi-analytic method due to equation (15) implies $y(x)=\sum_{n=i+1}^{\infty}Y(n)(x-x_0)^n$ is negligibly small. In fact i is decided by the convergence of natural frequency. DTM construct a series solution which is actually a truncated series solution. This series solution gives a good approximation to the true solution in a very small region. But it has slow convergent rate or completely divergent in the wider region.

Padé approximant is the approximation of a function by a rational function of given order. In this technique, we find a polynomial of degree $p+q+1$ as a rational function which is a polynomial of degree p divided by a polynomial of degree q. This gives often better approximation of the function than truncating its Taylor expansion, especially for a function has conditions at infinity.

Taking the differential transform of equations (9) and (10), we have

$$\begin{gathered}(n+3)(n+2)(n+1)F(n+3)-2\sum_{n_1=0}^{n}(n_1+1)(n-n_1+1)F(n_1+1)F(n-n_1+1)\\+\sum_{n_1=0}^{n}(n-n_1+1)(n-n_1+2)F(n_1)F(n-n_1+2)-M(n+1)F(n+1)=0\end{gathered} \tag{16}$$

$$\left(1+\frac{4}{3}R\right)(n+2)(n+1)T(n+2)+\Pr\sum_{n_1=0}^{n}(n-n_1+1)F(n_1)T(n-n_1+1)$$
$$-\Pr\sum_{n_1=0}^{n}(n_1+1)F(n_1+1)T(n-n_1)$$
$$+\Pr.\mathrm{Ec}.\sum_{n_1=0}^{n}(n_1+2)(n_1+1)F(n_1+2)(n-n_1+2)(n-n_1+1)F(n-n_1+2) \tag{17}$$
$$+\Pr.\mathrm{Ec}.M\sum_{n_1=0}^{n}(n_1+1)(n-n_1+1)F(n_1+1)F(n-n_1+1)=0$$

Where $F(n)$ and $T(n)$ are the differential transform of $f(\eta)$ and $\theta(\eta)$, respectively.

The boundary conditions are rewritten as initial conditions:

$$f(0)=0 \quad f'(0)=1 \quad \theta(0)=1 \quad f''(0)=\alpha \quad \theta'(0)=\beta. \tag{18}$$

The differential transformations of conditions in (18) are given as

$$F(0)=0 \quad F(1)=1 \quad T(0)=1 \quad F(2)=\frac{\alpha}{2} \quad T(1)=\beta \tag{19}$$

where α and β are to be determined.

Using the conditions (19) in the recurrence relations (16) and (17) leads to the solution of a system of algebraic equations. After finding the DTM solutions for velocity and heat transfer, the Padé approximant must be applied to obtain α and β. To find the values of α and β it is taken limit tends to infinity of the Padé approximant of DTM solution.

3 RESULTS AND DISCUSSION

In this study all computations are done with Maple. In the DTM solution 20 term are taken so that the order of Padé approximant is [10,10]. Skin friction coefficient and local Nusselt number are determined as $f''(0)=\alpha$ and $-\theta'(0)=-\beta$, respectively. The shear stress which is proportional to α and the values of heat transfer coefficient β are given in Table 1 and Table 2 respectively for various values of parameters.

Table 1, α values for different values of Magnetic parameter

	M=0	M=1/2	M=1	M=2
$\alpha = f''(0)$	-1.284989678	-1.466585678	-1.628654619	-1.912448326

Table 2, β values for different values of Magnetic parameter, Radiation Parameter and Eckert number

$\beta = \theta'(0)$	Ec	Pr	R	M=0	M=1/2	M=1	M=2
	0	1	0	-.9652361600	-.9120918338	-.8577114599	-.8073383447
			0.5	-.6765622412	-.6286223821	-.5881375790	-.5539571142
			1	-.5389970000	-.5149450441	-.4620199464	-.4068818952
	0.2	1	0	-.8712600289	-.7573788555	-.6713115526	-.5445636315
			0.5	-.6223039456	-.5471632796	-.4871645113	-.3895030707
			1	-.4937037509	-.4444687547	-.3832084466	-.3265448806

From the equation (16), it is seen that variations in Pr, R and Ec do not affect the value of α. It is observed from the Table 1 that surface shear stress (α) decreases as Magnetic Parameter increases. Table 2 indicates that an increase of parameters R, Ec and M increases the value of heat transfer coefficient β. To describe the flow behavior we consider the profiles of $f'(\eta)$ and $\theta(\eta)$. For the values of Pr=1, M=1,Ec=0.2 and R=0 ,the velocity profiles $f'(\eta), f(\eta)$ and the temperature profile $\theta(\eta)$ are shown in Figure 1. As seen that $f'(\eta)$ and $f(\eta)$ are inversely proportional to each other.

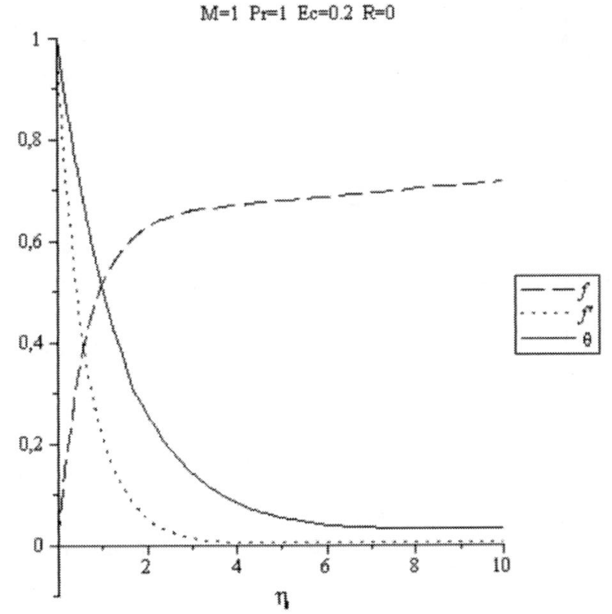

Figure 1. The velocity profiles $f'(\eta), f(\eta)$ and the temperature profile $\theta(\eta)$ at Pr=1, Ec=0.2, R=0, M=1.

The comparisons of the solutions of DTM with the DTM-Padé are shown in Figure 2. DTM is an analytical method to solve initial value problem in a small region. Figure 2 shows that the solution obtained from DTM is only valid for small values of η and does not satisfy the boundary condition at infinity. Further, DTM-Padé approximation is shown to yield a convergent solution at infinity and agree very well with literature [4-6].

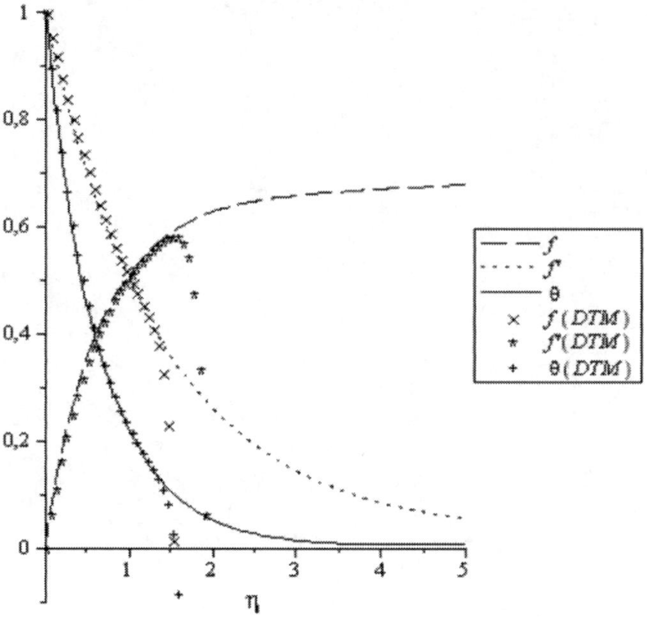

Figure 2. DTM and DTM-Padé solutions of velocity profile $f'(\eta)$ and $\theta(\eta)$ for Pr=1, Ec=0.2, R=0, M=1.

The influence of the magnetic parameter M on the velocity profile $f'(\eta)$ is shown in Figure 3. This figure indicates a decrease in the magnetic parameter M by choosing the fixed Pr=1, Ec=0 and R=0 that corresponds to an increase in the velocity boundary layer thickness. In figure 4, it is shown that the temperature profile $\theta(\eta)$ for various values of magnetic parameter M. The increase in magnetic Parameter M causes the increase in temperature profile.

DOI: 10.1051/itmconf/20171301025

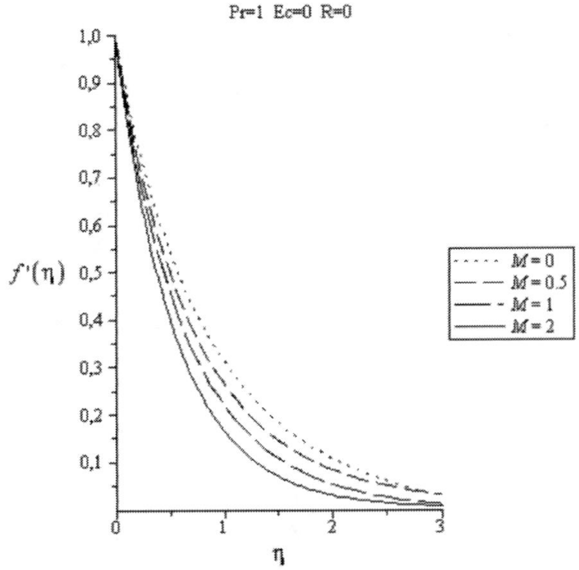

Figure 3. Velocity profile $f'(\eta)$ for different M values at R=0.

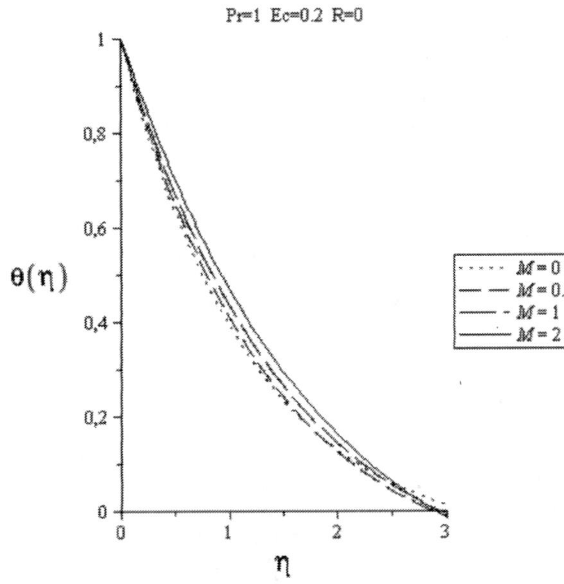

Figure 4. Temperature profile $\theta(\eta)$ for different M values at Pr=1,Ec=0.2,R=0.

From Figures 5and 6, it is seen that the effects of Radiation Parameter and Eckert Number on the temperature profile $\theta(\eta)$. The temperature profiles and the thermal boundary layer thickness increase with the increase of Ec and R.

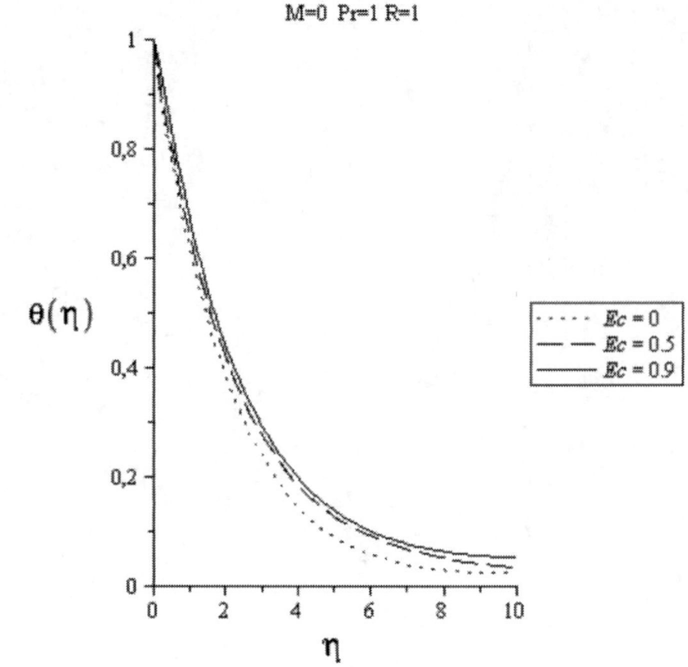

Figure 5. Temperature Profile $\theta(\eta)$ for different Ec values.

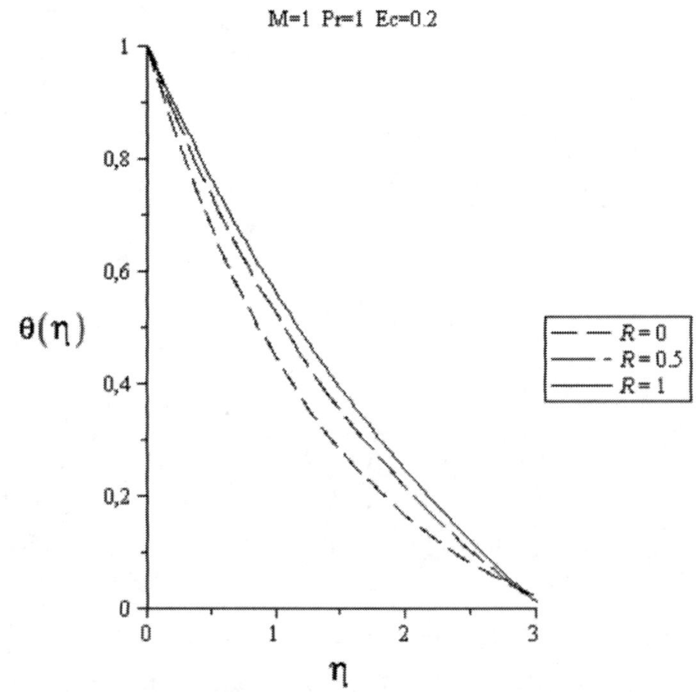

Figure 6. Temperature Profile $\theta(\eta)$ for different R values.

On the other hand, it has been observed in Figure 7 that the temperature profile decreases with an increase of the Pr. All these results are coincide with the literature.

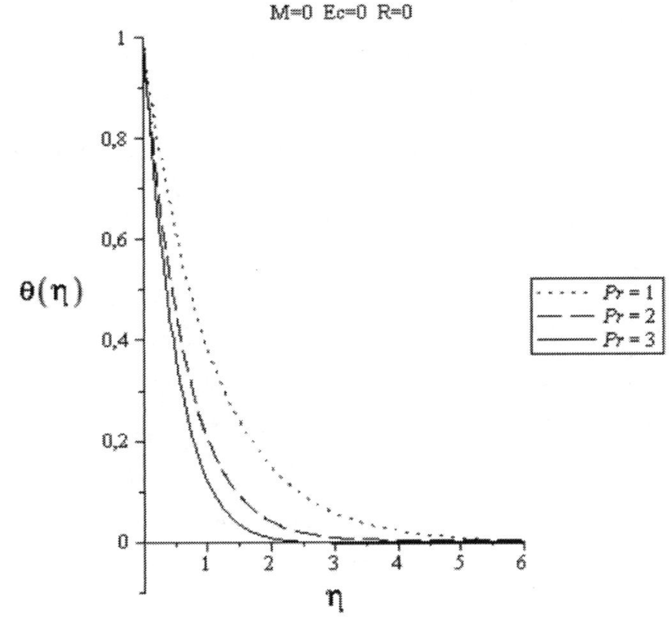

Figure 7. Temperature profile $\theta(\eta)$ for different Pr values.

4 CONCLUSION

In this work DTM-Padé is used to find analytical solution of boundary layer flow over an exponentially stretching sheet with thermal radiation. The comparison of the solution obtained from DTM and DTM-Padé are given by graphically. DTM is a convenient way to solve fluid equations because of the symbolic computation. But, for a wide region it is not enough to get a convergent solution. With using symbolic computation one can have a valid solution with DTM-Padé method. DTM-Padé is easy to implement and effective method for the problems on an unbound domain, when compared DTM.

REFERENCES

[1] B.C. Sakiadis, Boundary-layer behavior on continuous solid surfaces: I. Boundary-layer equations for two-dimensional and axi-symmetric flow, AIChE Journal, **7**, 26–28,(1961).

[2] L.J. Crane, Flow past a stretching sheet, JAMP, **21**, 645–647, (1970).

[3] P.S. Gupta, A.S. Gupta, Heat and mass transfer on a stretching sheet with suction or blowing, Can. J. Chem. Eng, **55**, 744-746, (1967).

[4] B. Bidin, R. Nazar, Numerical solution of the boundary layer flow over an exponentially stretching sheet with thermal radiation. Eur. J. Appl. Eng. Sci. Res., **33**(4), 710-717, (2009).

[5] A. Ishak, MHD boundary layer flow due to an exponentially stretching sheet with radiation effect, Sains Malays., **40**, 391-395, (2011).

[6] R. N. Jat, G. Chand, MHD Flow and Heat Transfer over an Exponentially Stretching Sheet with Viscous Dissipation and Radiation Effects, Applied Mathematical Sciences, 7, 4, 167 – 180,(2013).

[7] J.K. Zhou, Differential Transformation and its Applications for Electrical Circuits, Huazhong University Press, Wuhan, P. R. China, In Chinese.

[8] M.M. Rashidi, T. Hayat, M. Keimanesh, H. Yousefian, A Study on Heat Transfer in a Second-Grade Fluid Through a Porous Medium with the Modified Differential Transform Method, Heat Transfer—Asian Research, **42,** 31-45,(2013).

[9] M. Azimi, D. D. Ganji, F. Abbassi, Study on MHD Viscous Flow over a Stretching Sheet Using DTM-Pade' Technique, Mod. Mech. Eng., **2** ,126-129, (2012).

[10] M. Thiagarajan, K. Senthilkumar, DTM–Padé Approximants for MHD Flow with Suction/Blowing JAFM, **6,** 4, 537-543,(2013).

[11] O. Anwar Bég, M.M. Rashidi, M.T. Rastegari, T. A. Bég, S.S. Motsa and A. Halim, DTM-Padé Numerical Simulation of Electro hydrodynamic Ion Drag Medical Pumps with Electrical Hartmann and Electrical Reynolds Number Effects, J. Adv. Biotechnol. Bioeng., **1**, 62-79, (2013).

[12] M. Sajid, T. Hayat, Influence of thermal radiation on the boundary layer flow due to an exponentially stretching sheet, Int. Commun. Heat Mass Transfer, **3**,5,347-356, (2008).

 DOI: 10.1051/itmconf/20171301026

Continuous ultrasonic thickness measurement technology

Vitaliy Pronin[1], *Petr* Shkatov[1], *Anna* Sandulyak[1,*], and *Haci Mehmet* Baskonus[2]

[1]Moscow technological university, Moscow, RF
[2]Department of Computer Engineering, Munzur University, Tunceli, Turkey

Abstract. The article describes the technologies and means of ultrasonic thickness measurement for pipeline base metal and welded joints using phased array antennas with linear scanning by TOFD and DAF methods. According to the thickness measurement data provided through specialized software, maps are obtained of the control objects thickness. The measurement results are given using these technologies, as well as the prospects for these methods development.

1 Introduction

Pipelines in operation are subject to erosion-corrosion wear. Under the influence of the heat carrier, erosion of the protective oxide film destruction occurs on the inner surface, which contributes to the corrosion process, occurrence of corrosion cracks resulting in metal destruction of the pipeline wall.

To ensure reliable and safe pipelines operation, it is necessary to use modern achievements in ultrasonic non-destructive testing and diagnostics of pipeline walls metal and welded joints. One of these directions is continuous ultrasonic thickness measurement of the base metal and welded seams using antenna arrays (AA). Unlike selective ultrasonic thickness measurement, the use of automated scanning devices (SDs) allows identifying and capturing a complete representation of metal condition.

The following technologies and means of ultrasonic thickness measurement have been developed:

• Continuous ultrasonic thickness measurement of the pipelines base metal using antenna arrays.

• Ultrasonic thickness measurement of welded joints according to TOFD (Time-of-flight diffraction) scheme using antenna arrays.

• Ultrasonic thickness measurement of welded joints using digital antenna focusing method with the bottom surface profile plotting.

2 Continuous ultrasonic thickness measurement of the pipelines base metal

The control object (QA) is the base metal of monometallic pipelines, bends and conical transitions. The range of controlled pipeline wall thicknesses is 5.5 mm.

The relatively wide measurement strip of the pipeline thickness measurement is ensured by a control scheme where AA moves along the axis of the pipeline and oriented across the direction of scanning (Fig. 1). AA are used in the process of measurement, installed on flat-bent delays with the contact concave surface (Fig. 2). To carry out the thickness measurement of the pipelines base metal, a mechanized two-coordinate scanning device (SD) was developed on magnetic wheels according to the proposed control scheme (Fig. 3).

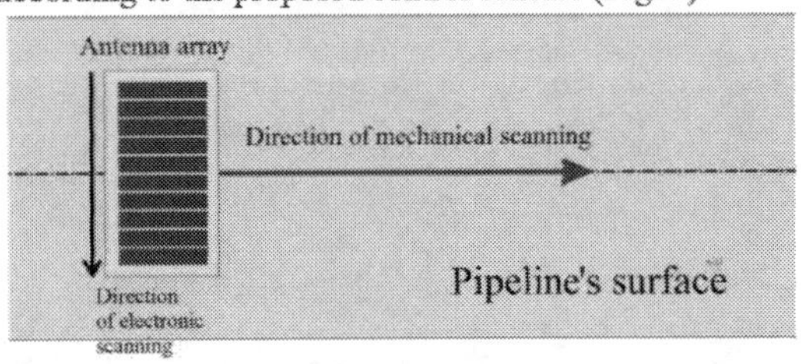

Fig. 1. Thickness measurement scheme for pipelines base metal.

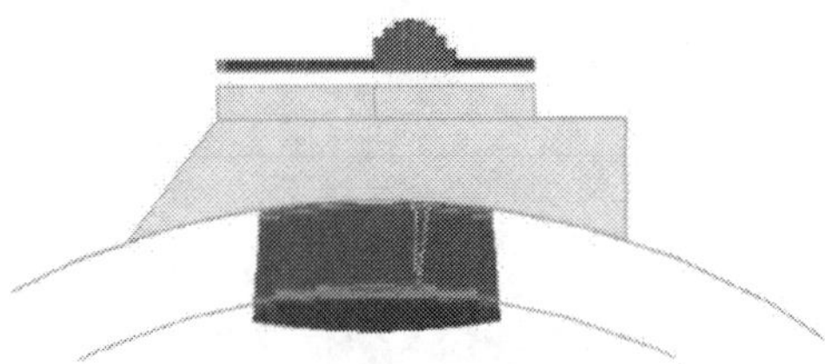

Fig. 2. Electronic scanning along AA aperture.

Specialized software (SW) was developed to handle the thickness measurement data. SW interface is shown in Fig. 4.

An example of measurements and resulting thickness map is shown in Fig. 5.

---

* Corresponding author: anna.sandulyak@mail.ru

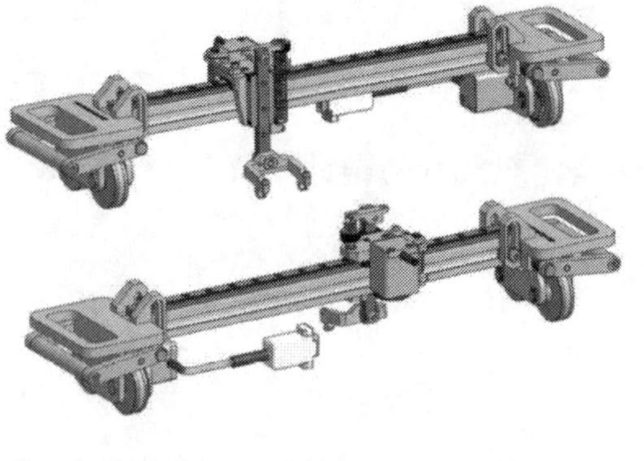

Fig. 3. Scanner for the base metal thickness measurement.

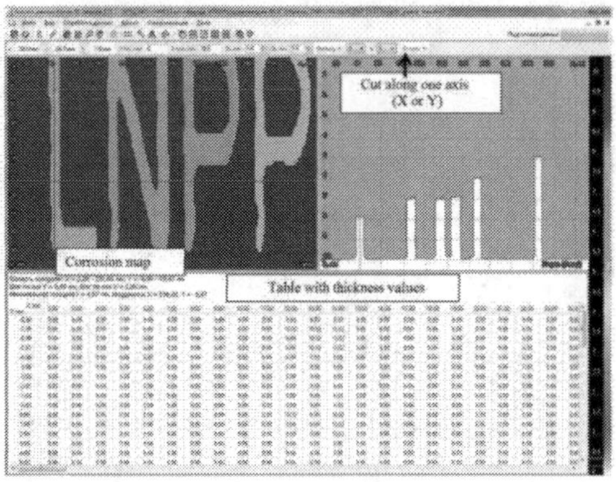

Fig. 4. Software interface for building and analyzing thickness maps.

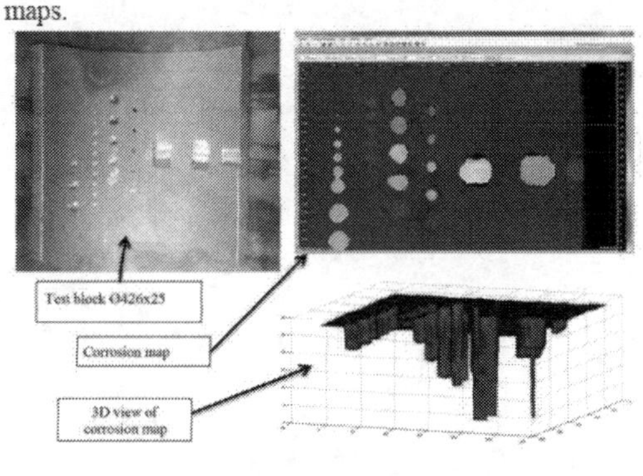

Fig. 5. Test specimen Ø426x25 mm with a thickness map built on the basis of thickness measurements using AA.

3 Ultrasonic thickness measurement of welded joints according to TOFD (Time-of-flight diffraction) scheme using antenna arrays

The control objects according to this method are welded joints of pipelines with external diameters from Ø159 mm and wall thickness of 5.5 mm.

The measurements are carried out using ultrasonic phased antenna arrays according to the ultrasonic time-of-flight diffraction method – TOFD (Fig. 6). At the same time, thickness measurements in the weld proximity are carried out using the same AAs operating in the electronic scanning mode with an input angle of 0°, based on the arrival time of the bottom signal (Fig. 7). Thus, a wide strip of thickness measurement is provided.

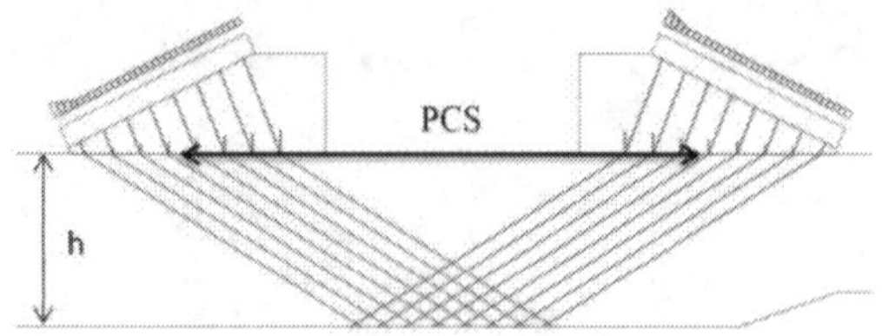

Fig. 6. Principle of thickness measurement in the area of welded joints according to TOFD scheme with AA and electronic scanning.

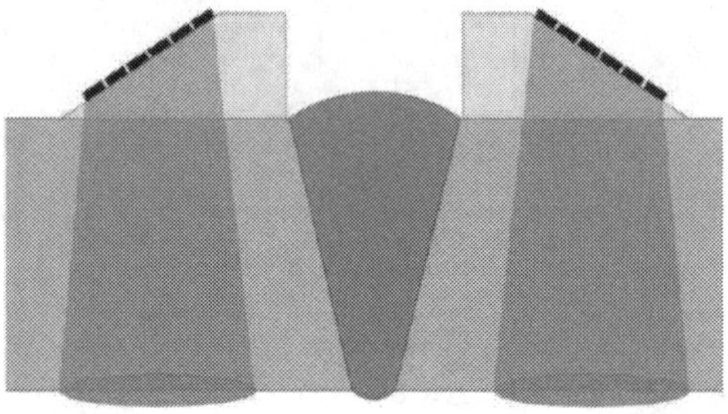

Fig. 7. Principle of thickness measurement in the weld proximity zone.

An unidirectional scanning device with a set of tracks for diameters Ø 159, Ø 219, Ø 273, Ø 325, Ø 377 and Ø 426 mm developed by SRC ECHO+ LLC (Fig. 8) is used to move the antenna arrays.

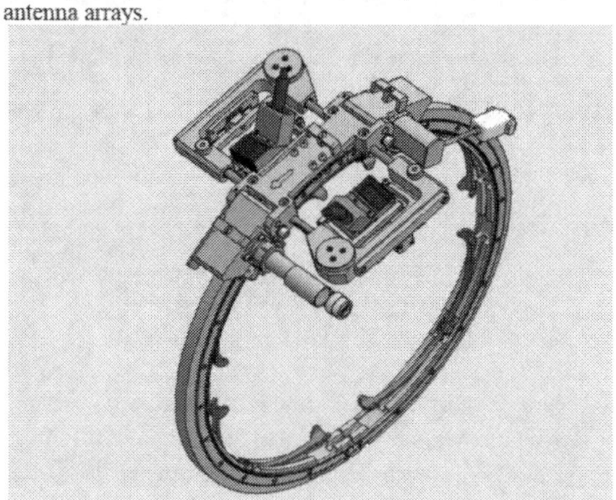

Fig. 8. Scanner SK-TD.160-426.DP-1 with track of Ø 325 mm.

An example of measurements and resulting thickness map is shown in Fig. 9.

DOI: 10.1051/itmconf/20171301026

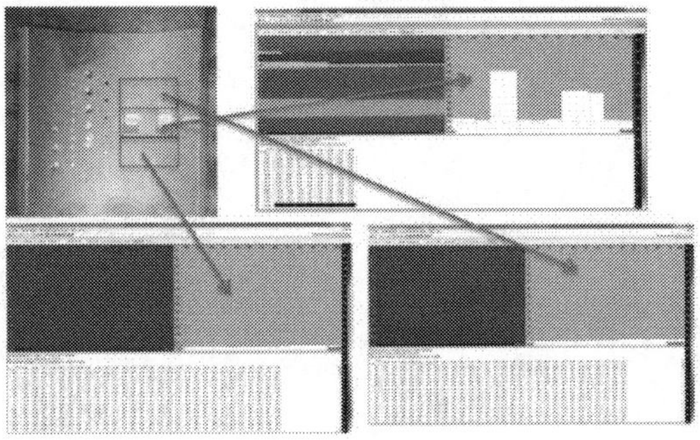

Fig. 9. Test specimen Ø426x25 mm with a thickness map built based on WJ thickness measurements according to TOFD scheme.

4 Ultrasonic thickness measurement of welded joints using digital antenna focusing method with the bottom surface profile plotting

To control the thickness of pipelines with non-equidistant surfaces, the [1] technology was developed for the bottom surface profile plotting based on an ultrasonic method with digital antenna focusing (DAF). The method is based on ultrasonic waves emission to CO and registration of these ultrasonic waves reflected from the internal boundaries of CO, as well as subsequent DAF digital processing [2] with generation of images from the registered signals. The technology is based on the use of two antenna arrays (AA) with two identical prisms according to two schemes: separate scheme and combined scheme (Fig. 10).

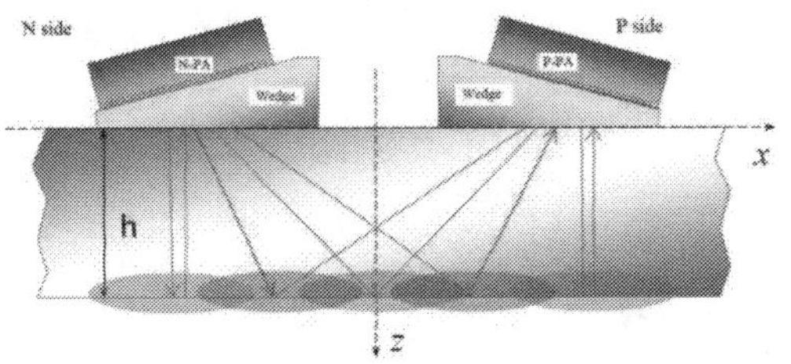

Fig. 10. DAF application for WJ thickness measurement. Violet color – longitudinal waves according to the combined scheme, red color – longitudinal waves according to the separate scheme, violet color – transverse waves according to the separate scheme.

Then, the received echo signals are processed according to DAF method and total reconstructed image with the bottom surface profile is determined (Fig. 11).

Fig. 12 shows examples of samples thickness measurement with real erosion-corrosive wear (ECW) with built thickness map.

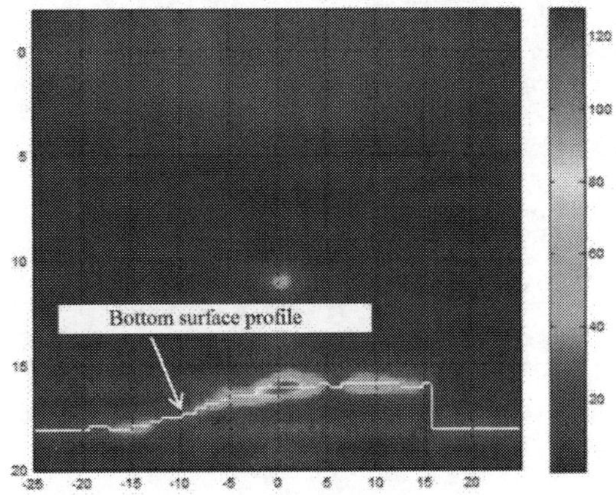

Fig. 11. Total reconstructed image with a bottom surface profile.

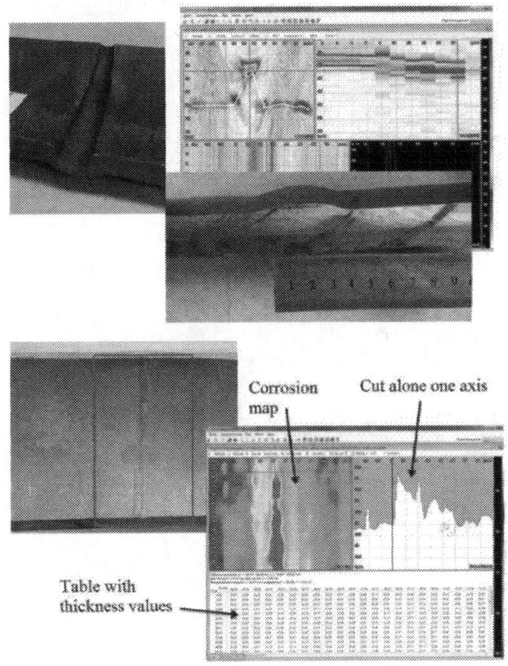

Fig. 12. Application of WJ ultrasonic thickness measurement technology based on DAF samples with ECW.

5. Conclusion

• The developed technologies make it possible to detect corrosive wear, edge displacement, thinning, sagging, various thicknesses, etc., in a timely manner when controlling objects with non-equidistant external and internal surfaces, and also to measure the geometrical parameters of the inner surface.

• The proposed technologies can be widely used in ultrasonic flaw detection of pipelines and other metal structures in nuclear power engineering, oil and gas transport infrastructure, chemical engineering,

DOI: 10.1051/itmconf/20171301026

shipbuilding with base metal control, as well as with non-equidistant surfaces of the control object.

• Internal surface recording will increase the reliability and efficiency of automated operational control for critical equipment by obtaining high-quality images of reflectors.

The bottom surface profile plotting technology will give impetus to the development of automated methods for recognizing and measuring discontinuities.

Acknowledgements

The research is conducted with financial support from RFFI within the frameworks of research project № 16-38-60034 mol_a_dk.

References

1. Way of ultrasonic control of a profile of an internal surface of a product with uneven surfaces. Patent RF № 2560754 (2014).
2. E.G. Basulin, S.A. Kokolev, A.S. Golubev. Defectoskopiya, **7** (2009).

ITM Web of Conferences **13**, 01027 (2017) DOI: 10.1051/itmconf/20171301027
CMES2017

Depth measurement of closely spaced surface cracks by the electric potential method

Petr Shkatov[1], *Irina* Lisitsyna[1], *Anna* Sandulyak[1, *] and *Hasan* Bulut[2]

[1] Moscow technological university, Moscow, RF,
[2] Department of Mathematics, University of Firat, Elazig, Turkey

Abstract. The study results are provided for the electropotential converter signals interacting with two closely spaced parallel surface cracks. The measurement scheme was studied with transducer electrodes placed at the vertices of the rectangle. Study results are important for depth measuring of the surface cracks of stress-corrosion origin.

1 Introduction

The depth of surface cracks refers to one of the most significant parameters that determines the technical condition of the metal objects of responsible use. To measure the depth of single cracks, the electropotential method has been successfully applied [1, p. 498-502]. This method is based on current passage through the defective area and recording voltage between the potential electrodes. The crack depth is evaluated by the relative change in voltage during transition from the defect-free area (voltage U_0) to the identical defective area (voltage U_r). The value $U^* = (U_r - U_0) / U_0$ depend on metal specific electric conductivity. To exclude the metal magnetic permeability effect on measurement result, the transmitted current is chosen to be constant [2]. In addition to the measured value – crack depth h, U^* value is also affected by such parameters as crack length ℓ, thickness T of the controlled section, surface curvature, crack plane deviation from the plane orthogonal to the surface of the controlled object. The influence of these factors and methods for measuring the crack depth h with the appropriate adjustment are considered in [3-6].

Known studies refer to measuring the depth of single cracks. At the same time, in practice, it is often necessary to measure the depth of surface cracks that form a network on the surface of a controlled area. This problem is typical for zones with cracks of stress-corrosion origin.

2 Selection of electropotential converter

When measuring the depth of surface cracks, the current and potential electrodes of the electropotential converter (EPC), usually [1, 2], are set along a single line (Fig. 1) perpendicular to the plane of the crack. Hereafter, we will call similar EPC as linear for short. It is not possible to apply linear EPCs in measurements within the crack network zone if the distance between adjacent cracks is comparable to the cracks depth. This is due to the fact that to ensure an acceptable measurement error, the distance between the current electrodes should be several times greater than the measured depth of the crack [3]. However, in this case there will be more than one crack between the current electrodes which makes the measurements impossible. Due to this limitation the measurements using electropotential method in the cracks network zone are not applied in practice.

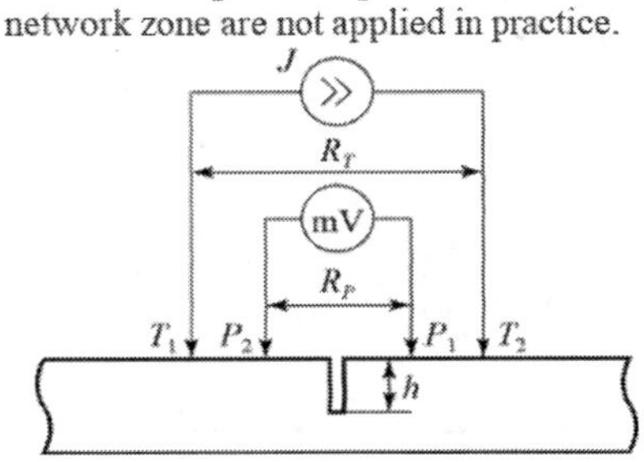

Fig. 1. Electropotential converter with placement of electrodes along the line ("linear" EPC).

To address this issue it is advisable to use the arrangement of the electrodes at the corners of the rectangle (Fig. 2). Hereafter, we will call for short the similar electropotential converters as rectangular. The use is known of the rectangular EPC for measuring the specific electric conductivity [2, 11], creep of metals in the welded joints zone [7] and depth of single cracks [8-10]. EPC advantages include a large relative sensitivity, which is important, for example, in the measurement of cracks of shallow depth [8], as well as the possibility of placing electrodes in areas with a cylindrical surface shape [9].

* Corresponding author: anna.sandulyak@mail.ru

 DOI: 10.1051/itmconf/20171301027

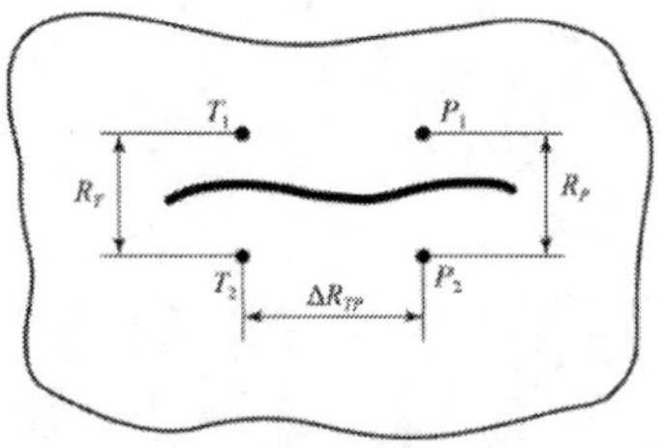

Fig. 2. Electropotential converter with placement of electrodes at the corners of the rectangle ("rectangular" EPC).

The rectangular EPC advantages with respect to the measurement of cracks network zone is that the pairs of current and potential electrodes can shift relative to each other without approaching adjacent cracks. In this case, as shown in [9-10], the range of possible measurement of the cracks depth is changing. In other words, increasing the distance ΔRTP between pairs of current and potential electrodes for a rectangular EPC (Fig. 2) leads to the same effect as an increase in RT distance for linear EPC (Fig. 1).

3 Study of the adjacent crack effect on stress introduced by a surface crack with a measured depth

The performed analysis shows the expediency of using rectangular EPC to measure the depth of surface cracks in the cracks network zone. To determine the possibility of measuring the depth of the surface crack in the cracks network zone by the finite elements method (FEM), the theoretical calculations were carried out in accordance with the geometric design model shown in Fig. 3.

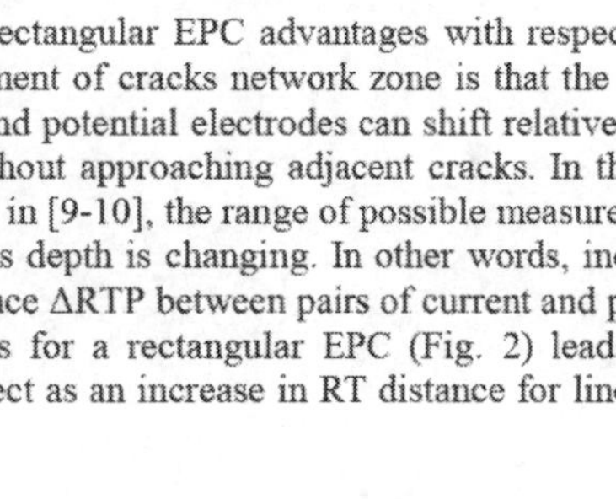

Fig. 3. Geometric design model of interaction of the rectangular EPC with two cracks parallel to each other.

To estimate the effect of an adjacent crack with a depth h^*_2 on signal recorded when placing a rectangular EPC above surface crack with a depth of h^*_1, the calculations were carried out for a variation of the distance b^*_{12} between the cracks planes and their depths. As an illustration Fig. 4 shows the voltage dependence U^*_1 EPC, located above the crack with a depth of h^*_1 in a function of depth h^*_2 of the adjacent crack for a number of b^*_{12} values for $\Delta R^*_{TP} = 1$, $\ell 1^* = \ell_2^* = 20$, $T \to$ □. In Fig. 4-a the value $h^*_1 = 0.8$, and in Fig. 4-b – $h^*_1 = 2$.

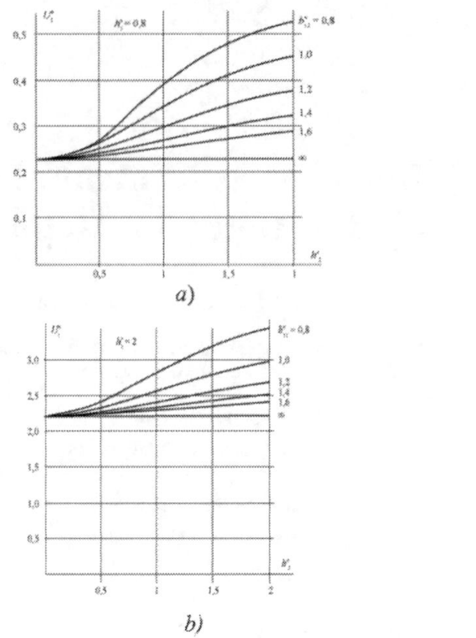

Fig. 4. Dependence $U^*_1 = U^*_1\,(h^*_2, b^*_{12})$ for a rectangular EPC placed above crack with a depth of h^*_1 and effect of an adjacent crack with a depth of h^*_2 for $\Delta R^*_{TP} = 1$, $\ell_1^* = \ell_2^* = 20$, $T \to \infty$: *a – at* $h^*_1 = 0{,}8$, *b* – at $h^*_1 = 2$.

It can be seen from these functions that the adjacent crack effect increases monotonically with its increasing depth and decreasing in distance between them. At the same time, with increasing in depth h^*_1 of the main crack, the effect of the adjacent crack increases in absolute value, and decreases on the relative crack.

Calculations show that for distances b^*_{12} comparable to ΔR^*_{TP} between close-in-depth cracks, the effect of the adjacent crack is large enough. For example, at a value of $b^*_{12} = 1$ and $h^*_1 = h^*_2 = 0.8$, the voltage U^*_1 increases by 50% due to the effect of the adjacent crack. Taking into account the non-linearity of the dependence $U^*_1 = U^*_1$ (h^*_1 this will lead to more than 100% error in measuring the depth h^*_1 of the crack.

Thus, for a reliable measurement of the crack depth in the cracks network zone, it is necessary to take into account the mutual effect of adjacent cracks. This effect depends both on the parameters that are known or can be measured: lengths ℓ_1 and ℓ_2 of each crack, distance b^*_{12} between them, metal thickness T in the controlled section, inter-electrode distances, and unknown depths h^*_1 and h^*_2 of adjacent cracks.

The cracks depth of can be determined by measuring the voltages U^*_1 and U^*_2 on each of the cracks taking into account all the above-mentioned influencing parameters. This statement can be illustrated by the diagrams shown in Fig. 5-6.

 DOI: 10.1051/itmconf/20171301027

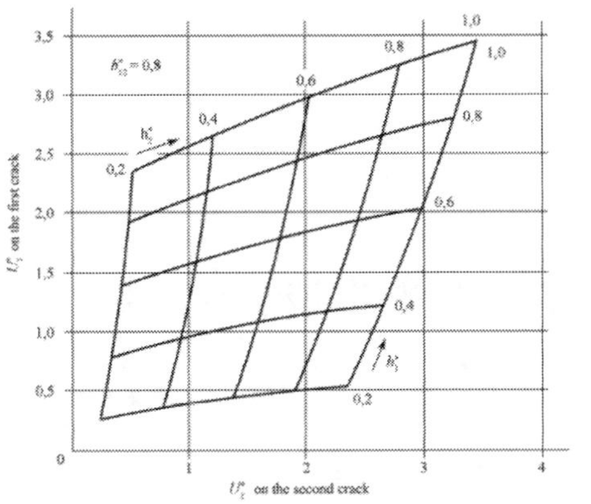

Fig. 5. Functions $U^*_1 = U^*_1\,(U^*_2)$ for rectangular EPC interacting with two cracks at $b^*_{12} = 0.8$, $\Delta R^*_{TP}=1$, $\ell_1^* = \ell_2^* = 20$, $T\to\infty$.

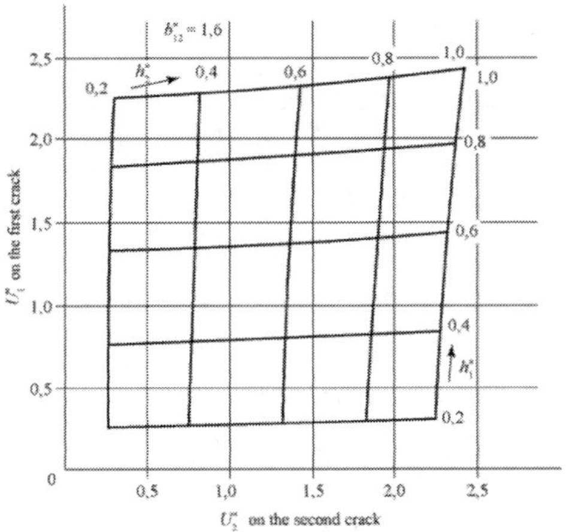

Fig. 6. Functions $U^*_1 = U^*_1\,(U^*_2)$ for rectangular EEC interacting with two cracks at $b^*_{12} = 1.6$, $\Delta R^*_{TP}=1$, $\ell_1^* = \ell_2^* = 20$, $T\to\infty$.

They show the measurement results for voltages in U^*_1 and U^*_2 axes for a set of crack depths h^*_1 and h^*_2 with other fixed parameters indicated under the figures. The diagram in Fig. 5 is constructed at a normalized distance $b^*_{12} = 0.8$ between the cracks, and diagrams in Fig. 6 – at $b^*_{12} = 1.6$. The given diagrams show that with increasing distance b^*_{12} between cracks, their mutual effect decreases, which leads to the fact that the lines of the diagram become more linear, and their slope relative to the corresponding axis decreases.

4. Conclusions

To measure the depth of closely spaced parallel cracks, it is advisable to use an electropotential converter with electrodes placed in the corners of a rectangle (rectangular EEC).

To determine the depth of each of the two cracks effecting each other, it is necessary to measure the input voltage when placing the electropotential converter over each of the cracks, measure the distance b_{12} between them and their lengths ℓ_1 and ℓ_2, and use the previously calculated multidimensional dependencies when interpreting.

Multidimensional dependencies should be calculated for a set of depths h^*_1 and h^*_2, as well as the effecting factors b_{12}, ℓ_1 and ℓ_2, determined by direct measurements.

In a particular and practically important case of measuring the depth of long cracks, the recorded signals, with an acceptable error, can be interpreted using the pre-calculated diagrams $U^*_1 = U^*_1\,(U^*_2)$ calculated for a number of values of b^*_{12}.

Acknowledgements

The research is conducted with financial support from RFFI within the frameworks of research project № 16-38-60034 mol_a_dk.

References

1. V.P. Vavilov, K.V. Podmasteriev, F.R. Sosnin, et al. *Non destructiv control.* Under redaction of V.V.Kliuev. Moscow (2006).
2. E.I. Brainin. *Control of elements of electrical machines and devices by an electropotential method,* Moscow, Energiya (1980).
3. P.N. Shkatov. *Research of an error of measurement of depth of cracks by an electropotential method.* Proceedings of the 14th Russian science conference «Nondestructive control and diagnostics», Moscow (1996).
4. P.N. Shkatov. *Increased preciseness of deep defects defecting through electropotential method.* 7th European Conference on NDT, Copenhagen (1998).
5. P.N. Shkatov. *The solution of the return problem of an electropotential defektometry for a superficial crack of final length.* Proceedings of the 3rd international science conference "Computer methods and the return tasks in nondestructive control and diagnostics", Moscow (2002).
6. P.N. Shkatov. Mech. Eng. **2** (1998).
7. S. Prajapati, P.B. Nagy, P. Cawley. *Potential drop detection of creep damage in the vicinity of welds.* AIP Conference Proceedings **1430** (2012).
8. G. Gille. NDT, **4** (1971).
9. P.N. Shkatov. A.A. Elisov. Devices, **4** (2013).
10. P.N. Shkatov. I.O. Lisitsyna. Fundamental and applied problems of the equipment and technology, **1**, 315 (2016).

ITM Web of Conferences **13**, 01028 (2017) DOI: 10.1051/itmconf/20171301028
CMES2017

Benefits of reverse engineering technologies in software development makerspace

M.H. Aabidi[1,*], *Bouziane El Mahi*[1], *Chafik Baidada*[1], *Abdeslam Jakimi*[1], *and Hany Ammar*[2]

[1]Software Engineering & Information Systems Engineering Team, Computer Sciences Department, Moulay Ismaïl University, FST Errachidia, Morocco
[2]West Virginia University, Lane Department of Computer Science and Electrical Engineering, Morgantown, USA

Abstract. In the recent decades, the amount of data produced by scientific, engineering, and life science applications has increased with several orders of magnitude. In parallel with this development, the applications themselves have become increasingly complex in terms of functionality, structure, and behavior. In the same time, development and production cycles of such applications exhibit a tendency of becoming increasingly shorter, due to factors such as market pressure and rapid evolution of supporting and enabling technologies. As a consequence, an increasing fraction of the cost of creating new applications and manufacturing processes shifts from the creation of new artifacts to the adaption of existing ones. A key component of this activity is the understanding of the design, operation, and behavior of existing manufactured artifacts, such as software code bases, hardware systems, and mechanical assemblies. For instance, in the software industry, it is estimated that maintenance costs exceed 80% of the total costs of a software product's lifecycle, and software understanding accounts for as much as half of these maintenance costs. To facilitate the software development process, it would be ideal to have tools that automatically generate or help to generate UML (Unified Modeling Language) models from source code. Reverse engineering the software architecture from source code provides a valuable service to software practitioners. Case tools implementing MDA and reverse-engineering constitute an important opportunity of software development engineers. So MDA and reverse engineering is an important key witch make makerspace more productive and more efficient.

1. Introduction

The architecture of a system is a specification of software components, interrelationships, and rules for component interactions and evolution over time. In 2001 OMG adopted an architecture standard, the Model Driven Architecture (MDA). With the emergence of internet applications, the interoperability problem moved from the integration of platforms and programming languages on a company intranet to the integration of different middleware on the Internet. In this situation, the middleware is part of the problem itself [1]. The original inspiration around the definition of MDA had to do with this internet middleware integration problem. Apart from interoperability reasons, there are other good benefits to use MDA such as to improve the productivity, code and processes quality and, software maintenance costs. MDA is an architectural framework for improving portability, interoperability and reusability through separation of concerns. It uses models to direct the complete lifecycle of a system; all artifacts such as requirement specifications, architecture descriptions, design descriptions and code, are regarded as models. MDA provides an approach for specifying a system independently of the platforms that it supports, specifying platforms, selecting a particular platform for the system, and transforming the system specification into one implementation for the selected particular platform. It distinguishes Computation Independent Model (CIM), Platform Independent Model (PIM), Platform Specific Model (PSM) and Implementation Specific Model (ISM). The Unified Modeling Language (UML) [2,3] combined with the Object Constraint Language (OCL) [4] is the most widely used way for writing either PIMs or PSMs. Model Driven Development (MDD) refers to a range of development approaches that are based on the use of software models as first class entities. MDA is the specific realization of MDD proposed by OMG. It is carried out as a sequence of model transformations: the process of converting one model into another one of the same system preserving some kind of equivalence relation between them. The idea behind MDA is to manage the evolution from CIMs to PIMs and PSMs that can be used to generated executable components and applications. The high-level models that are developed independently of a particular platform are gradually

DOI: 10.1051/itmconf/20171301028

transformed into models and code for specific platforms. The concept of metamodel, an abstract language for describing different types of models and data, has contributed significantly to some of the core principles of the emerging MDA. The Meta Object Facility (MOF), an adopted OMG standard, (latest revision MOF 2.0) provides a metadata management framework, and a set of metadata services to enable the development and interoperability of model and metadata driven systems [5]. MDA reverse engineering can be used to recover architectural models of legacy systems that will be later used in forward engineering processes to produce new versions of the systems. OMG is involved in a series of standards to successfully modernize existing information systems. Modernization supports, but are not limited to, source to source conversion, platform migration, service oriented architecture migration and model driven architecture migration. Architecture Driven Modernization (ADM) is an OMG initiative related to extending the modeling approach to the existing software systems and to the concept of reverse engineering [6]. One of ADM standards is Knowledge Discovery Metamodel (KDM) to facilitate the exchange of existing systems meta-data for various modernization tools [7]. The following section presents the concepts of model, metamodel, transformation, MDA and reverse engineering in more details and finally presents our study and how to share it in makerspaces.

2. MDA and Reverse Engineering

To understand automated reverse engineering, we must first understand model driven development/architecture [8] [9] and the transformation framework. Model driven development and code generation from models is called in literature forward engineering. In a model driven development approach, given two meta-models, i.e., a source meta-model and a target meta-model and the transformation rules that can transform the source meta-model into the target meta-model, any given platform independent model that adheres to the source meta-model can be translated into a platform specific model (PSM) that adheres to the target meta-model. The resulting PSM can then be translated into various implementation artifacts on the target platform. This is called forward engineering. By reversing this approach, platform independent models can be extracted from platform specific models and implementation artifacts. Extraction of models from existing artifacts of a business application is termed reverse engineering.

Figure 1 shows forward engineering transformation approach and reverse engineering transformation approach. The figure represent software transformations that automatically translate artifacts on the left to the artifacts on the right of the arrows they reside. Model transformation is the process of converting one model into another model of the same system preserving some kind of equivalence relation between both of these models.

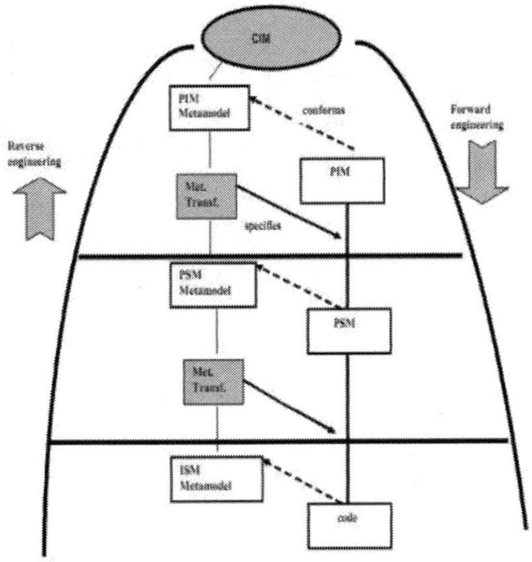

Fig. 1. Model, metamodels and transformations

The idea behind MDA is to manage the evolution from CIMs to PIMs and PSMs that can be used to generate executable components and applications. The high-level models that are developed independently of a particular platform are gradually transformed into models and code for specific platforms. The transformation for one PIM to several PSMs is at the core of MDA. A model-driven forward engineering process is carried out as a sequence of model transformations that includes, at least, the following steps: construct a CIM; transform the CIM into a PIM that provides a computing architecture independent of specific platforms; transform the PIM into one or more PSMs, and derive code directly from the PSMs.

The success of MDA depends on the existence of CASE tools that make a significant impact on software processes such as forward engineering and reverse engineering processes, however all of the MDA tools are partially compliant to MDA features. The major developments taking place in the framework of the Eclipse project [10]. For instance, the Eclipse Modeling Framework (EMF) was created for facilitating system modeling and the automatic generation of Java code. The major developments taking place in the framework of the Eclipse project [10]. For instance, the Eclipse Modeling Framework (EMF) was created for facilitating system modeling and the automatic generation of Java code. EMF started as an implementation of MOF resulting Ecore, the EMF metamodel comparable to EMOF. EMF has evolved starting from the experience of the Eclipse community to implement a variety of tools and to date is highly related to Model Driven Engineering (MDE). For instance, ATL (Atlas Transformation Language) is a model transformation language in the field of MDE that is developed on top of the Eclipse platform [11]. Commercial tools such as IBM Rational Software Architect, Spark System Enterprise Architect or Together are integrated with Eclipse-EMF [12] (figure 2).

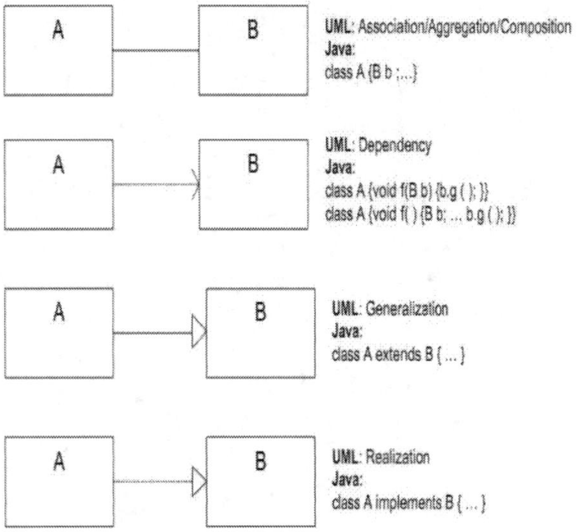

Fig. 2. ISM Java constructs versus PSM Java constructs

Reverse engineering is the process of analyzing a subject system to identify the system's components and their interrelationships and create representations of the system in another form or at higher levels of abstraction [13]. Reverse engineering is needed when the process to understand a software system would take a long time due to incorrect, out of date documentation, complexity of the system and the insufficient knowledge of the maintainer of the system. Reverse Engineering is used to recover lost information, to improve or provide documentation, to detect side effects, to reuse the components and to reduce the maintenance effort.

This technique is widely used in several disciplines including new technologies such as mechanics, electronics, etc. In computer science, Reverse engineering dynamic models is to convert the code to models such as UML sequence diagram, state diagram, etc, the goal is to increase the level of abstraction to better understand the behavior of software.

Reverse engineering dynamic UML models is a very efficient way to understand complex software whose source code is absent or documentation is outdated. It is also used in software engineering activities such as testing and validation.

In this paper, we consider UML diagrams as target design models for reverse-engineering. This means that we focus only on RE of UML diagrams (i.e. the process that analyzes the execution of the system and creates UML diagrams that depicting its structure and its behavior). UML is an object-oriented language for software system modeling [2,3]. It is composed of a set of diagrams which allow specifying the several aspects of a system. The two main aspects are: static (structural diagrams) and behavior (dynamic) aspects. The static aspect of system can be specified using class or component diagrams, while the behavior aspect can be specified using sequence diagrams, state machines and activity diagrams...etc.

In the earlier approaches, reverse engineering of object oriented code resulted in generation of class diagrams, interaction diagrams and use case model in general. However they represent only the static nature of the object oriented systems.

3. Related Works

Several studies have been performed on the reverse engineering of UML diagrams [14,15,16,17,18,19,20,21]. We distinguish two categories in existing approaches: static and dynamic. Static analysis is to use the code structure to generate the sequence diagram. One of the main works based on static analysis is that Rountev et al. [14]. They proposed an approach for the extraction of UML sequence diagrams from code through building the control flow graphs. The dynamic analysis is to analyze the performance of the application. Several studies try to generate the sequence diagram by analyzing the execution traces. In [15] is proposed an approach to build a high-level sequence diagram incrementally from basic diagram using the operators introduced by UML 2.0. In [16] they try to build a high-level sequence diagram from combined fragment using the state vector describing the system. In [17], it is proposed an approach completely dynamic based on the LTS (labeled transition system) for merger traces collected and generate a high-level sequence diagram.

These approaches have succeeded in generating representative UML behavior. But they recognize some limitations. These limitations include the information filtering problem. Thus, the resulting sequence diagram contains a lot of useless information that does not help to understand the software.

Figure 3 shows an MDA-based Reverse Engineering Framework for reverse engineering that distinguishes three different abstraction levels linked to models, metamodels and formal specifications.

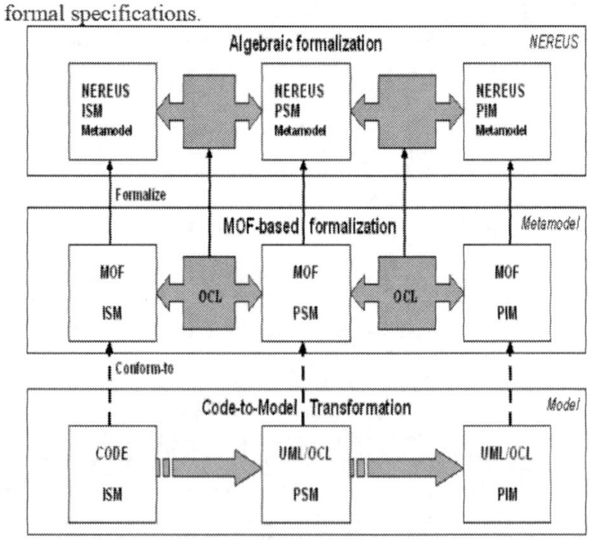

Fig. 3. MDA-based Reverse Engineering Framework

The model level includes code, PIMs and PSMs. A PIM is a model with a high-level of abstraction that is independent of an implementation technology. A PSM is a tailored model to specify a system in terms of specific platform such J2EE or .NET. PIMs and PSMs are expressed in UML and OCL [2,3,4]. The subset of UML diagrams that are useful for PSMs includes class diagram, object diagram, state diagram, interaction diagram and package diagram. On the other hand, a PIM can be expressed by means of use case diagrams, activity diagrams, interactions diagrams to model system processes and state diagrams to model lifecycle of the system entities. An ISM is a specification of the system in source code.

At model level, transformations are based on static and dynamic analysis. The metamodel level includes MOF metamodels that describe the transformations at model level. Metamodel transformations are specified as OCL contracts or QVT transformations [30] between a source metamodel and a target metamodel. MOF metamodels "control" the consistency of these transformations. The level of formal specification includes specifications of MOF metamodels and metamodel transformations in the metamodeling language NEREUS that can be used to connect them with different formal and programming languages. This framework could be considered as an MDA-based formalization of the process described by [22]. In this chapter we exemplify the bases of our approach with Class Diagram reverse engineering. However, our results include algorithms for extracting different UML diagrams such as interaction diagram, state diagram, use case diagram and activity diagram [23,24,25,26].

4. Methodology

Our approach consists on converting platform specific artifacts into a platform independent model (Figure 4). Platform specific code, artifacts, UI elements and schema are processed in a Model Generator Module to generate a platform specific model. The platform specific code, artifacts, UI elements and schema could be present in many forms and formats including code written in programming languages such as Java, or C, or C++ and schema and other artifacts represented as xml files or other files. A Model Generator Module processes the platform specific artifacts in their various formats and extracts a platform specific model from them. In order to do this, it has to know the metamodel of the underlying platform. If one exists, then the implementation artifacts can be mapped to such a platform specific model. But in cases where one does not exist, we use a semi-automated approach to derive metamodels from specific platforms. In general, extracting the meta-models for non-standards based and proprietary platforms is an engineering challenge. Depending on the platform, varying amounts of manual effort may be required to extract the meta-modal of the platform. If the meta-models are not published or not accessible, then one may have to resort to manual observation of exemplars to derive the meta-model from the exemplar. This means an exemplar with all possible types of elements needs to be constructed. An exemplar contains the implementation artifacts which include code, schemas, xml files etc. The meta-model extraction may be automated using exemplar analysis tools available in vendor tools such as IBM's Rational Software Architect (RSA).

However, an exemplar must be created first to conduct the exemplar analysis. In our work, for the two vendor platforms chosen, we were able to obtain the metamodels for one of the vendor platforms while we had to manually create the other using exemplar creation and exemplar analysis.

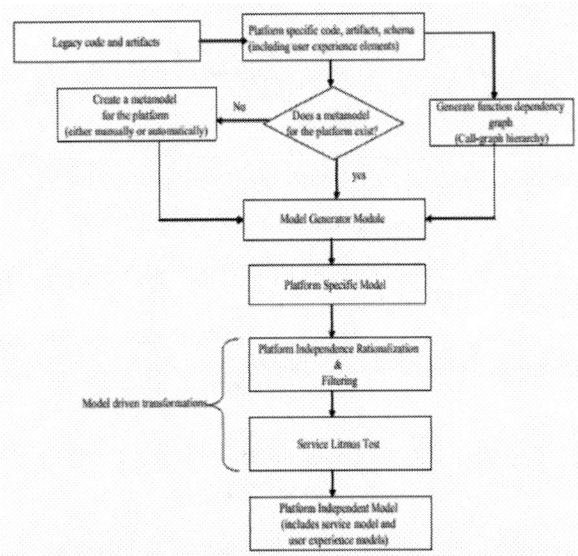

Fig. 4. Our approach to deriving platform independent models from implementation artifacts

This metamodel is then used by the Model Generator Module to generate a platform specific model for specific model instances. Then, filtering is performed to extract only those elements that would be of 'value' at platform independent level in an SOA environment. The rationalization and filtering mechanism can employ predefined rules to perform this. For example, models of artifacts such as factory classes for business objects, and auxiliary data structures and code that setup environment variables and connectivity with legacy systems etc need not be translated onto platform independent models. These types of business objects, data structures, application services, their operations are cleansed and filtered at this stage. Then from the platform specific model, we extract service models and apply a service litmus test as given in IBM's SOMA method [27] to categorize services as process services, information services, security services, infrastructure services etc. SOMA method defines these categories of services. Each service along with its ecosystems of services can be examined in detail to derive this information either automatically or manually. Once done, additional tagging is done on services to note which ones are exposed externally and which ones are internal implementations. The litmus test can be administered manually or can be automated if there is enough semantic information about the code/artifacts to know about the behavior and characteristics.

In our work, we used a user- directed mechanism for doing this filtering. A tool has been developed to enable a developer to conduct the filtering. This along with the user experience elements and models are all extracted into a platform independent model via model-driven transformations. In addition one can use code analysis tools to understand the call-graph hierarchy to retrieve an ecosystem of mutually dependent services. Several vendor tools are available for doing this for various programming languages. We use IBM's Rational Software Architect (RSA) [28] tool to do code analysis [29].

This information is captured and reflected in a platform specific model which then gets translated into a platform independent model via model driven transformations. This helps generate a service dependency model at the platform independent model. The service model and the service dependency information together provide static and the dynamic models at the platform independent level.

5. Reverse Engineering in Makerspace

This study can be shared through experimental workshops working on concrete case studies using one of the MDA software, these workshops should simply be equipped with a few computers or a single computer with a video projector can do the trick. The sharing can also be done by videoconference or by recorded videos illustrating the use of these tools by demonstrations or by other means allowing sharing like a local network.

6. Case Study

Transformations create elements in a target model (domain) based on elements from a source model' [6]. A model driven transformation is a set of mapping rules that define how elements in a given source model map to their corresponding elements in a target domain model. These rules are specified between the source and target platform metamodels. Depending on what need to be generated there could be multiple levels of transformations such as model-to-model, model-to-text, model-to-code and code-to-model. Also, depending on the domain and the desired target platform multiple levels of transformations might be required to transform a PIM into implementation artifacts on a target platform in the case of forward engineering and vice versa for reverse engineering. For example, transformations may be required across models of the same type such as a transformation from one PSM to another PSM to add additional levels of refinement or across different levels of abstraction such as from PIM to PSM or from one type of model to another such as from PSM to code or even PIM to code. In our case (Figure 5), we use the traditional PIM-to-PSM and PSM-to-code transformations for forward transformations and code-to-PSM and PSM-to-PIM transformations for model extraction or reverse engineering. Operationally, multiple levels of transformations can be chained so that the intermediate results are invisible to the consumer of the transformations. The proposed case tool to prove our approach is Rational Software Architect or Eclipse-EMF (the Eclipse Modeling Framework).

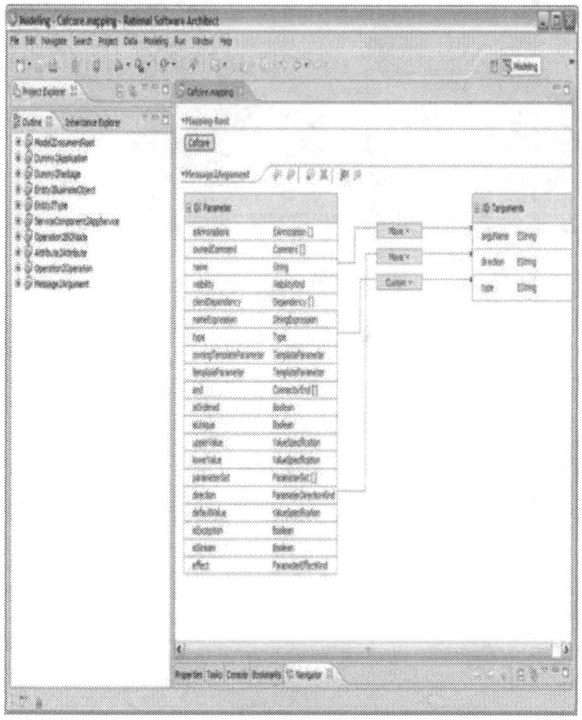

Fig. 5. Case tool for deriving platform independent models

7. Results/Discussion

Reverse engineering is one the essential parts of software maintenance. It involves high risk especially when the maintenance and development teams are different. Extracting the static model from the source code may not be motivating factor for software maintenance. Many reverse engineering tools and approaches are proposed in literature. However each represents only a subset of operational requirements. The process of reverse engineering is resulting in models consisting of only the static parts of design. Though the communication among objects is transformed, but reverse engineering of the internal state of the object is not presented. Static models are limited in their usefulness. It is important to realize that quality attributes such as performance and reliability can be predicted from the dynamic behavioral models of the system.

In this paper we proposed a new approach to extract UML dynamic behavior diagrams from the Java source code using both the static and dynamic analysis.

There are two main categories of existing UML behavior reverse engineering approaches: the first category refers to approaches which are based on static analysis while the second concerns dynamic analysis based approaches. Static analysis is done on static information which describes the structure of the software as it is written in the source code. However, dynamic analysis is based on the system runtime behavior information which can be captured by separated tools as in [17], by instrumentation techniques as in [18], or by debugging techniques. There approaches [31,32, 33] which combine both static and dynamic approaches for more efficiency.

We show in this paper the importance of the reverse engineered static and dynamic views of the system architecture in order to predict the system quality attributes and analyze possible plans of the system evolution.

Acknowledgments

We would like to thank the FST Errachidia for providing us with assist us in the development part of our research. My sincere thanks also goes to Moulay Ismail University for its financial support of the project "Algorithmic and software support for transformational approaches to UML".

8. Conclusion

Reverse-engineering aims to provide design models from existing software. Hence, this can facilitate program comprehension and by consequence maintenance and evolution of systems.

In this paper, we have presented an overview on the reverse engineering of static and behavioral diagrams. Our approach helps software development engineers to reverse engineer object-oriented software. This paper gives also manners of how to share and benefit of the reverse engineering technologies in software development Makerspace.

A Makerspace is a shared community space where people can come together and collaborate while sharing tools, resources and knowledge. Makerspaces can include a range of equipment and instructions from industrial arts, electronic & robotic technologies and 3D prototyping to software development and digital arts. A Makerspace represent an opportunity to share our software development approach. So, in this paper we presented our approach to porting software solutions on multiple software middleware platforms. We propose the use of model-driven transformations to achieve cross-platform portability. We propose approaches for two scenarios. First, in cases where no software solution exists on any of the desired target middleware platforms, we advocate developing a platform independent model of the software solution in a formal modeling language such as UML and then applying model-driven transformations to generate implementation artifacts such as code and schemas from the models on the desired target platforms. Second, if a software solution already exists on one specific middleware platform, we propose applying reverse transformations to derive a platform independent model from the implementation artifacts and then applying forward transformations on the derived model to port that software solution on to a different target platform. We advance the traditional model-driven technique by presenting a service-oriented approach to deriving platform independent models from platform specific implementations. Reverse-engineering aims to provide design models from existing software. Hence, this can facilitate program comprehension and by consequence maintenance and evolution of systems. Our approach helps software development engineers to reverse engineer object-oriented software.

Our future work is to evaluate our approach on more complex system such as embedded systems

References

1. The Model Driven Architecture. www.omg.org/mda. MDA (2005).
2. Unified Modeling Language: Infrastructure. Version 2.3. OMG Specification formal/ 2010-05-03. www.omg.org. UML (2010a).
3. UML: Unified Modeling Language: Superstructure. Version 2.3. OMG Specification: formal/2010-05-05. www.omg.org. UML (2010b).
4. OCL: Object Constraint Language. Version 2.2. OMG: formal/2010-02-01. www.omg.org. OCL (2010).
5. MOF: Meta Object Facility (MOF ™) 2.0. OMG Specification formal/2006-01-01. www.omg.org/mof. MOF (2006).
6. Standards Roadmap. ADM Task Force. adm.omg.org. ADM (2010).
7. Knowledge Discovery Meta-Model, Version 1.3-beta 2, March 2011. OMG specification formal 2010-12-12. http://www.omg.org/spec/kdm/1.3/beta2/pdf. KDM (2011).
8. Mellor, S. J., Clark, A. N., and Futagami, T. Model-driven development. IEEE Software, Vol-**20**, Issue-5, pp. 14–18, (2003).
9. Frankel, David S.: Model Driven Architecture: Applying MDA to Enterprise Computing. OMG Press: 2003. OMG Unified Modeling Language Specification, Object Management Group, (2003).
10. The eclipse modeling framework. http://www.eclipse.org/emf/. Eclipse (2011).
11. ATL Documentation. www.eclipse.org/m2m/atl/documentation. ATL (2011).
12. CASE MDA. www.case-tools.org. (2011).
13. Elliot j. Chikofsky and James H. Cross II. Reverse engineering and design recovery : a taxonomy. IEEE Software, (1990).
14. A. Rountev, O. Volgin, and Miriam Red-doch. Control flow analysis for reverse engineer-ing of sequence diagrams. Technical Report OSU-CISRC-3/04-TR12, Ohio State University, (March 2004).
15. M. K. Sarkar , T. Chaterjee ; Reverse Engineering : An Analysis of Dynamic Behavior of Object Oriented Programs by Extracting UML Interaction Diagram, International Journal of Computer Technology & Applications, Vol **4** (3),378-383, (2013).
16. R. Delamare, B. Baudry, Y.L/ Traon "Reverse-engineering of UML 2.0 Sequence Diagrams from Execution Traces" Workshop on Object-Oriented Reengineering at ECOOP 06, Nantes, France, (July 2006).
17. T. Ziadi, M. Aurélio A. da Silva, L. Hillah, M. Ziane, "A Fully Dynamic Approach to the Reverse Engineering of UML Sequence Diagrams", 16th IEEE International Conference on Engineering of Complex Computer Systems, 2011, pp. 107-116, (IEEE Computer Society). http://www.ptidej.net/material/inanutshell. (August 2009).
18. L. C. Briand, Y. Labiche, J. Leduc, "Towards the Reverse Engineering of UML Sequence Diagrams for Distributed Java Software", IEEE Transactions on Software Engineering, vol. **32** (9), pp. 642-663, (2006).
19. R. Zhao, and L. Lin, "An UML Statechart Diagram- Based MM-Path Generation Approach for Object-Oriented Integration Testing",Enformatika, Vol. **16**, p259, (2006).
20. D.H.A. van Zeeland, "Reverse-engineering state machine diagrams from legacy C-code", proceedings of 12th Conf. on Entity-Relationship Approach - Arlington-Dallas, (Dec. 1993).
21. S. Christian, "Extracting n-ary relationships through database reverse engineering", B. Thalheim (Ed.), Proc. 15th Internat. Conf. on Conceptual Modeling, Cottbus, Germany, pp. 392–405, (1996).
22. Tonella, P., & Potrich, A. Reverse Engineering of Object-oriented Code. Monographs in Computer Science. Heidelberg: Springer-Verlag, (2005).
23. Favre, L. (2010). Model Driven Architecture for Reverse Engineering Technologies: Strategic Directions and System Evolution. Engineering Science Reference, IGI Global, USA.
24. Favre, L., Martinez, L., & Pereira, C. MDA-based Reverse Engineering of Object- oriented Code. Lecture Notes in Business Information Processing (Vol **29**, pp. 251-263). Heidelberg: Springer-Verlag, (2009).
25. Pereira, C., Martinez, L., & Favre, L. Recovering Use Case Diagrams from Object- Oriented Code: an MDA-based Approach. In Proceedings ITNG 2011, 8th. International Conference on Information Technology: New Generations (pp. 737-742), Los Alamitos: IEEE Computer Press, (2011).
26. Martinez, L., Pereira, C., & Favre, L. Recovering Activity Diagrams from Object- Oriented Code: an MDA-based Approach. In Proceedings

 DOI: 10.1051/itmconf/20171301028

2011 International Conference on Software Engineering Research and Practice (SERP 2011)(Vol. **1**, pp. 58- 64), CSREA Press, (2011).

27. Arsanjani A.: Service Oriented Modeling and Architecture (SOMA). http://www.ibm.com/developerworks/webservices/library/ws-soa-design1/

28. IBM Rational Software Architect http://www.ibm.com/developerworks/rational/products/rsa/

29. http://www.eclipse.org/articles/Article-JET/jet_tutorial1.html Java Emitter Templates (JET).

30. QVT: MOF 2.0 Query, View, Transformation. Formal/2008-04-03. www.omg.org. QVT (2008).

31. A. Jakimi, L. Elbermi and M. El Koutbi, "Software Development for UML Scenarios: Design, fusion and code generation". International Review on Computers and Software, Vol. **6**. n. 5, pp. 683-687, (2011).

32. M. H. Abidi, A. Jakimi, E. H. El Kinani. A New Approach the Reverse Engineering UML State Machine from Java Code. International Conference on Intelligent Systems and Computer Vision (ISCV'2015), Fes, Morocco, (March 25-26 2015).

33. Chafik Baidada, El Mahi Bouziane and Abdeslam Jakimi, An Analysis and New Methodology for Reverse Engineering of UML Behavioral, International Journal of Advanced Engineering, Management and Science (IJAEMS), Vol-2, Issue-4, (April- 2016).

DOI: 10.1051/itmconf/20171301029

Two-exponential model of ferroimpurities extraction during magnetic inspection of these impurities content in various media

Anna Sandulyak[1,*], *Alexander* Sandulyak[1], *Maria* Polismakova[1], and *Darya* Sandulyak[1]

[1]Moscow technological university, Moscow, RF

Abstract. Illustrative examples of process media for which the key mass-operational characteristics of magnetophoretic control of ferroimpurities (represented in test semilogarithmic coordinates) are not subject to the basic (one) exponential model are presented and analyzed. The concept is described and implemented of a two-exponential model that assumes the desuperposition of the key control characteristic (for two subfractions of ferroimpurities). The possibilities are demonstrated of performing practical mass calculations of ferroimpurities in a process media sample and corresponding values of their concentration in this media.

1 Introduction. The main provisions of (one)-exponential model of the step-by-step extraction of ferroimpurities

Many liquid and free-flowing media in industry (fuel and lubricants, raw materials in the production of glass and ceramics, molding mixture for casting molds, food products, etc.) contain a fraction of impurities of almost any kind, as inclusions in the form of ferroparticles. The metal (structural steels) wear and corrosion in various equipment are the main sources of their formation and entry into operating environments.

To ensure the necessary quality (purity) of certain media, which in turn requires professional choice and application of appropriate purification methods and devices, the target content of ferroimpurities is of fundamental importance. Such control is performed via various methods (depending on specific media), including the polyoperational magnetophoresis method [1-4]. In particular, in this case a sample of the liquid media is passed through a sectional magnetic analyzer with the number of sections-operations n or n operations of "magnetic scanning" of a granular medium thin layer [1,2]. This approach leads to one of the conceptual provisions of the physical control model of this type: the analytic form appears of the decrease in mass m of ferroimpurities, extracted from the sample while operations, desirable for the extrapolation analysis, and this form is usually an exponential one [1-4]:

$$m = A \cdot \exp(-k \cdot n), \qquad (1)$$

where A and k – empirical parameters. Then, based on this actual section of experimentally obtained mass-operational dependence at the disposal of the experimenter (operator), it is possible to set the parameter values A and k.

In addition to the exponential regularity (1) itself, the conceptual provisions of the corresponding physical model of this control method include the possibility of an experimental calculation of the total mass of released ferroimpurities. Thus, in analytical terms, the dependence (1) is discrete, representing a quantitative series, the terms of which, beginning with the first one $m_1=A\cdot\exp(-k\cdot 1)$, decrease in a geometric progression with the geometric ratio $q=\exp(-k)$. Therefore, given the functional extrapolation (in order to obtain objective predictive data) beyond the actual experimental operations of magnetophoresis, they can also be summed up to any amount value of n, up to $n\to\infty$.

Hence, a reliable determination of the total (both, actually separated and residual) mass of ferroimpurities of the analyzed medium sample amount to finding the sum $m_{1...\infty}$ of an infinite number of terms ($n\to\infty$) of the given progression, i.e.

$$m_{1...\infty} = \sum_{n=1}^{\infty} m_n = \frac{m_1}{1-q} = \frac{A}{\exp k - 1}. \qquad (2)$$

Noticed, when using the calculated expression (2) we are taking into account not only the mass values of the actually separated ferroparticle (as a result of n operations of their magnetophoretic separation), but also the data of the residual mass of such particles, which has not been separated yet from the analyzed medium sample after actual n operations. To calculate $m_{1...\infty}$ according to (2), it is only necessary to use the values of empirical parameters A and k previously established (as a result of performing magnetophoresis operations). It should be noted that in the absence of such an original experimental-analytical approach, the obtaining of such total ferroimpurities mass ($m_{1...\infty}$) is hardly possible: it could be obtained only under hypothetically unlimited

* Corresponding author: anna.sandulyak@mail.ru

(not feasible in practice, of course) number of magnetophoresis operations.

It is clear, the established value of the sediment total mass $m_{1...\infty}$ of ferroimpurities in the analyzed medium volume V and mass M is the key for determining the true values of the desired ferroimpurities concentration $c = m_{1...\infty} / V$ and/or its analogue – mass fraction of these ferroimpurities: as $c = m_{1...\infty} / M$, including in %, i.e. $c = (m_{1...\infty} / M) \times 100\%$.

2 Simulation results of magnetic inspection characteristics atypical behavior (two-exponential model). Desuperposition concept

In terms of the above-mentioned conceptual position of the basic physical model (polyoperational magnetophoresis), the experimentally obtained dependence of the separated masses of ferroimpurities particles m from the ordinal number of magnetophoresis operations n should correspond to the regularity (1). Testing the statement of such correspondence is easy to implement, in particular, with the convenient semi-logarithmic coordinates, where this dependence should become quasilinear.

At the same time, along with numerous data [1-4], which showed the specified correspondence, a number of data revealed, which obviously do not correspond to the regularity (1). Among them – the data of polyoperational magnetophoresis, obtained for the molding mixture, sand sugar, feldspar samples. Thus, the expanded mass-operational dependences for these media (Fig.1*a*, 2*a*, 3*a*), being represented with semilogarithmic coordinates (Fig. 1*b*, 2*b*, 3*b*), are not quasi-linearized (by a single line), and therefore do not follow the classic exponent (1).

Herewith, it cannot go unnoticed that for such atypical mass-operational dependences, as if having a break in semilogarithmic coordinates (Fig. 1*b*-3*b*), the propensity to quasilinearization (and, hence, to exponential form) clearly preserves their tail sections. This circumstance directly indicates the presence in the composition of the ferroimpurities fraction (each of the media analyzed here) of a relatively wide range of coarseness and/or magnetic susceptibility, i.e. the presence of the ferroimpurities "subfractions". For example, a similar fact ("break" of a specific ξ-characteristic) was noted for [5] in magnetic-filtration purification of liquids from ferroparticles of different kinds with a change in the length of the magnetic-capture zone.

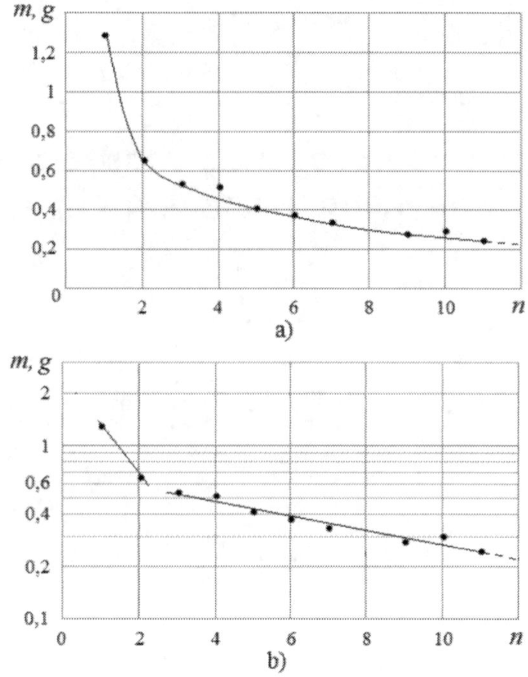

Fig. 1. Dependence of the operational (separated through polyoperational magnetophoresis) fraction mass of the molding mixture ferroimpurities; *a* and *b*) – in ordinary and semilogarithmic coordinates. Judging from the fact that the data in figure "*b*" are not quasilinearizable, this dependence is the result of the probable superposition of the ferroimpurities subfractions operational masses.

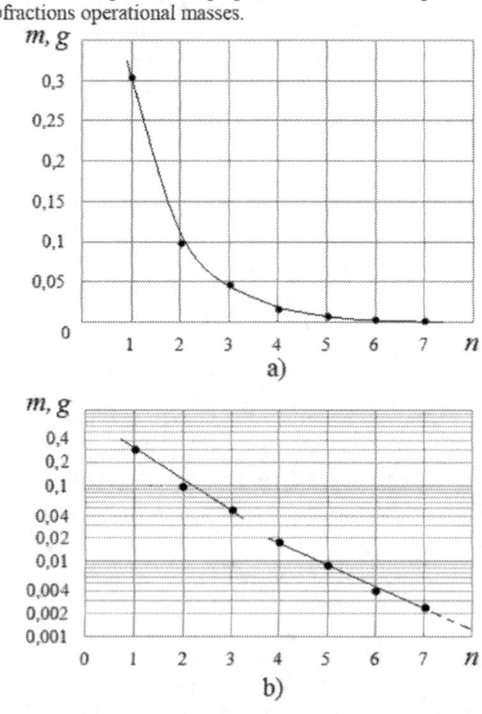

Fig. 2. Same as in Fig. 1, but for the sand sugar ferroimpurities.

DOI: 10.1051/itmconf/20171301029

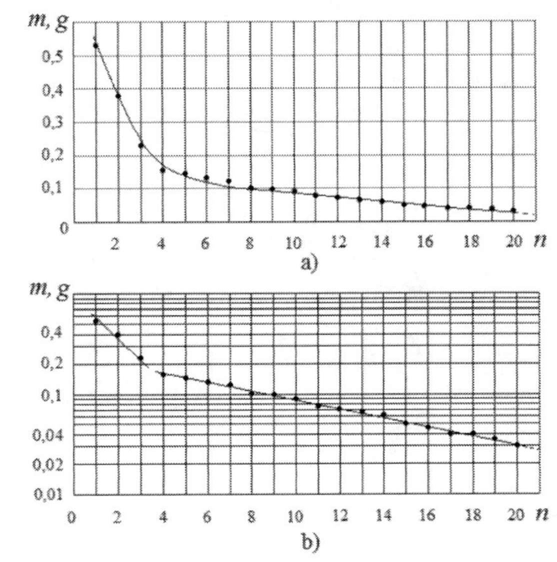

Fig. 3. Same as in Fig. 1, but for the feldspar ferroimpurities.

Consequently, same as above [5], it is expedient to resort to a justified quasi-separation of impurities ferroparticles, for example, into two subfractions, which differ considerably from one another, in the whole fraction of the ferroimpurities. This follows from Fig.1*b*-3*b* in terms of indirect indications, namely, by the presence of two distinct sections of the mass-operational dependence, which is, please note, characteristic of not an "alternating", but simultaneous precipitation of both subfractions. The first subfraction is conditionally easily extracted (*A*), consisting of larger particles and/or possessing certain magnetic properties. The second subfraction is conditionally difficult to extract (*B*), consisting of particles with greater degree of dispersion and/or possessing comparatively (in comparison with the first one) deteriorated magnetic properties.

Hence, same as in [5], it will be fair not to completely abandon the exponential model, but its in-depth analysis. Herewith, it is enough to assume that such an atypical dependence (Fig.1*b*-3*b*) – is the result of a superposition, for example, of two exponential types (characteristic of the magnetophoretic extraction of each of the two subfractions). Then, it is appropriate to raise the question of these "autonomous" dependencies legalization, i.e. question of a peculiar desuperposition of the resulting dependence (some of them are shown in Fig. 1-3).

To develop the corresponding physical model (in order to obtain the necessary calculation dependences) of the two-exponential model, in essence, the independence principle should be used, here – the principle of independent (from each other) magnetophoretic behavior of the ferroparticles subfractions *A* and *B*. Then, we can state the obvious fact that for each of the ferroparticles subfractions *A* and *B*, similar to (1), the individual characteristics (m_A and m_B) of the exponential operational masses loss are valid:

$$m_A = A_A \cdot \exp(-k_A \cdot n),$$

$$m_B = A_B \cdot \exp(-k_B \cdot n), \quad (3)$$

in this case, the alternative to the superposition expression (1) for the operational mass m is the detailed (as $m = m_A + m_B$) expression:

$$m = m_A + m_B = A_A \cdot \exp(-k_A \cdot n) + A_B \cdot \exp(-k_B \cdot n), \quad (4)$$

where A_A and A_B, k_A and k_B – the corresponding empirical parameters for *A*–subfraction and *B*–subfraction. Obtaining their values (as the required initial step in the implementation of the proposed desuperposition concept) is as follows.

One or another experimental operational dependence of m on n (Fig. 4, shaded points), which is subject, as mentioned above, to the partial linearization (for the tail part) in the semi-logarithmic coordinates used here, must be decomposed using graph-analytic method into dependency components (by fractions *A* and *B*). The "break point" (for $n=[n]$) of the experimentally obtained dependence (Fig. 4, shaded points) can serve as a reference mark for such a decomposition.

Obviously, the area of this dependence at $n>[n]$ (Fig. 4, shaded points), i.e., namely the tail section, which is subject to quasilinearization, is practically characteristic of the fraction *B* only – under conditions of almost complete absence of fraction *A* (separated to $n=[n]$). In order to fully identify the component *B* the actual section *B* follows the dependence of m on n (its tail section in Fig. 4: solid line, shaded points) to extrapolate to the region of small values of n (dashed line, ushaded points).

To identify the component *A* (at $1 \le n \le [n]$ – see Fig. 4), a simple mutual subtraction of the corresponding ordinates is performed of the actual main here section (*A*+*B*) of m on n dependence and extrapolated section *B* (dashed line *B*). This component (*A*) in Fig. 4 is also illustrated by the dashed line *A* (unshaded enlarged points).

Extrapolation of the identified dependences *A* and *B* (dashed lines in Fig. 4) to the formal area up to $n \to 0$ directly provides values A_A and A_B, and the necessary calculations using these obtained values A_A and A_B, expressions (3) and specific dependences *A* and *B* data in Fig. 4 allow finding also the values of the parameters k_A and k_B. Of course, using the corresponding graph-analytical identification data for dependencies *A* and *B* (Fig. 4), we can obtain similar results for determining the values A_A, A_B, k_A and k_B using one of the known programs, in particular, Excel.

On the basis of (3), similarly to (2), it is also possible to obtain calculated expressions for the total masses of the ferroparticles of each of the subfractions:

$$(m_{1...\infty})_A = \frac{A_A}{\exp k_A - 1},$$

$$(m_{1...\infty})_B = \frac{A_B}{\exp k_B - 1}. \quad (5)$$

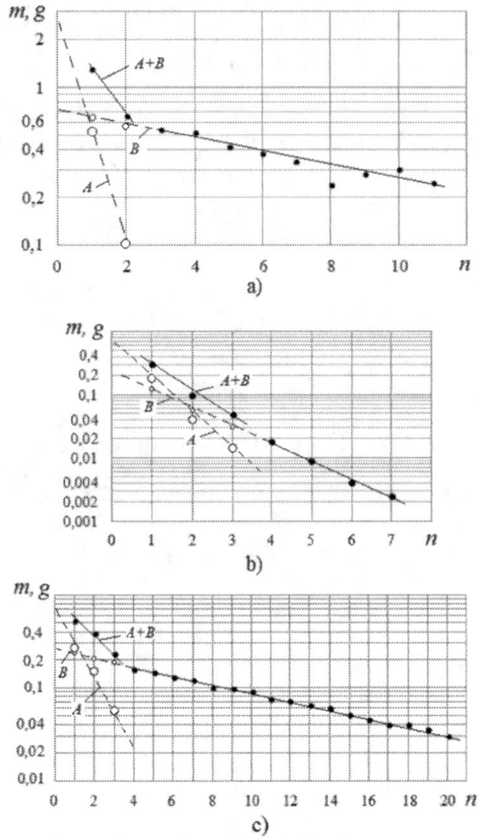

Fig. 4. Illustration of graph-analytic desuperpositions of the mass-operational dependencies shown in Fig. 1-3 (shaded points), on individual mass-operational dependencies for *A*–subfraction (ushaded enlarged points) and *B*-subfraction (unshaded reduced points first, and then, shaded ones): *a*) molding mixture, *b*) sand sugar, *c*) feldspar.

Hence, based on the concept of desuperposition proposed here, an alternative to the superposition expression (2) for the total mass $m_{1...\infty}$ of ferroimpurities contained in the medium, should be the detailed (by subfractions *A* and *B*) expression:

$$m_{1...\infty} = (m_{1...\infty})_A + (m_{1...\infty})_B = \frac{A_A}{\exp k_A - 1} + \frac{A_B}{\exp k_B - 1} \qquad (6)$$

An important consequence of the desuperposition concept considered herein is that it becomes possible to determine individual share value of each subfraction in the general fraction of ferroimpurities:

$$\lambda_A = (m_{1...\infty})_A / [(m_{1...\infty})_A + (m_{1...\infty})_B], \quad \text{x100\%} \qquad (7)$$

$$\lambda_B = (m_{1...\infty})_B / [(m_{1...\infty})_A + (m_{1...\infty})_B] = (1-\lambda_A).\text{x100\%} \qquad (8)$$

moreover – clear quantitative values (identified by the experimental data, with their inherent difference parameters), in addition to the qualitative ones declared at the stage of this concept definition.

Corresponding values of the magnetophoretic control parameters of the conventionally easily extracted (*A*) and difficult-to-extract (*B*) fractions of the ferroimpurities are given in the table.

Table 1. The parameters of polyoperational magnetophoretic control obtained for the subfractions *A* and *B* of the ferroimpurities fraction (as shown in Fig. 1-4, as a result of desuperposition).

Sample of analyzed media	Values of parameters			
	A_A, g; A_B, g	k_A; k_B	$(m_{1...\infty})_A+(m_{1...\infty})_B=$ $= m_{1...\infty}$, g	λ_A, %; λ_B, %
Molding mixture	2,6; 0,7	1,63; 0,094	0,63+7,1= **=7,73**	8,2; 91,8
Sand sugar	0,8; 0,21	1,37; 0,64	0,273+0,234 **=0,507**	53,8; 46,2
Feldspar	0,7 0,27	0,84 0,12	0,532+2,118 **=2,65**	20; 80

3 Conclusion

The main provisions are declared of a well-proven exponential model of the mass-operational (prone to linearization in semi-logarithmic coordinates) characteristics of magnetophoretic control of ferroimpurities in liquid and free-running media. Attention is drawn to those characteristics that are not typical, i.e. formally do not follow the exponential model. It has been established (and shown by an appropriate theoretical-experimental data analysis) that an atypical form of such characteristics is not a result of removing from the exponential regularity of the operational masses loss of the extracted (under polyoperational magnetophoresis) ferroimpurities. It is caused by a superposition of at least two exponential regularities – for two mutually different subfractions of ferroimpurities. The concept is proposed of desuperposition of the atypical mass-operational characteristic of the magnetophoretic control of ferroimpurities resulting in particular (for each of the subfractions) exponents, particular expressions for the total masses and, accordingly, information on the fractional presence of each subfraction.

Acknowledgements

The research is conducted with financial support from RFFI within the frameworks of research project № 16-38-60034 mol_a_dk and from Russian Federation Ministry of Education and Science №9.9626.2017.

References

1. A.A. Sandulyak, M.N. Polismakova, D.V. Ershov, et al. Meas. Tech. **53 (8)**, 914–918 (2010).
2. A.A. Sandulyak, A.V. Sandulyak, V.A. Ershova, et al. Journal of Magnetism and Magnetic Materials, **426**, 714–720 (2017).
3. A.A. Sandulyak, A.V. Sandulyak, H. Bulut, et al. Web of Conferences 76, 02045 (2016)

4. A.A. Sandulyak, D.A. Sandulyak, A.E. Ablaeva, et al. Proceedings of the International. Conference on «Advances in Environmental and Agricultural Science». Dubai. Feb. 135–138 (2015).
5. A. V. Sandulyak. Magnetic and filtration purification of liquids and gases. Moscow: Chemistry (1988).

Preparation of Improved Turkish DataSet for Sentiment Analysis in Social Media

Semiha Makinist[1,*], *İbrahim Rıza* Hallaç[2,*], *Betül* Ay Karakuş[2,*], and *Galip* Aydın[2,*]

[1]Sentis Software
[2] Department of Computer Engineering, Firat University, Elazig, Turkey

Abstract. A public dataset, with a variety of properties suitable for sentiment analysis [1], event prediction, trend detection and other text mining applications, is needed in order to be able to successfully perform analysis studies. The vast majority of data on social media is text-based and it is not possible to directly apply machine learning processes into these raw data, since several different processes are required to prepare the data before the implementation of the algorithms. For example, different misspellings of same word enlarge the word vector space unnecessarily, thereby it leads to reduce the success of the algorithm and increase the computational power requirement. This paper presents an improved Turkish dataset with an effective spelling correction algorithm based on Hadoop [2]. The collected data is recorded on the Hadoop Distributed File System and the text based data is processed by MapReduce programming model. This method is suitable for the storage and processing of large sized text based social media data. In this study, movie reviews have been automatically recorded with Apache ManifoldCF (MCF) [3] and data clusters have been created. Various methods compared such as Levenshtein and Fuzzy String Matching have been proposed to create a public dataset from collected data. Experimental results show that the proposed algorithm, which can be used as an open source dataset in sentiment analysis studies, have been performed successfully to the detection and correction of spelling errors.

Keywords. *Sentiment Analysis, Natural Language Processing, Hadoop, Turkish Data Set*

1 Introduction

Today, most of the social media sources we meet frequently encounter in a very wide range of fields are generate by video sharing (YouTube), photo sharing (Instagram), and location based applications (Foursquare), blogs, microblogs (Twitter), social networks (Facebook)... Social media, made up of these different digital media, has become an important source of information to many researchers in the form of constantly updated information on many issues such as popular events, politics, products, or services. For example, a politician can use this data to make a conclusion about what people think about him/her when conducting an election campaign. Similarly, a company may collect data from social media for analyzing the opinions of their customers on their products.

According to March 2015 data, Facebook has more than 1 billion registered users, and Twitter has 288 million active users per month [4]. Linkedin, Tumblr, Instagram and many other social networks are increasing their number of users every day. The need to store, analyze and visualize data in social networks is not only a problem of these social networking companies. In many different areas, people and organizations want to access social media data and analyze it from various perspectives.

The collection, management and analysis of these data is a major big data problem [5]. Many studies have been done in this field in the literature and usually Twitter is preferred as the social media platform. One of the most important reasons for this is that Twitter users share short and many messages. By benefitting from these social media posts one can come up with the distribution of positive and negative thoughts, the tendencies of the targeted customer groups, the reputation and influence status on social media for people or companies. These type of analyses are mainly carried out by applying machine learning techniques on large volumes of data. Different methods are used during the collection of data such as data can be collected by using the application programming interface

* Semiha Makinist: semihamakinist@gmail.com
* İbrahim Rıza Hallaç: irhallac@firat.edu.tr
* Beetül Ay Karakuş: betulay@firat.edu.tr
* Galip Aydın: gaydin@firat.edu.tr

(Twitter API, Facebook API, etc.) of the platform, can be saved manually, or a problem specific data collection method can be used.

In the literature, Pak and Paroubek [6] created a data structure for automatic sentiment analysis to predict if a newly added data is positive, negative, or neutral. Albert and Eibe [7] proposed a sliding window Kappa statistic for evaluation in time-changing data stream and they dynamically classify the instant data flows on Twitter. Yerve et al. [8] detects in microblogs whether there were posts related to difficult companies, such as "Apple, BlackBerry". Roberto, Smaranda and Nina [9] analyzed tweets in their study to analyze existence of sarcasm. Agarwal et al. [10] performed sentiment analysis of tweets using a tree kernel method. Saif et al. [11] obtained quite good results by adding additional semantic features (eg, the feature of the device on which the tweet is inserted) to the classification methods to identify the sentiment of the tweets that are posted by the companies or the individuals. Spencer and Uchyigit [12] used a web-based tool which is called as "Sentimator" to collect data from Twitter in real-time according to specified search terms, and they developed a sentiment analysis tool using the Naive Bayes classification algorithm and natural language processing libraries. Kouloumpis, Wilson and Moore [13] performed a classification task by using the hashtags and emoticons written in the tweets along with the consideration of the linguistic features a language. Saif and his friends [14] formed two different clusters in their studies to reduce the density of the Twitter data which is to be used for sentiment analysis. Agarwal and Sabharwal [15] proposed a sentiment analysis technique for Twitter texts which uses an end-to-end pipeline method to group tweets into four groups; positive, negative, objective and neutral categories. Cody et al. [16] tried to identify climatic changes by analyzing data collected from Twitter. Paltoglou [17] conducted a study on emotion-based event detection. Singhal and his friends [18] used Twitter data to predict which party would win the elections.

In the mentioned studies, sentiment analysis and other prediction based applications were performed using textual data. The quality of the data sets used in these studies is highly critical and it affects the success of the results at high rates. In addition, the size of the word-vector-matrices, which are input data of the text-based classification and clustering algorithms such as TF-IDF, are greatly increased as a result of deriving different words because of the wrong spellings of the same word. This increase, significantly reduces the success of the machine learning algorithm. In addition, every word added will increase the required processing capacity and reduce the computational performance. In this study we propose a method for creating a data set in Turkish to improve the performance of sentiment analysis studies which are based on textual data. Various improvement studies have been carried out in order to increase the data quality and different methods have been tried. Some problems have been encountered in the experiments and some solutions have been proposed for solving these problems. In the first part of the article, the introduction is given, in the second part how the data is gathered, in the third part how the data set is formed, in the fourth section the proposed algorithm is explained. In the fifth part we compare our method with other methods in the literature. In the sixth part we share the experimental results.

2 Data Collection

Obtaining the data covers a significant part of this work. User comments in a movie site are preferred as data source. There are also ratings on the movie site, with comments from users and ratings of 1 and 5 points. The reason for choosing film critics as datasets is that there are numerical rating values given along with the user comments.

Apache ManifoldCF (MCF) was used to obtain the data. MCF technology is an application that transfers data contained in data sources to target resources. In the general use of MCF, data mapping is done by establishing a bridge to content indexing systems from enterprise content management systems (also applying content security policies during the process).

In this study, MCF was used as a web browser and user comments in the designated movie site were drawn to the file system. The first step in collecting data with MCF is to set the source from which the data will be pulled. In MCF, this is called repository connector and there are many repository connectors in MCF. One of these is the web connector. The web connector passes the data of the web pages to the target sources, and if any authorization system is used before this connector is created, an appropriate idle authorization connector must be created. Using this technology, the data is drawn with the following sequence:

1. Create an empty authorization group.
2. Create a Web adapter.
3. A time adapter is created to configure which time intervals the data will be retrieved.
4. Create a file adapter to save the data.
5. After all the necessary adapters have been created, finally a transfer job needs to be created. In other words, which site to connect to and the index area (user comments) is determined.
6. Once all the work is done, the created job is started.

Once all the configurations are complete, Apache ManifoldCF scans all web pages and saves the pages with user comments to the local file system in HTML format. In this way, 2983 movie reviews were collected for use as sample data sets. In order to extract the comments out of this data, the Groovy script language which can work on the JVM was used and the edited data was saved in JSON format.

3 Data Set Creation

The step of generating words or groups of words is a very important step in text-based analysis where natural language processing methods are used. While there are many resources and software libraries for creating these data sets in the literature, most of them are for the English language. There are very few methods and software libraries developed for Turkish. In this study, language processing and word base generation techniques have been proposed based on Turkish language [19].

The primary aim of the work done is to clear the comments correctly from unnecessary words and to create correct word lists from 1, 2 and 3 grams according to the scores given by the users. Another goal is to identify strong negative and strong positive word and word groups in word lists.

For this purpose, comments are first cleared of unnecessary words (conjunctions, prepositions, adverbs, special names, place names and country names), punctuation marks, and numbers. In the next step, we use the Zemberek library to check whether each word is in Turkish (language detection). If a word is misspelled in Turkish, the proposed method corrects the word and the word is converted to the correct one. As a result of these operations, 3 different files are obtained for each rating score between 0 and 5 points. A total of 15 files will be created in this way. From all these operations, lists of words in different grams in different types of ratings are obtained and they are saved in a file that is not in any other file, ie only in the file it is in. The main purpose of this step is to identify the most powerful words and phrases, especially those for the rating values of 0 and 5. Some of the results from the comments of 0 and 5 points of ratings for 1 gram is shown in Fig.1. Results are obtained by checking the spelling of the words with Zemberek and without making any additional change in the text. As you it can be seen in the figure, using the text without editing process is performed is not generating a very effective word list.

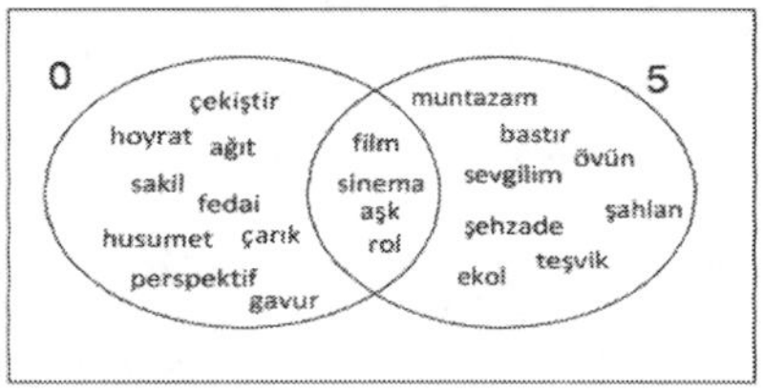

Fig. 1. 1-gram Turkish words from movie reviews with a star rating on 0-5 scale.

Some of the results that have been come into prominence on the list of words used and not used by Zemberek in 1 gram, which are obtained after correcting the movie reviews of 0 and 5 stars are shown in Fig. 2.

0	0&5	5
iğrenç nefret gereksiz izlemeyin rezalet sıkıcı etmiyorum berbat	film sinemada iyi çok kötü kesinlikle güzel tavsiye	muhteşem mükemmel harika etkileyici süper efsane gerçekten ederim

(a)

0	0&5	5
iğrenç nefret gerek rezalet sıkıcı berbat	film sinema iyi çok kötü kesin güzel tavsiye	muhteşem mükemmel harika etki süper efsane gerçek

(b)

Fig. 2. The results from the comments of 0 and 5 stars of ratings for 1 gram (a) The results from the comments of 0 and 5 stars of ratings for 1 gram with Zemberek (b).

Some of the results that have been come into prominence on the list of words used and not used by Zemberek in 2 grams (Fig. 3) and 3 grams (Fig. 4).

0	0&5	5
en kötü berbat film berbat ötesi saçma sapan tavsiye etmiyorum vasat film en berbat yazıklar olsun vakit kaybı hak etmiyor	kötü film çok kötü son zamanlarda tek kelimeyle filmin sonu kesinlikle tavsiye	çok güzeldi film iyi görsel şölen mutlaka izleyin muhteşem film süper film baş yapıt hak ediyor çok beğendim gerçekten harika en iyi çok kaliteli

(a)

0	0&5	5
en kötü berbat film berbat öte saçma sapan vasat film en berbat yazık ol vakit kap	kötü film çok kötü son zaman tek kelime film son kesin tavsiye hak et tavsiye et	çok güzel film iyi görsel şölen mutlaka izle muhteşem film süper film baş yapıt çok beğen gerçek harika

(b)

Fig. 3. The results from the comments of 0 and 5 stars of ratings for 2 grams (a) The results from the comments of 0 and 5 stars of ratings for 2 grams with Zemberek (b).

0	0&5	5
izlediğim en kötü en berbat film en kötü film puanı hak etmiyor verdiğim paraya acırım berbat ötesi film gördüğüm en berbat izlediğim en berbat saçma sapan film hayatımda böyle saçma rezalet ötesi film	hayatımda izlediğim en	izlediğim en iyi en iyi film kesinlikle izlenmesi gereken çok iyi film şiddetle tavsiye ederim herkese tavsiye ederim mutlaka izlenmesi gereken gerçekten çok güzel tek kelimeyle muhteşem tam baş yapıt

(a)

0	0&5	5
izle en kötü en berbat film en kötü film saçma sapan film ver para acı berbat öte film gör en berbat tek kelime berbat izle en berbat hayat böyle saçma rezalet öte film	hayat izle en puan hak et herkes tavsiye et	izle en iyi en iyi film kesin izle gerek çok iyi film şiddet tavsiye et mutlaka izle gerek gerçek çok güzel tek kelime muhteşem tam baş yapıt

(b)

Fig. 4. The results from the comments of 0 and 5 stars of ratings for 3 grams (a) The results from the comments of 0 and 5 stars of ratings for 3 grams with Zemberek (b).

It was found that the best results obtained from the studies were 2 and 3 grams word lists and it was observed that the word lists created by the proposed method compared to the word list created by using only Zemberek are more successful.

4 Proposed Method

In this study, some of the algorithms that have emerged in the literature such as Levenshtein Distance and Fuzzy String Matching have been tested. However, a new study was conducted because these methods did not achieve the desired success. The main goal of this study is to make:

- The words which are not written correctly in Turkish such as «cok, saglam, simarik, guzel» and
- Words with repetitive letters like «şoook, eeeen»,
- Abbreviations used in social media such as «kib, grrrz, slm, mrb, aeo, tmm» and
- To identify and correct spelling mistakes such as «gelrm, biliyrm, sonclr»

All of these steps are given below, respectively:

1. In the first step, it is checked whether there are any unnecessary repetitive characters in the word, if any, they are cleared. For example, a word "çoookkkk" is corrected to "çok".
2. In the next step, words with misspelled letters, such as "cok", "saglam" words, are converted into the correct Turkish writing in the form of "çok" and "sağlam". Different algorithms have been tried for this transformation process. For this purpose, *Levenshtein Distance*, *Fuzzy String Matching* and proposed word correction algorithm are used. The best results are obtained with the newly developed word correction algorithm.
3. Fuzzy String Matching Algorithm is used to detect letter errors. The test results are shown in Table 1 (the threshold value for all data is 0.8).
4. The Levenshtein Distance Algorithm is used to determine the word to be corrected according to the obtained results.

Table 1. Correcting letter errors with Fuzzy String Matching Algorithm.

Word to fix	Correct Word	Results	
		Output	Prob.
gelirm	gelirim	geldim	0.83
		gelirim	0.92
		geliyorum	0.8
		germ	0.8
biliyrm	biliyorum	bildirim	0.8
		bilim	0.83
		biliyorum	0.88
sonuşlar	sonuçlar	sonuçlar	0.88

In Table 1, a word list was used for the purpose of checking and this list was taken from "mythes-tr \ data" addressed by Anıl Özbek's GitHub account [20].

4.1 Word Correction Algorithm

A simple probability equation is used when the word correction algorithm is developed. With this algorithm, new words are derived by determining the letters that are likely to be corrected. For example, "cok, cök, çok, çök" words were expected to be derived from "çok" word, and it was confirmed that these words were derived as a result of the studies done. This process can be performed for more than one letter, and successful results up to 6 letters have been obtained. The formula for this algorithm is given in Eq. (1).

$$\text{power} = 2^n \quad (1)$$

Table 2 describes the variables used in the equation used in the algorithm.

Table 2. Variable Definitions.

Variables	Description
n	Number of elements to fix
2	Letter change status
power	Derived word count

The operating principle of the proposed algorithm is shown in Fig. 5.

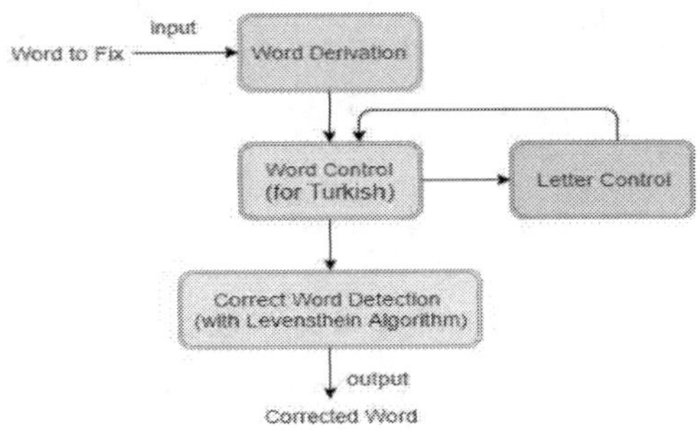

Fig. 5. The working principle of the word correction algorithm.

DOI: 10.1051/itmconf/20171301030

For example, the number of letters "2" to be changed in "çok" word (that is n = 2), the probability of each word is 2 since each word has two states, and the number of letters to be derived is calculated as 4 using the equation in (1).

As a result of the tests performed, it was observed that over 90% of the corrected words were achieved using the developed algorithm. The only significant disadvantage of this method is that the system slows down as the number of letters to be corrected increases. For example, if there are 18 letters in a word, this process takes approximately 1 hour. The reason for this is that (that is 262144) derivations of the word are done.

5 Experimental Results

The experimental results of the methods used to correct the words are given comparatively in the table context in this section. Several similarity algorithms have been used in the literature for this process, but in accordance with this study, tests have been performed using Levenshtein, Fuzzy String Matching and the proposed algorithm. The comparison of these algorithms as a result of the tests performed is given in Table 3. The threshold value for all tests performed in Table 3 was set at 0.6. In the Levenshtein Algorithm, it is assumed that the cost is closer to 0, so the similarity is closer to 1 in the Fuzzy String Matching Algorithm. A threshold value is not used for the word correction algorithm.

Table 3. The Experimental Results of the Methods.

Algorithms	Test Data	Results	
		Output	**Prob.**
Levenshtein	cok	cık	0.33
		cop	0.33
		cuk	0.33
		çok	0.33
Fuzzy String Matching	cok	cık	0.66
		cokey	0.75
		coşarak	0.66
		coşku	0.75
		coşkulu	0.6
		coşkun	0.66
		coşmak	0.66
		cuk	0.66
		çok	0.66
Proposed Method	cok	çok	1

As a result of observations made according to the test results, It has been found that using only Levenshtein and Fuzzy String Matching Algorithms is not sufficient. It is observed that these algorithms are better used in the correct place in the word correction algorithm.

6 Conclusion

Although data warehouses and open-source libraries for many languages are included in the literature, a full Turkish data set and library are not open source. Therefore, open source data sets and libraries are needed in Turkish sentiment analysis and other natural language processing studies. Moreover, the storage of social media data and the high cost of calculation make it a major problem. These problems can be eliminated through frameworks that support distributed programming such as Hadoop, Spark. For this purpose, the paper presented an improved Turkish dataset with an effective spelling correction algorithm with Hadoop Distributed File System and MapReduce programming model.

This study was carried out to create a dataset in the project "Cloud Based, Distributed Reputation Analysis Systems for Social Networks", code 2140281, which is supported by Tübitak 1512 Entrepreneurship Progressive Support Program. In the next study, sentiment analysis studies will be done using this dataset.

References

1. B. Pang, and L. Lee, Opinion mining and sentiment analysis, Foundations and Trends® in Information Retrieval, 1-135 (2008)
2. W. Tom, *Hadoop: The definitive guide*, O'Reilly Media, Inc. (2012)
3. https://manifoldcf.apache.org/ (2015)
4. http://www.statista.com/statistics/272014/global-social-networks-ranked-by-number-of-users/, (2015)
5. G. Aydin, and İ. R. Hallac, Distributed NLP, ISITES2015 (2015)
6. A. Pak, and P. Paroubek, Twitter as a Corpus Sentiment Analysis and Opinion Mining, In Proceedings of the Seventh Conference on International Language Resources and Evaluation, 1320-1326 (2010)
7. A. Bifet, and E. Frank, Sentiment Knowledge Discovery in Twitter Streaming Data, 13th International Conference, DS 2010, Canberra, Australia, pp 1-15 (2010).
8. S. R. Yerva, Z. Mikl´os, and K. Aberer, It was easy, when apples and blackberries were only fruits, Working Notes for CLEF 2010 Conference, **1176** (2010)
9. R. González-Ibáñez, S. Muresan, and N. Wacholder, Identifying Sarcasm in Twitter: A Closer Look, The 49th Annual Meeting of the Association for Computational Linguistics: Human Language Technologies, 581-586 (2011)
10. A. Agarwal, B. Xie, I. Vovsha, O. Rambow, R. Passonneau, Sentiment Analysis of Twitter Data, Proceedings of the Workshop on Language in Social Media (LSM 2011), 30–38 (2011)
11. H. Saif, H. Yulan, and H. Alani, Semantic sentiment analysis of twitter, The 11th International Semantic Web Conference (ISWC 2012), Boston, MA, USA (2012)
12. J. Spencer, and G. Uchyigit, Sentimentor: Sentiment Analysis of Twitter Data, Proceedings of the 1st International Workshop on Sentiment Discovery from Affective Data (SDAD 2012), 56-66 (2012)

DOI: 10.1051/itmconf/20171301030

13. E. Kouloumpis, T. Wilson, and J. Moore, Twitter Sentiment Analysis: The Good the Bad and the OMG!, 5. International AAAI Conference on Weblogs and Social Media, 538-541 (2011)
14. H. Saif, H. Yulan, and H. Alani, Alleviating data sparsity for twitter sentiment analysis, CEUR Workshop Proceedings (2012)
15. A. Agarwal, and J. S. Sabharwal, End-to-End Sentiment Analysis of Twitter Data, 24th International Conference on Computational Linguistics (2012)
16. E. M. Cody, A. J. Reaga, L. Mitchell, P. S. Dodds, and C. M. Danforth, Climate change sentiment on Twitter: An unsolicited public opinion poll, 2015.
17. G. Paltoglou, Sentiment-based event detection in Twitter, Journal of the Association for Information Science and Technology (2015)
18. K. Singhal, B. Agrawal, and N. Mittal, Modeling Indian General Elections: Sentiment Analysis of Political Twitter Data, In Information Systems Design and Intelligent Applications, 469-477 (2015)
19. https://github.com/ahmetaa/zemberek-nlp (2015)
20. https://github.com/maidis/mythes-tr/tree/master/veriler (2015)
21. J. Lin, and A. Kolcz, Large-scale machine learning at twitter, Proceedings of the 2012 ACM SIGMOD International Conference on Management of Data, 793-804 (2012)
22. A. Pak, and P. Paroubek, Twitter based system: Using Twitter for disambiguating sentiment ambiguous adjectives, Proceedings of the 5th International Workshop on Semantic Evaluation. Association for Computational Linguistics (2010)
23. S. Mukherjee, A. Malu, B. AR, and P. Bhattacharyya, TwiSent: a multistage system for analyzing sentiment in twitter, Proceedings of the 21st ACM international conference on Information and knowledge management (2012)
24. J. P. Carvalho, and L. Coheur, Introducing UWS-A fuzzy based word similarity function with good discrimination capability: Preliminary results, Fuzzy Systems (FUZZ), 2013 IEEE International Conference on. IEEE (2013)
25. A. Kumar, and T. M. Sebastian, Sentiment analysis on twitter, IJCSI International Journal of Computer Science, 372-378 (2012)
26. B. O'Connor, R. Balasubramanyan, B. R. Routledge, and N. A. Smith, From Tweets to Polls: Linking Text Sentiment to Public Opinion Time Series, ICWSM 11., 122-129 (2010)

A study on the improved tan($\phi(\xi)/2$) –expansion method

Berat KARAAGAC[1,⋆], *N. Murat* YAGMURLU[2,⋆⋆], and *Alaattin* ESEN[2,⋆⋆⋆]

[1] *Department of Mathematic Education,Adıyaman University , Adıyaman, Turkey*
[2] *Department of Mathematics,Inonu University , Malatya, Turkey*

Abstract. In this study, the improved tan($\phi(\xi)/2$)-expansion method (ITEM), one of the improved expansion methods, has been applied to (3+1)- dimensional Jimbo Miwa and Sharma-Tasso-Olver equations using symbolic computation. With the aid of the method, many new and abundant analytical solutions have been obtained. The newly obtained results show that ITEM is a new and significant technique for solving nonlinear differential equations which plays an important role on fluids mechanics, engineering and many physics fields.

1 Introduction

Scientists have used mathematics to describe the physical properties of the universe for many years by modelling. Many fundamental phenomena observed in ecology,physics, finance, data science, mechanical engineering etc. can be formulated using differential equations. Therefore obtain analytical and numerical solutions for differential equations is a significant part of scientific studies. So, over the years a wide range of efficient methods have been suggested and improved for solving those equations such as Refs. [1–6] and many other techniques in their references. The aim of the current study is to derive new analytical solutions of (3+1)-dimensional Jimbo Miwa (JM) and Sharma-Tasso-Olver (STO) equations which are physical model by using an improved expansion method called as improved tan($\phi(\xi)/2$)-expansion method. The rest of the study is organized as follows, in the second section is a description of improved tan($\phi(\xi)/2$)-expansion method (ITEM) and third section consists of an application of the method to the STO and JM equations and newly obtained results. At the end of the study, conclusion and references are laid out, respectively.

2 Algorithm of the improved tan($\phi(\xi)/2$)-expansion method

In this section of the study, we are going to present the algorithm of ITEM for constructing analytical solutions of nonlinear differential equations succesfully. The algortihm has four steps given as follows:

Step 1*:* The main idea of this step is to reduce the nonlinear partial differential equation to the nonlinear ordinary differential equation. First of all, let us symbolize the NPDEs with polynomial R of $u(x, y, z, t)$ as

$$R\left(u, u_x, u_y, u_z, u_t, u_{xx}, u_{yy}, ...\right) = 0 \tag{1}$$

⋆e-mail: b.karaagac@adiyaman.edu.tr
⋆⋆e-mail: murat.yagmurlu@inonu.edu.tr
⋆⋆⋆e-mail: alaattin.esen@inonu.edu.tr

where x, y, z and t are independent variables and u is an unknown function. R consist of partial derivatives, the highest order derivative and nonlinear term. For reduction process, combining four independent variables x, y, z and t into one parameter ξ, i. e, using traveling wave transformation given as

$$u(x,t) = u(\xi) \qquad \xi = \alpha x + \beta y + \gamma z - \mu t. \tag{2}$$

Eq. (1) reduces to nonlinear ODE follows

$$R'\left(u, \alpha u', \beta u', \gamma u', -\mu u', \alpha^2 u'', \beta^2 u'', ...\right) = 0 \tag{3}$$

where α, β, γ and μ are real parameters.

Step 2*:* Assuming that the wave solution of Eq. (3) can be expressed as

$$u(\xi) = \sum_{k=0}^{m} A_k [p + \tan(\phi(\xi)/2)]^k + \sum_{k=1}^{m} B_k [p + \tan(\phi(\xi)/2)]^{-k} \tag{4}$$

where A_0, A_k, B_k, $(1 \leq k \leq m)$ and p are constants to be determined by symbolic computation, m is a positive integer to be calculated using homogenous balance between the highest order derivative and nonlinear term. Also $\phi(\xi)$ is the general solution of certain Riccati equation with constant coefficients given as

$$\phi'(\xi) = a\sin(\phi(\xi)) + b\cos(\phi(\xi)) + c. \tag{5}$$

Manafian, obtained special solutions of Eq. (5) and expressed in the following nineteen solution families, one can see in Ref. [7]

Step 3*:* In this step, substitution of derivatives for $u(\xi)$ respect to ξ into (3) is going to be carried out. After this proceeding, a polynomial yields in term of $\tan(\phi(\xi)/2)^k$ and $\cot(\phi(\xi)/2)^k (k = 1, 2, ...)$ function. Collecting coefficients of the newly polynomial according to the power of trigonometric functions and setting to zero, we can get an algebraic equation system in terms of A_0, A_k, B_k, a, b, c and p.

Step 4*:* At the final step, after solving the algebraic equation systems which are obtained in Step 3 for A_0, A_k, B_k and μ and using these values in (4) with Families 1 ~ 19 seen in Ref.[7] , the desired solutions are obtained.

3 Applications

We are going to now exert ITEM described in Section 2 to solve Jimbo Miwa and Sharma-Tasso-Olver equations.

3.1 Jimbo-Miwa equation

Consider the following nonlinear partial differential equation of which name is Jimbo Miwa equation

$$u_{xxxy} + 3u_y u_{xx} + 3u_x u_{xy} + 2u_{yt} - 3u_{xz} = 0. \tag{6}$$

Here we apply traveling wave transformation to Eq.(6) given by (2) with $\alpha = \beta = \gamma = 1$ and as a result of the first step, we obtain the following nonlinear ODE:

$$u^{iv} + 6u'u'' - (2\mu + 3)u'' = 0 \tag{7}$$

During the second step, using homogeneous balance between the highest order derivative u^{iv} and nonlinear term $u'u''$, solution of Eq. (7) expressed for $n = 1$ is obtained by

$$u(\xi) = A_0 + A_1 [p + \tan(\phi(\xi)/2)] + B_1 [p + \tan(\phi(\xi)/2)]^{-1} \tag{8}$$

At the last for steps 3 and 4, when we substitute the first, second and fourth order derivatives of $u(\xi)$ given in (8) into (7), we hold an over determined algebraic polynamial in terms of $A_0, A_k, B_k, p, a, b, c$ and μ. Then collecting coefficients of the newly obtained polynomial as trigonometric functions and equate to zero, and solving the algebraic system for A_0, A_k, B_k and p by symbolic computation software Maple, the following solution set is obtained.

$$\text{set 1:}\quad \{A_0 = A_0,\quad A_1 = b - c,\quad B_1 = 0,\quad \mu = \tfrac{a^2+b^2-c^2-3}{2}.\}$$
$$\text{set 2:}\quad \{A_0 = A_0,\quad A_1 = b - c,\quad B_1 = \tfrac{b^2+a^2-c^2}{b-c},\quad \mu = \tfrac{2(b^2+a^2-c^2)-3}{2},\quad p = \tfrac{a}{c-b}.\}$$

To show power and effect of the improved method, using the solution set 2 with Families 1, 2, 3, 4 and 8 in Ref. [7], we get the following solutions;

- For $a^2 + b^2 - c^2 < 0$ and $b - c \neq 0$,

$$u_1 = A_0 + \left(c^2 - a^2 - b^2\right)^{1/2}\left(\cot\left(\frac{\sqrt{c^2 - a^2 - b^2}}{2}(\xi + C)\right) - \tan\left(\frac{\sqrt{c^2 - a^2 - b^2}}{2}(\xi + C)\right)\right)$$

- For $a^2 + b^2 - c^2 > 0$ and $b - c \neq 0$,

$$u_2 = A_0 + \left(a^2 + b^2 - c^2\right)^{1/2}\left(\tanh\left(\frac{\sqrt{a^2 + b^2 - c^2}}{2}(\xi + C)\right) + \coth\left(\frac{\sqrt{a^2 + b^2 - c^2}}{2}(\xi + C)\right)\right)$$

- For $a^2 + b^2 - c^2 > 0$ and $b \neq 0$ and $c = 0$,

$$u_3 = A_0 + \left(a^2 + b^2\right)^{1/2}\left(\tanh\left(\frac{\sqrt{a^2 + b^2}}{2}(\xi + C)\right) + \coth\left(\frac{\sqrt{a^2 + b^2}}{2}(\xi + C)\right)\right)$$

- For $a^2+b^2-c^2 < 0$ and $c \neq 0$ and $b = 0$, $u_4 = A_0 - \left(c^2 - a^2\right)^{1/2}\left(\tan\left(\frac{\sqrt{c^2-a^2}}{2}(\xi + C)\right) - \cot\left(\frac{\sqrt{c^2-a^2}}{2}(\xi + C)\right)\right)$
- For $a^2 + b^2 = c^2$, $u_8 = A_0 + \frac{2}{\xi+C}$

In the rest of the subsection, we depict graphical simulation of the considered equation. In Fig 1, we plot 3D graphs of the solutions given above. plotting graphics, we have chosen $x, y \in [-15, 15]$, $t = 1, C = 1$ for all figures and $a = 2, 2, 1, 1$ $b = 3, 3, 3, 0, 1$ and $c = 4, 2, 0, 3, \sqrt{2}$ for u_1, u_2, u_3, u_4 and u_8, respectively.

3.2 Sharma-Tasso-Olver equation

In this subsection, we will next obtain analytical solutions of the following Sharma-Tasso-Olver equation through the ITEM

$$u_t + \alpha\left(u^3\right)_x + \frac{3}{2}\alpha\left(u^2\right)_{xx} + \alpha u_{xxx} = 0 \tag{9}$$

To start, let us transform Eq.(9) into nonlinear ODE by $u(x,t) = u(\xi)$, $\xi = x - \mu t$ where with, $\beta = \gamma = 0$ and α, μ is an arbitrary parameters and μ symbolizes wave speed. Thus, we get the following equation

$$-\mu u' + 3\alpha u\, u' + 3\alpha (u')^2 + 3\alpha u\, u'' + \alpha u''' = 0. \tag{10}$$

DOI: 10.1051/itmconf/20171301031

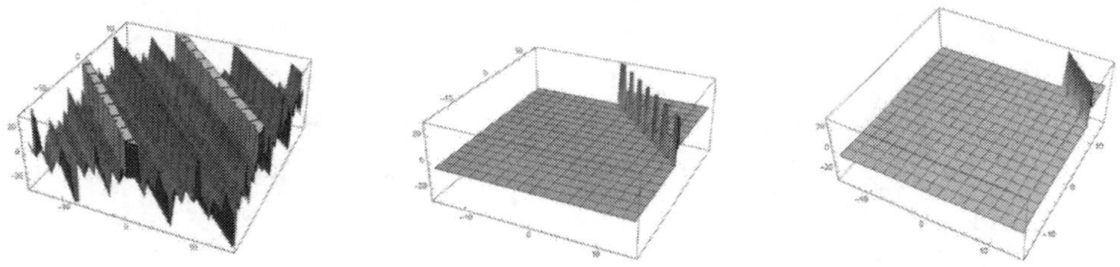

(a) u_1 (b) u_2 (c) u_3

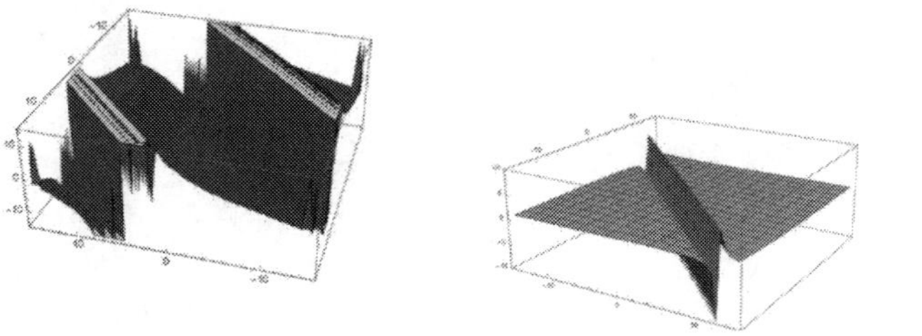

(d) u_4 (e) u_8

Figure 1: JM: The simulations of u_1, u_2, u_3, u_4 and u_8 for *set*2, respectively.

Considering the balance between u''' and $(u')^2$, we find $n + 3 = 2n + 2$ so that $n = 1$. Now we can build solution structure of Eq. (10) as

$$u(\xi) = A_0 + A_1 \left[p + \tan(\phi(\xi)/2)\right] + B_1 \left[p + \tan(\phi(\xi)/2)\right]^{-1}. \tag{11}$$

Substituting solution (11) into (10) and setting each coefficients of $\tan(\phi(\xi)/2)$ and $\cot(\phi(\xi)/2)$ functions to zero, we obtain an algebraic equation system for $A_0,\ A_k, B_k, p, \alpha, a, b, c$ and μ. After solving these system, we attain the values of $A_0,\ A_k,\ B_k$ and μ as follow:

$$\begin{array}{llll} Set1: A_0 = \frac{p(c-b)-a+1}{3} & A_1 = \frac{b-c}{3} & B_1 = 0 & \mu = \alpha \\ Set2: A_0 = \frac{1}{3} & A_1 = \frac{b-c}{3} & B_1 = \frac{b^2+a^2-c^2}{3(b-c)} & p = \frac{-a}{b-c} \end{array}$$

Finally, substituting obtained parameters in set 2 to the solution families 1,2, 8 and 11 into (11), we get new analytical solutions of STO equation as follow

- For $a^2 + b^2 - c^2 < 0$ and $b - c \neq 0$, $u_1 = \frac{1}{3} + \frac{\sqrt{c^2-a^2-b^2}}{3}\left(\cot\left(\frac{\sqrt{c^2-a^2-b^2}}{2}(\xi + C)\right) - \tan\left(\frac{\sqrt{c^2-a^2-b^2}}{2}(\xi + C)\right)\right)$
- For $a^2 + b^2 - c^2 > 0$ and $b - c \neq 0$, $u_2 = \frac{1}{3} + \frac{\sqrt{a^2+b^2-c^2}}{3}\left(\tanh\left(\frac{\sqrt{a^2+b^2-c^2}}{2}(\xi + C)\right) + \coth\left(\frac{\sqrt{a^2+b^2-c^2}}{2}(\xi + C)\right)\right)$
- For $a^2 + b^2 = c^2$, $u_8 = \frac{1}{3} + \frac{2}{(\xi+C)}$
- For $c = a$, $u_{11} = \frac{1}{3} + \frac{b-a}{3}\left(-\frac{a}{b-a} - \frac{(a+b)e^{b(\xi+C)}-1}{(a-b)e^{b(\xi+C)}-1}\right) + \frac{b^2+a^2-c^2}{3(b-c)}\left(-\frac{a}{b-a} - \frac{(a+b)e^{b(\xi+C)}-1}{(a-b)e^{b(\xi+C)}-1}\right)^{-1}$

Then, 3D graphical simulations of u_1, u_2, u_8 and u_{11} are depicted in Figure 3 using Set 2 in Fig. 2. For plotting, the range of x and t values are chosen the same with previous subsection. In Figures, for u_1, u_2, u_8 and u_{11}, we chose $a = 1, 2, 1, 2$, $b = 1, 3, 1, 1$ and $c = 2, 3, \sqrt{2}, 2$, respectively. In the literature, there are many studies on obtaining analytical solutions of Jimbo Miwa equation [8, 9] and Sharma-Tasso-Olver equation [10, 11]. To the best knowledge of the authors of this article there is not any study in the literature about to obtain analytical solutions of the mentioned equations using ITEM and these new results.

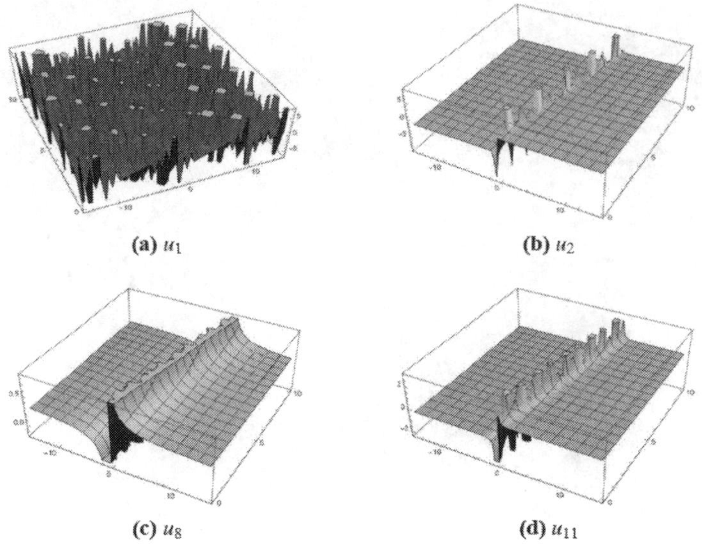

(a) u_1 (b) u_2

(c) u_8 (d) u_{11}

Figure 2: STO: The simulations of u_1, u_2, u_8 and u_{11} for *set*2, respectively.

4 Conclusions

In summary, the aim of this study is to obtain new analytical solutions of Jimbo Miwa and Sharma-Tasso-Olver equations with the aid of ITEM and symbolic programming Maple. The newly obtained results are expressed as hyperbolic, trigonometric, exponential and rational functions. Evidently, it can be concluded that ITEM is an effective, powerful and concise mathematical tool for finding analytical solutions of other nonlinear evolution equations in science and physical nature.

References

[1] H. Bulut, Y. Pandir, S. T. Demiray ,IJMO, **4:4**, 315-320, (2014).
[2] K. Hosseini, R. Ansari, Journal Waves in Random and Complex Media ,**1-9**, (2017).
[3] J. Manafiana, M. F. Aghdaei, M. Khalilianc, R. S. Jeddi,Optik, **135**, 395–406, (2017).
[4] G. Ebadi, A. Biswas, Acta Math. Appl. Sin. Engl. Ser., **32:3** , 623–630, (2016).
[5] S. Kutluay, Y. Ucar, N. M. Yagmurlu, BMMSS, **39:4**, 1603-1614, (2016).
[6] S. Kutluay, A. Esen, Mathematical Problems in Engineering,**2006**, 1-14,(2006).
[7] J. Manafian,Optik, **127**,4222–4245, (2016).
[8] E.M.E. Zayed, Journal of Math. Resear. and Appl., **2 : 2**, 5-13, (2014).
[9] R. Abazari, Çankaya University Jour. of Sci. and Engineering, **9 : 1**, 59-67, (2012).
[10] O. T. Kolebaje, Int. Jour. of Mod. Math. Sci., **5(3)**: 144-155, (2013).
[11] Y. Zhou , F. Yang , Q. Liu, Commun. Nonlinear. Sci. Numer. Simulat. **16**, 641–646, (2011).

ITM Web of Conferences **13**, 01032 (2017) DOI: 10.1051/itmconf/20171301032
CMES2017

On numerical solution of fractional order boundary value problem with shooting method

Hüseyin Demir[1,*], *Yücel* Baltürk[2]

[1] Ondokuz Mayıs University, Institute of Science, Department of Mathematics, 55139, Samsun, Turkey
[2] Ondokuz Mayıs University, Institute of Science, Department of Mathematics, 55139, Samsun, Turkey

Abstract. In this study the shooting method is used for calculation of the second order boundary value problem with fractional order. This method is found to be useful for the application. The accuracy of the shooting method is tested. Some examples are given to illustrate the efficiency of the method with respect to different value of fractional orders.

1. Introduction

The history of fractional calculus and fractional differential equations date back to 1695 (refer to [6] for a brief history). Studies on fractional calculus goes back to studies on calculus [11]. Numerous publications relevant with fractional differential equations have been published in time. Some of them are [1]-[4], [7], [8], [10]. Also some important books on fractional field can remark as [3] and [7]. Related with our subject, Mohamed and Mahmoud have investigated fractional Euler's method and modified trapezoid rule in ([7]). Additionally Tong, Feng and LV have also investigated the error analysis of the fractional Euler's method in ([9]).

In this study a second order boundary value problem (9) is considered. First of all this BVP has been solved analytically and evaluated the value of this exact solution at the certain h points. Then the shooting method has been used to evaluate the numerical solution of this boundary value problem by using the formula (7)-(8) which is given in ([7]). During the application of the shooting method, for predictor the fractional Euler method and for corrector the modified trapezoid rule for systems has been applied to solve the system of IVP (10)-(11). This solution has implemented by taking $m=1$, $n=1$ and we have seen that the solution values of the shooting method are compatible with exact solution values at the same h point. Then the calculations have been repeated with different m, n values which are fractional orders. Consequently the results was have been submitted in a table.

* Corresponding author: hdemir@omu.edu.tr

2. Shooting method

For the use of the shooting method, firstly need the changing of the boundary value problem to system of initial value problems and then need to assume a guess value at the lower bound of the interval. After these, the solution goes on like the solution of system of differential equations.

Let us consider a second order boundary value problem

$$p'' = f(t, p, p'), \ \phi \le t \le \psi \tag{1}$$

with

$$p(\phi) = \gamma, \ y(\psi) = \eta \tag{2}$$

boundary conditions.

At first boundary value problem, which is given, is converted to two first order differential equation:

$$\begin{cases} \dfrac{dp}{dt} = p' = r \\ r = g(t, p, r) \end{cases}, \qquad p(\phi) = \gamma \tag{3}$$

$$\frac{dr}{dt} = k(t, p, r), \qquad r(a) = \omega \tag{4}$$

After deciding the h step size we use one of the known methods like Euler, Runge-Kutta, etc. to solve these equations. Then we gain the first numerical result of the BVP that we consider subject to the first guess value ω. The solution is valid if the solution is enough closer to the upper bound η, else the procedure will repeat with another guess.

After two guesses, instead of making the third guess arbitrary, we can use the interpolation to obtain the third and if necessary the next guesses [5].

3. For predictor the Fractional Euler method and for corrector the Modified Trapezoid Rule formula for systems

In this section the basic formula for the numerical solution of the system of fractional differential equations will be given. Let us consider the system of

$$D_*^m p(x) = g(x, r(x)), \qquad p(x_0) = p_0, \tag{5}$$

$$D_*^n r(x) = k(x, p(x)), \quad r(x_0) = r_0, \ 0 < m, n < 1 \tag{6}$$

fractional differential equations. To solve this system numerically the basic formula that we use is:

$$p(x_s)=\frac{h^m}{\Gamma(m+2)}\left((s-1)^{m+1}-(s-m-1)s^m\right)g\left(x_0,r(x_0)\right)+p(x_0)+$$
$$\frac{h^m}{\Gamma(m+2)}\sum_{q=1}^{s-1}(\,(s-q+1)^{m+1}-2(s-q)^{m+1}+(s-q+1)^{m+1})g\left(x_q,r(x_q)\right)+ \quad (7)$$
$$\frac{h^m}{\Gamma(m+2)}g\left(x_s,r(x_{s-1})+\frac{h^m}{\Gamma(m+1)}g\left(x_{s-1},r(x_{s-1})\right)\right)$$

$$r(x_s)=\frac{h^n}{\Gamma(n+2)}\left((s-1)^{n+1}-(s-n-1)j^n\right)k\left(x_0,p(x_0)\right)+r(x_0)+$$
$$\frac{h^n}{\Gamma(n+2)}\sum_{q=1}^{s-1}(\,(s-q+1)^{n+1}-2(s-q)^{n+1}+(s-q+1)^{n+1})k\left(x_q,p(x_q)\right)+ \quad (8)$$
$$\frac{h^n}{\Gamma(n+2)}k\left(x_s,p(x_{s-1})+\frac{h^n}{\Gamma(n+1)}k\left(x_{s-1},p(x_{s-1})\right)\right)$$

[7].

4. Numerical example

Example: Let us consider the

$$\frac{d^2p}{dt^2}-p=0 \qquad (9)$$
$$p(0)=20,\ p(1)=100$$

boundary value problem. Solve this problem with shooting method.

Solution: The exact solution of the boundary value problem is $p=39,416e^t-19,416e^{-t}$. At first we consider two initial value problems below.

$$\frac{dp}{dt}=r=g(t,p,r)\ \ ,\ p(0)=20 \qquad (10)$$

$$\frac{dr}{dt}=p=k(t,p,r)\ \ ,\ r(0)=\omega=50 \qquad (11)$$

The step size is $h=\frac{1-0}{4}=0.25$ and $m=1$, $n=1$. On the other hand the first guess value is obtained arbitrary as $\omega=50$. The result subject to first guess is 89.590 and so the solution is not correct. Next guess is the second arbitrary guess is $\omega=80$ and the result subject to second guess is 124.871 so the solution is not correct.

The third guess which is obtained by the interpolation of first and two guesses is $\omega=58.851$. The result subject to third guess is 99.999 and the solution is correct.

Under these circumstances for numerical solution, the formula for predictor the fractional Euler method and for corrector the modified trapezoid rule for systems is used, by taking $m=1$, $n=1$. Further the results that obtained with the same procedure for the same example with different m,n values were given in Table 1.

x	Exact Solution	Numerical solution for $m=1$, $n=1$	Numerical solution for $m=0.6$, $n=0.25$	Numerical solution for $m=0.75$, $n=0.25$
0.25	35.490	36.551	26.67655019	28.81987638
0.5	53.209	53.550	48.74752752	49.05826313
0.75	74.272	74.432	65.72867265	68.53210261
1	100	99.999	100.0000003	100.0000000

Table 1. Exact and numerical solution values of the example.

5. Conclusion

In this study a boundary value problem, which has exact solution is considered. The shooting method is used for the numerical solution of this boundary value problem. To solve the two initial value problems which arise when applying shooting method, the formula for predictor the fractional Euler method and for corrector the modified trapezoid rule for systems is used. Because of the incorrect solution values of the boundary value problem subject to arbitrary guesses the third guess is made by using the interpolation formula. The solution was repeated with this guess value. The results that obtained were compared with exact solution values and approximated the target value (the other boundary condition of boundary value problem) with high sensivity. Additionally, for different m, n values, numerical solution values at certain h points are in line with the exact solution values of the boundary value problem. Finally, the shooting method can be used for the numerical solution of the fractional order boundary value problem as a result of this consistence of the values in the Table 1.

References

1. T.S. Alaroev, H.T. Alaroeva, N.M. Nie and Y.F. Tang, Mem. on Diff. Eq. and Math. Phys. *Boundary Value Problems For Differential Equations Of Fractional Order,* **49**, 21-82 (2010)
2. M. Benchohra, S. Hamani, and S.K. Ntouyas, Surv. in Math. and its App. *Boundary Value Problems for Differential Equations with Fractional Order,* **3**, 1-12 (2008)
3. K. Diethelm, *The Analysis of Fractional Differential Equations*, Berlin: Springer, (2004)
4. A. Doğan, *"Kesirli Mertebeden Diferansiyel Denklemler ve Sayısal Çözümleri Üzerine Bir Çalışma"*,Yüksek Lisans Tezi. Ankara: Gazi Üniversitesi, (2011)
5. İ. Karagöz, *Sayısal Analiz Ve Mühendislik Uygulamaları*, Ankara: Nobel, (2011)
6. A. Loverrro, *Fractional Calculus: History, Definitions and Applications for the Engineer,* Department of Aerospace and Mechanical Engineering University of Notre Dame, (2004).
7. D.A.S. Mohamed, and R.A. Mahmoud, Int. Jou. of Comp. App. *An Algorithm For The Numerical Solution of System of Fractional Differential Equations*, ,**65**, 27-31 (2013).
8. I. Podlubny, *Fractional Differential Equations*, San Diego, California: Academic Press, (1999).

9. P. Tong, Y. Feng, H. LV, Wseas Trans. on Math., *Eulers Method for Fractional Differential Equations,* **12**, 1146-1153, (2013).
10. V. Turut, *"Kesirli Diferansiyel Denklemlerin Nümerik Çözümleri"*, Doktora Tezi. İstanbul: Yıldız Teknik Üniversitesi, (2010).
11. M. Weilbeer, *Efficient Numerical Methods for Fractional Differential Equations and their Analytical Background,* Von der Carl-Friedrich-Gauß-Fakultät für Mathematik und Informatik der Technischen Universität Braunschweig, (2005).

Analysis of Output Voltage Harmonics of Voltage Source Inverter used PI and PID Controllers Optimized with ITAE Performance Criteria

Ozan Gül[1,*], and *Nusret* Tan[2]

[1]Bingol University, Department of Electrical and Electronics Engineering, Bingol, Turkey
[2]Inonu University, Department of Electrical and Electronics Engineering, Malatya, Turkey

Abstract. In this article, PI and PID controllers are designed using ITAE (Integral Time Absolute Error) Performance Criteria in order to obtain the controller parameters assuring improved response at selected load. The voltage source inverter (VSI) including PI and PID controllers whose parameters are estimated by minimizing errors using ITAE performance criteria are modeled in MATLAB environment. The stability analysis of the control system will be presented. VSI controlled with the PI-ITAE and the PID-ITAE controller are simulated for various loads and the results are analyzed using FFT analysis for observing and comparing the total harmonic distortion (THD) of the output voltage. The quality of the sinusoidal waveform is more important than the quantity and in order to achieve that, we need to reduce the harmonic content in the output.

1 Introduction

Minimization of the total harmonic distortion (THD) at the output of an inverter is essential to obtain a pure sinusoidal wave. Various methods and strategies to obtain püre sinusoidal voltage or current waveforms at the output of a DC/AC inverter [1-4], are intensely studied with the aim of eliminating dominant harmonics. Conventional proportional-integral (PI) and proportional-integral-derivative (PID) controllers that are widely used in the voltage source inverter (VSI) can reduce such harmonic distortion when operated with nonlinear or periodic loads, the output voltage of these systems contains significant harmonic distortion. In PI and PID controllers design, an optimization algorithm is commonly employed to search for the optimal controller parameters. The most commonly used functions are the time domain integral error performance criteria [2], which are based on calculating the error signal between the system output and the input reference signal. In this study, the VSI including PI and PID controller whose parameters are estimated by minimizing errors using integral of time multiplied by absolute error (ITAE) performance criteria are simulated in MATLAB simulink for resistive loads and nonliner loads and the results are analyzed for observing the total harmonic distortion

* Corresponding author: ogul@bingol.edu.tr

(THD) of the output voltage. The comparison of the PI-ITAE and the PID- ITAE, is presented by taking into consideration the THD at the inverter output voltage under the same conditions for different types of loads.

2 Minimum error integral criteria tuning methods

The shape of the complete closed loop response, from time t=0, until steady state has been reached, could be used for the formulation of a dynamic performance criterion. Unlike the simple criteria of this category are based on the entire response of the process .
The most often used are,
Integral of the Time -Weighted Absolute Error (ITAE) where

$$ITAE = \int_0^\infty t.|e(t)|.dt \tag{1}$$

Integral of the Absolute Error (IAE) where

$$IAE = \int_0^\infty |e(t)|.dt \tag{2}$$

Integral of the Square Error (ISE) where

$$ISE = \int_0^\infty e(t)^2.dt \tag{3}$$

Integral of the Time-Weighted Square Error or Mean Square Error (MSE) where

$$MSE = \int_0^\infty t.e(t)^2.dt \tag{4}$$

Where e(t) = Ysp(t) – Y(t) is the deviation (error) of the response from the desired set point.

3 Design of controller

3.1 Voltage source of inverter (VSI)

A single phase full-bridge inverter shown in Fig.1 [4]

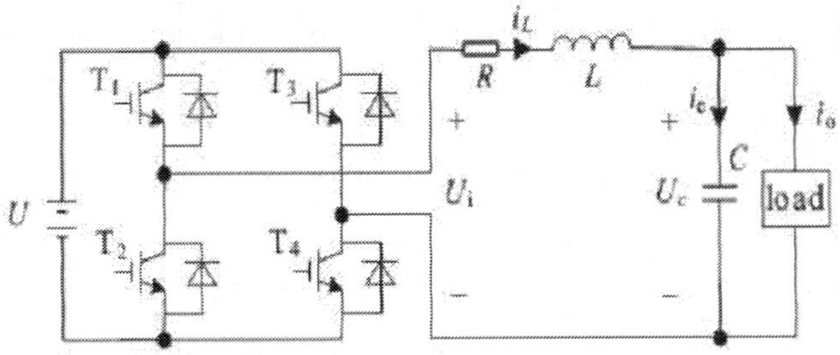

Fig. 1. Single phase voltage source inverter

By applying KVL and KCL (9),

$$\frac{dv_C}{dt} = \frac{1}{C} i_L - \frac{1}{C} i_o \qquad (5)$$

$$\frac{di_L}{dt} = \frac{1}{L} v_i - \frac{1}{L} v_C - \frac{R}{L} i_L \qquad (6)$$

The inverter transfer function with no load can be derived from above equations as,

$$P(s) = \frac{U_C(s)}{U_i(s)} = \frac{1}{LCs^2 + RCs + 1} \qquad (7)$$

where, U_c is the voltage of filter capacitance C and U_i is the inverter bridge output voltage. By putting values , L=0.552 mH, R=0.3Ω, C=140μF, the transfer function of VSI is found as;

$$P(s) = \frac{1}{77.28*10^{-9}s^2 + 42*10^{-6}s + 1} \qquad (8)$$

3.2 Optimization with ITAE performance criteria for PI and PID controller

Kp, Ki, Kd values of PI and PID controls designed using ITAE performance criterion for VSI system are shown in Table 1.

Table 1. Kp, Ki, Kd values of PI-ITAE and PID-ITAE controller

	Kp	**Ki**	**Kd**	**ITAE**
PID	1,69	3,13	0,002	0,5931
PI	10	1110	--------	0,022804

4 Result and conclusion

A MATLAB simulation is carried out for VSI controlled with the PI-ITAE and the PID-ITAE controller for observing the total harmonic distortion (THD) of the output voltage. To demonstrate the advantages, the PI-ITAE and the PID-ITAE are applied to the same VSI for comparative purposes. The tracking performance is tested with both the resistive load and the nonlinear load. Simulation results are given in Table 2.

Table 2. THD output of voltage under different values of resistive and nonlinear loads

		R (50k Watt)	**R (100k Watt)**	**R (150k Watt)**	**RLC(50 k Watt, 5k Var, 5k Var)**	**RLC(100 k Watt, 10k Var, 10k Var)**	**RLC(150 k Watt, 15k Var, 15k Var)**
PI	THD(%)	6.57	5.64	5.69	5.01	4.14	4.55
PID	THD(%)	7.17	7.22	7.5	2.79	3.23	4.05

In Matlab Simulink, a network is created, VSI is used with voltage of 440V, frequency of 50Hz, the RLC filter having the value of L=0.552 mH, C =140μF, R=0.3Ω.
From the simulation results, it can be found that resistive loads cause high THD value both systems. PID-ITAE system show better THD restraint under lighter load. The PID-ITAE control will improve the THD performance of the system for nonlinear loads, but PI-ITAE control is better for resistive load.

References

1. L. Guo, J. Y. Hung, R. M. Nelms, Proc. IEEE Appl. Power Electron. Conf. 383. (2002)
2. D. Maiti, A. Acharya, M. Chakraborty, A. Konar, R. Janarthanan, ICIAFs, 457. (2008)
3. B. Husodo, S. Ayop, ICCAIE, (2011)
4. P. C. Buchade, V. A. Vyawahare, V. V. Bhole, CSCITA, (2014)

Mathematical Modelling of the Synthesis of New Materials

Valery Antonov[1], *Roman* Davidov[1], *Nikolay* Kalinin[2] and *Alexey* Ryakhovskiy[2,*]

[1]Peter the Great Saint-Petersburg Polytechnical University, 195251 Saint-Petersburg, Russia
[2]Ioffe Physical-Technical Institute of Russian Academy of Sciences, 194021 Saint-Petersburg, Russia

Abstract. The paper focuses primarily on a magnetohydrodynamical approach to solving the numerical problems arising in studying the high energy input technologies of the synthesis of new materials. The underlining mathematical model combines the description of both electromagnetic and hydrodynamic phenomena for the metals and surrounding media in supercritical conditions. A detailed specification of this model is provided. The crucial role of the correct choice of equation of state is outlined. The results section includes the latest finding regarding the numerical simulation of underwater electrical explosion.

1 Introduction

Powerful electrical discharges in the plasma channels produced in liquids are widely used in scientific research and development of modern advanced technologies. A large part of applied research is directed towards the use of hydrodynamic pulses for the treatment of materials. High efficiency energy conversion, introduced into the plasma, to the fluid motion, provides a variety of technological applications of electrical discharges [1].

In this regard, a series of simplified discharge models useful for solving a given problem have been developed [2]. Some of the most accuratemathematical description of the discharge areradiative magnetohydrodynamics (RMHD) models [3]. These models allow analyzing the change in the physical condition of plasma and liquid based on the current pulses, generated by the discharge circuit.

It is equally important to determine the role of the chemical reactions as effective method for phase formation in polycrystalline systems. This approach can create models adequately describing the process of changing the state of matter as a result of the high energy supply in a short period of time.

Our scientific interest towards the powerful electrical discharge in the water is caused by the aim of getting supercritical water conditions under which it acquires properties of effective solvent. The study of the reaction gives the possibility to get physical-chemical condition of fluid required for the implementation in various technological processes [4].

* Corresponding author: alexey.i.ryakhovskiy@mail.ioffe.ru

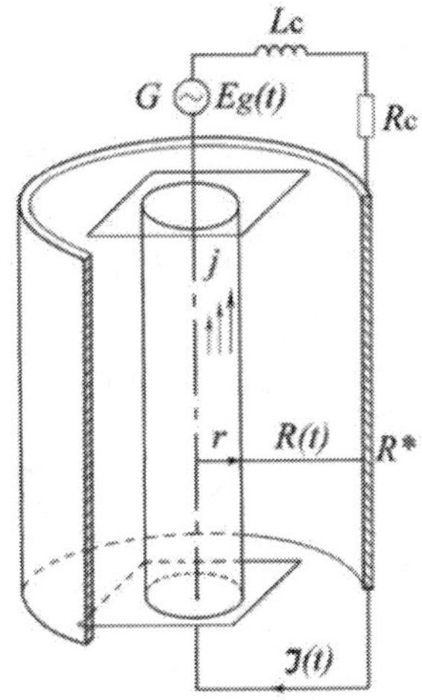

Fig. 1. Scheme of an electrical explosion setup.

Discharge power supply system consists of a high-voltage pulse generator *Eg(t)*, exacerbating fuses, inductor L_c and an electric circuit of the transmission line, which is represented in the scheme by a wave resistance R_c. The plasma channel can be regarded as a conductor, having a variable radial conductivity and inductance, and its capacity can be neglected. Pulse load — high-current discharge — represented in the diagram variable as an impedance plasma channel *R(t)* and the inductance of the area enclosed between the plasma channel and the reverse current lead $L_d(t)$.

2. Mathematical model

The aim of this study is developing a mathematical model of the sharp changes in the phase state of a substance. To construct a mathematical model, a phenomenological method of thermal dynamics, RMHD and statistical physics has been created [5]. This approach allows developing a model which can efficiently describe the processes of heat and mass transfer during the inelastic deformation of materials.

An important step in the work is an application of the aformentioned models to describe the phase formation.

Electrical breakdown of the discharge gap initiates the discharge, forming a plasma channel in water. The phenomenon of electrical breakdown of water is not considered. The radius of the channel and the resulting plasma properties are treated as parameters of the problem and must be specified.

The plasma temperature in the discharge channel can reach tens of thousands of degrees, and heat radiation in the energy balance is beginning to play a significant role. To calculate the characteristics of the radiation in the continuous spectrum used a simplified model based on the approach of "forward - backward" (Schuster-Schwarzschild approximation) with only two spectral bands [6].

2.1 Basic equations

The basic set of MHD equations is as follows [5]:

$$\frac{\partial}{\partial t}\rho + \frac{1}{r}\frac{\partial}{\partial r}(r\rho u) = 0 \tag{1}$$

$$\frac{\partial}{\partial t}\ \rho u) + \frac{1}{r}\frac{\partial}{\partial r}[r(\rho u^2 + \tau)] + \frac{\partial}{\partial r}(P + P_\omega) + jB = 0 \tag{2}$$

$$\frac{\partial}{\partial t}W + \frac{1}{r}\frac{\partial}{\partial r}\left[ru(\varepsilon + P + P_\omega + \tau) + r\frac{\partial T}{\partial r}\right] = \sigma E^2 - Q_R \tag{3}$$

$$\frac{1}{r}\frac{\partial}{\partial r}(rB) = \mu_0 j; j = \sigma E \tag{4}$$

$$W = \rho\left(\varepsilon + \frac{1}{2}u^2\right); \tau = \left(\frac{3}{4}\frac{\partial u}{\partial r} - \frac{2}{3}\frac{u}{r}\right) \tag{5}$$

$$P = P(T,\rho), \varepsilon = \varepsilon(T,\rho) \tag{6}$$

2.2 Electrical circuit equation

$$U_G(t) = L_0\frac{dI}{dt} + R_0 I + \frac{d}{dt}(L_D I) + R_p I + E_s \tag{7}$$

$$R_D(t) = I(t)^{-2}\int_0^{r_{pl}(t)} \sigma E^2 r dr \tag{8}$$

$U_G(t)$ - strength of the electric field; L_0, R_0- inductance and resistance in an electrical circuit; $R\rho$-impedance transporting line; L_D - inductance of discharge plasma- reverse conductor of electric current; E_s-electrical field strength on the border out of the discharge.

2.3 Lagrangian mass variables

$$s = \int_{x_0}^{x} \rho(y,t)dy \tag{9}$$

The system of equations of gas dynamics for a one-dimensional plane nonstationary flow in Lagrangian mass coordinates takes the form

$$\frac{\partial}{\partial t}\left(\frac{1}{\rho}\right) = \frac{\partial v}{\partial s}; \frac{\partial v}{\partial t} = \frac{-\partial p}{\partial s}; \frac{\partial x}{\partial t} = v, \tag{10}$$

$$\frac{\partial}{\partial t}\left(\varepsilon + \frac{v^2}{2}\right) = \frac{-\partial}{\partial s}\ pv) - \frac{\partial W}{\partial t}, \tag{11}$$

$$p = p(\rho,T); \varepsilon = \varepsilon(\rho,T). \tag{12}$$

2.4 Radiative gas

$$\frac{\partial r}{\partial t} = u, \rho r dr = dm, \tag{13}$$

$$\frac{\partial u}{\partial t} = -r\frac{\partial}{\partial m}\ p + \omega) + \frac{F}{\rho}, F = \frac{1}{c} j \times H, \tag{14}$$

$$\frac{\partial \varepsilon}{\partial t} = -(p+\omega)\frac{\partial}{\partial t}\left(\frac{1}{\rho}\right) - \frac{\partial}{\partial m}\ W + W_T) + \frac{q}{\rho}, q = \frac{j_z^2}{\sigma}, \tag{15}$$

$$\frac{\partial}{\partial t}\left(\frac{H_\varphi}{r\rho}\right) = \frac{\partial E_z}{\partial m}, E_z = \frac{\rho}{4\pi\sigma}\frac{\partial}{\partial m}(rH_\varphi), j_z = \sigma E_z, \tag{16}$$

$$W_T = -\chi\rho\frac{\partial T}{\partial m}, W = \sum_{k=1}^{n} r\int_{-1}^{1}\frac{\mu d\mu}{\sqrt{1-\mu^2}}\int_{-1}^{1}\sqrt{1-\gamma^2} I_k d\gamma \tag{17}$$

2.5 Resolving the shockwave

Tocalculatetheshockwave, additionaldissipativeterms (artificialviscosity) areintroducedintothefinite difference equations. Instead of the wave front, its spectral representation is obtained. The equations of gas dynamics are written as follows

$$\frac{\partial}{\partial t}\left(\frac{1}{\rho}\right)=\frac{\partial v}{\partial s},\frac{\partial v}{\partial t}=\frac{-\partial}{\partial s}\ p+\omega),\frac{\partial \varepsilon}{\partial t}=-(p+\omega)\frac{\partial v}{\partial s},\omega=-\nu\rho\frac{\partial v}{\partial s} \tag{18}$$

Artificial viscosity is an additive to pressure. It is calculated at half-integer grid points (at the boundary of the reference volume) [5].

2.6 Equations of state

An important step in modelling is to create an equation of state - a thermodynamic equation describing the state of matter. This equation provides a mathematical relationship between state functions such as its temperature, pressure, volume, or internal energy.

The most promising for this class of problems is to use the free energy of the system, taking into account various interaction factorsbetween the particles. It is also possible to use simpler equations of state for the internal energy of the system.

Table 1. Thermodynamic variables

Name	Symbol	Formula	Natural variables
Internal energy	U	$\int(TdS-pdV)+\sum_i \mu_i dN_i)$	$S,V,\{N_i\}$
Hemholtz free energy	F	$U-TS$	$T,V,\{N_i\}$
Enthalpy	H	$U+pV$	$S,p,\{N_i\}$
Gibbs free energy	G	$U+pV-TS$	$T,p,\{N_i\}$

DOI: 10.1051/itmconf/20171301034

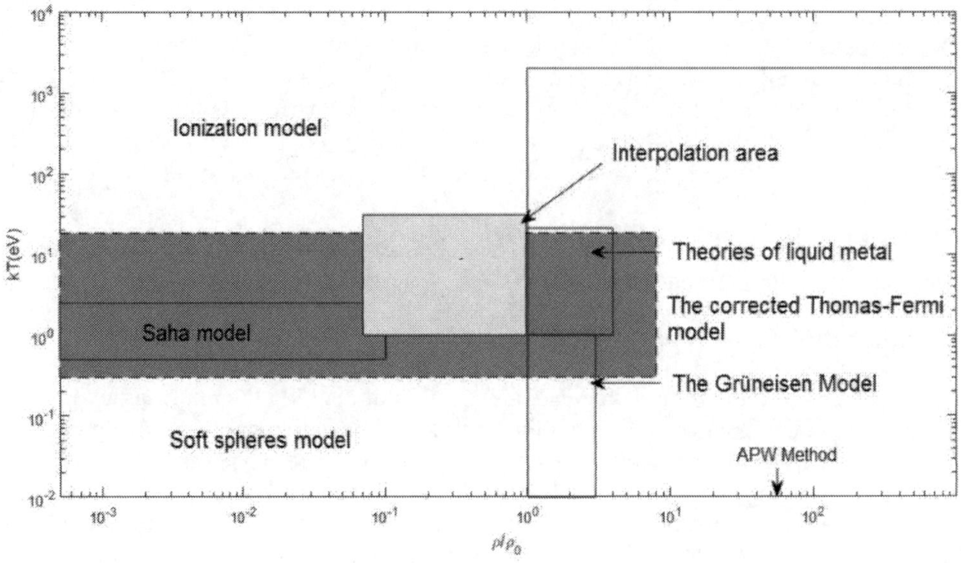

Fig. 2. Diagram of different equation of state approaches. In dark grey is an area of interest for electrical explosion of conductors.

An equation of state can be applied to either vapour-liquid or supercritical phenomena without any conceptual difficulties. Many equations of state have been proposed in the literature with either an empirical, semi-empirical or theoretical basis. Comprehensive reviews can be found in the works of Martin (1979), Gubbins (1983), Anderko (1990), Sandler (1994), Economou and Donohue (1996), Wei and Sadus (2000) and Sengers et al. (2000).

The free Helmholtz energy F used the elastic part of the interaction (F_c) [8]:

$$F = F_e + F_a + F_c \tag{19}$$

For example,

$$F_a = \frac{3T}{2m_a} ln\left(\frac{\theta^2}{T^2} + \frac{\alpha(V_0/V)^{2/3}}{T}\right) \tag{20}$$

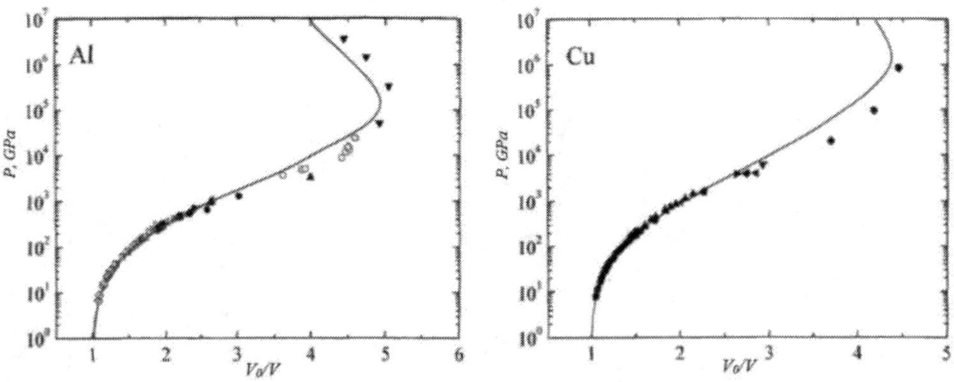

Fig. 3. Shock adiabats for Al and Cu.

2.6.1 Kolgatin equation of state

There are restrictions for the applicability of this equation on the temperature, which should be less than 10^5K and for the pressure (P≤10^4 MPa) (includes supercritical water conditions).

For gas:the pressure is created by molecules, atoms and ions

$$p = 300[(1-\beta) + 2\beta(1+Z)]\rho T \quad (21)$$

The internal energy is composed of the thermal energy of molecules and atoms, ions, electrons motion, ionization and dissipationterms

$$\varepsilon = \varepsilon_m + \varepsilon_{a,i,e} + \varepsilon_I + \varepsilon_d \quad (22)$$

For water:the pressure consists of elastic and thermal components:

$$p = p_{el} + p_T \quad (23)$$

Internal energy

$$\varepsilon = \varepsilon_{el} + \varepsilon_T + \varepsilon_I + \varepsilon_d \quad (24)$$

3. Results

The results below highlight the recent finding regarding the numerical simulation of underwater electrical explosions. We consider two real-life experiments with a different timescale: microsecond [9] and nanosecond [10]. On the graphs below circles indicate experimental data, cross and stars – numerical data from the original source, and point-dash line – our numerical results.

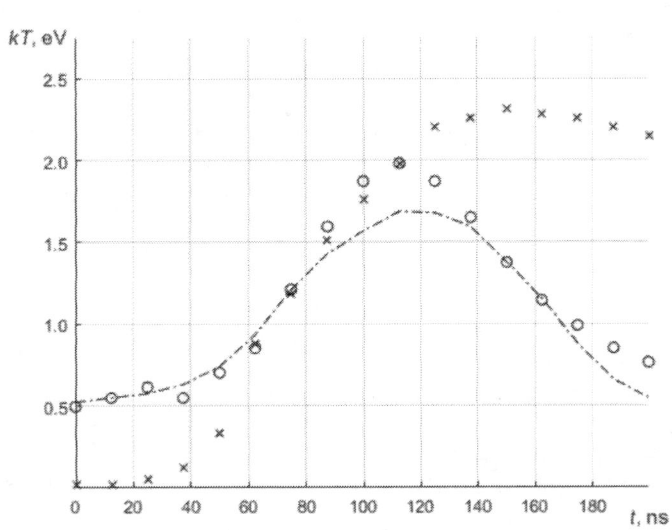

Fig. 4. Comparison of temporal temperature distributions on the interface between the metal vapor and water for the nanosecond explosion [9].

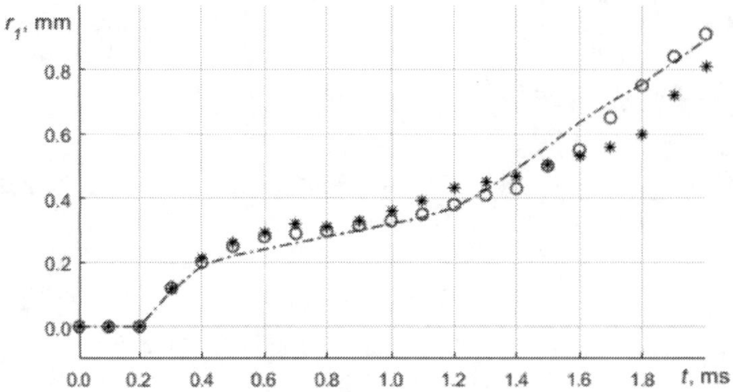

Fig. 5. The dynamics of metal vapor boundary for the microsecond explosion [10].

As the graphs indicate, we have been able to improve the numerical estimation for the thermodynamical evolution of the studied system. This was in part achieved by a careful selection of the auxiliary models, such as equations of state.

4. Conclusion

In this paper, we present a mathematical model for a part of the process that results in obtaining materials with programmable properties. The main emphasis is placed on the analysis of various equations of state, allowing calculating sharp changes in the properties of matter in the investigated interval of physical parameters and time intervals. It is obtained that the most rational approach is based on the use of internal or free energy of the thermodynamic system. Test calculations have shown that even simple equations of state give quite satisfactory results.

Further development of the work will go in two directions.

1. Continue testing the mathematical model on more complex equations of state. Ideally, it is desirable to use one or more combined equations that allow to significantly expand the range of the process parameters.
2. Carry out modeling of physical and chemical processes with the use of various metallic conductors as a source of reaction.

References

1. Pozdeev V A. Applied fluid dynamics of the electric discharge in liquid. (Kiev: Naukova Dumka,1980)
2. Zeldovich, Ya B., and Yu P. Raizer. Physics of shock waves and high-temperature hydrodynamic phenomena. No. FTD-MT-64-514. (Foreign Technology Div Wright-Patterson AFB OH, 1965.)
3. Chetverushkin B N. Mathematical modeling of dynamic problems of radiating gas. (Moscow: Nauka, 1985).
4. Naugolnykh K A and Roy N A. Electrical discharges in water.(Moscow: Nauka, 1971)
5. Samarsky A A and Popov Y P. Difference methods for solving problems of gas dynamics. (Moscow: Nauka, 1980)
6. Siegel, R., and C. M. Spuckler. "Approximate solution methods for spectral radiative transfer in high refractive index layers." International journal of heat and mass transfer 37 (1994): 403-413.
7. Kolgatin S N.J.Technical Physics. B 65 (7) 10 (1995)

8. R. V. Davydov, V. I. Antonov. Simulation of femtosecond pulsed laser ablation of metals. Journal of Physics: Conference Series, Volume 769 (2016), 012060
9. Sasaki, Toru, et al. "Electrical conductivities of aluminum, copper, and tungsten observed by an underwater explosion." Physics of Plasmas 17.8 (2010): 084501.
10. Krasik, Yakov E., et al. "Underwater electrical wire explosion and its applications." IEEE Transactions on Plasma Science 36.2 (2008): 423-434.

On Inverse Sturm- Liouville Problems With Symmetric Potentials

Münevver Tuz[1]
[1]Department of Mathematics Firat University,
23119 Elazığ,Türkiye
mtuz@firat.edu.tr

Abstract

In this paper an inverse problem

$$-y'' + p(x)y = \lambda y$$
$$y'(0) - hy(0) = y'(1) + hy(1) = 0$$

where p are integrable on $[0,1]$ and satisfy the symmetry conditions $p(x) = p(1-x)$ almost everywhere in the interval $0 \le x \le 1$. We reconstruct the potential of the inverse Sturm Liouville problem's solution is obtain for finite interval with symmetric potential.

Keywords: Inverse problem symmetric potential, fixed point theorem,second-order differential equation.

Introducton

In general, an inverse eigenproblem measures the frequencies of a vibrating system and tries to obtain some physical properties of the system. Difficulties in achieving higher eigenvalues in the field of practice exist, but only in a limited amount of data. However, the solution may have been predicted beforehand. The chronological problem, for example, indicates whether the first guess is consistent with the data and how it can be changed if it does not. The results to be found are very suitable to answer such a question.
The reverse Sturm-Liouville problem is examined in at least four different ways. The most common is the problem investigated by Gel'fand and Levitan [6], in which the potential and boundary conditions are uniquely determined by spectral function. This situation has also been investigated by [13,16,22]. Second, the potential and boundary conditions are uniquely determined by two spectra. This situation can be reduced to the first case by examining [5,15,16,22]. The third is uniquely determined by potential, boundary conditions and two possible reduced spectra. This is also discussed in [3,8,14]. Knowing the boundary conditions implies that the lowest eigenvalue in one of the spectra is neglected. The latter has been shown to be potentially specific, provided that boundary conditions and a possible reduced spectrum are given, as long as it is an equally functioning function in the middle of the potential. The study [3,8,14] In this article we will introduce a constructive method by dealing with an inverse sturm liouville problem for the fourth stage
The main situation is similar to the formula obtained by Hochstadt for two potential

differences and is based on the technique developed by Hochstadt [8]. This formula leads to uniqueness theories, as shown in [3,7,8,14].
However, taking into account the relationship between the lowest eigenvalue and the boundary conditions, it is clearly stated that if the boundary conditions are given and mixed, the lowest eigenvalue is neglected. In addition, the inverse Sturm-Liouville problem leads to a natural generalization of the Borg original formula and establishes a connection between the four above mentioned conditions.

1. Statement of the main results

In this section we will consider two Sturm-Liouville problems with different potentials

Theorem 1. 1 Consider the eigenvalue problems

$$-y'' + p(x)y = \lambda y$$
$$y'(0) - hy(0) = y'(1) + hy(1) = 0 \qquad (1)$$
$$-y'' + \tilde{p}(x)y = \lambda y$$
$$y'(0) - \tilde{h}\, y(0) = y'(1) + \tilde{h}\, y(1) = 0 \qquad (2)$$

where p and $\tilde{p}$ are integrable on [0,1] and satisfy the symmetry conditions $p(x) = p(1-x)$ and $\tilde{p}(x) = \tilde{p}(1-x)$ almost everywhere in the interval $0 \le x \le 1$

Let λ_j and $\tilde{\lambda}_j$, be the eigenvalues of (1) and (2). Let $\tilde{u}_j$ and $\tilde{v}_j$ be the solution of

$$y'' + (\lambda - \tilde{p})y = 0 \qquad (3)$$
$$y(0) = 1 \quad , y'(0) = \tilde{h} \qquad (4)$$
$$v(1) = 1 \quad , v'(1) = -\tilde{h} \qquad (5)$$

with $\lambda = \lambda_j$,. Define the functions $\tilde{y}$ by

$$\tilde{z}_j = 2\frac{\tilde{v}_j - k_j \tilde{y}_j}{\omega'(\lambda_j)} \qquad (6)$$

Here $\frac{k_j}{\omega'(\lambda_j)} = \frac{1}{\int_0^1 y_j^2 dx}$ where $k_j = (-1)^j$ and $y_j(x)$ are the eigenfunctions of (1) normalized such that $y_j(x) = 1$. If

$$\sum_j \left|\lambda_j - \tilde{\lambda}_j\right| \le \infty \qquad (7)$$

then

$$p - \tilde{p} = \sum_j \left(\tilde{z}_j y_j\right)' \qquad (8)$$

Proof: Then y satisfies the Volterra integral equation

$$y(x)=\cos\sqrt{\lambda}x+\frac{h}{\sqrt{\lambda}}\sin\sqrt{\lambda}x+\frac{1}{\sqrt{\lambda}}\int_0^x\sin\sqrt{\lambda}(x-t)p(t)y(t)dt \qquad (9)$$

Let $y(x)$ and $v(x)$ be the solutions of eq. (3) with initial conditions (4) and (5), where $\tilde{h}$ and $\tilde{p}$ are replaced by h and p.

Also $\lambda = s^2$ where $s=\sigma+i\tau$. From eq. (1.9) follows that for each $x, y(x)$ is an entire function of λ of order $\frac{1}{2}$ and asymptotically we have

$$y(x,\lambda)=\cos sx+O\left(\frac{e^{|\tau|x}}{|s|}\right) \qquad (10)$$

$$y'(x,\lambda)=-s\sin sx+O\left(e^{|\tau|x}\right) \qquad (11)$$

Titchmarsh [19]. Since $p(x)=p(1-x)$ a.e. we find that $v(x)=y(1-x)$ and thus e (10) and (11) provide the asymptotic expansions for v and v' as well. We introduce now the Wronskian

$$\omega(\lambda)=-hy(1,\lambda)+y'(1,\lambda) \qquad (12)$$

and note that λ is an eigenvalue of (1) iff $\omega(\lambda)=0$. From (10) and (11) follows that the asymptotic expansion for $\omega(\lambda)$ is

$$\omega(\lambda)=s\sin s\pi+O\left(e^{|\tau|\pi}\right) \qquad (13)$$

f is continuous and that the square of f is integrable and assume that f is meromorphic

$$\Omega(x,\lambda)=\frac{\tilde{v}\int_0^x yfdz+\tilde{y}\int_x^1 vfdz}{\omega(\lambda)}$$

Here $\tilde{y}$ and $\tilde{v}$ are the solutions of (3) with initial conditions (4) and (5), and have the same asymptotic expansions as y and v. We will integrate Ω along a large contour in the λ plane.

In the s plane we let R be the rectangle with vertices at $\pm d+iO$ and $\pm d+id$ where $d=n+\frac{1}{2}$ and we let Γ be the contour in the λ plane which corresponds to the points of

R for which $\tau>0$. By using the asymptotic estimates for y, $\tilde{v}$ and ω we find that

$$\frac{\tilde{v}(x,\lambda)y(z,\lambda)}{\omega(\lambda)}=\frac{\cos s(1-x)\cos z}{s\sin s\pi}+O\left(\frac{e^{\tau(z-x)}}{|s|^2}\right)$$

This is precisely what is needed to make Titchmarsh's [19], and we conclude that

$$\frac{1}{2\pi i}\int_{\Gamma}\Omega(x,\lambda)d\lambda-\frac{1}{2\pi i}\int_{\Gamma}\int_{0}^{x}\frac{\cos s(1-x)\cos z}{s\sin s\pi}f(z)dzd\lambda-\frac{1}{2\pi i}\int_{\Gamma}\int_{x}^{1}\frac{\cos sx\cos s(1-z)}{s\sin s\pi}f(z)dzd\lambda$$

converges uniformly to zero on the interval $0 \leq x \leq 1$ as n $\to \infty$ .It follows from the residue theorem that the sum of the last two terms is the first $n+1$ terms of the Fourier cosine expansion of the function f. It is therefore natural to extend f as an even, 2π periodic function and since f is square integrable and $f(-\pi) = f(\pi)$ we know that the Fourier series for f converges uniformly, Zygmund [22]. By using the residue theorem to evaluate the first term in the above expression and letting $n \to \infty$ we obtain

$$f(x)=\sum_{j=0}^{\infty}\frac{\tilde{v}_j\int_0^x y_j f dz+\tilde{y}_j\int_x^1 v_j f dz}{\omega'(\lambda_j)}$$

We note that y_j and v_j represent the same eigenfunction, whereas fit y_j and v_j are just solutions of (3) with $\lambda = \lambda_j$ Since $p(x) = p(1-x)$ we see that $v_j = k_j y_j$ where $k = (-1)^j$.

If p= p and $h = \tilde{h}$ then (13) reduces to the Sturm-Liouville expansion and consequently $\frac{k_j}{\omega'(\lambda_j)} = \frac{1}{\int_0^1 y_j^2 dx}$. Let now f be equal to the first eigenfunction y_0 of (1). From (13) and definition (6) follows that

$$y_0=\tilde{y}_0+\frac{1}{2}\sum_j\tilde{z}_j\int_0^x y_j y_0 dt \qquad (14)$$

We can now obtain the results stated in the theorem by differentiating eq. (14) formally. To realize that let $f_j = \tilde{z}_j \int_0^x y_j y_0 dt$ dır. Thus f(0)=0 and f(0)= $\tilde{y}_j(0)$. Since y_j and y_0 are eigenfunctions of (1) and $\tilde{z}_j$ is a solution of (3) with $\lambda = \lambda_j$ we find by differentiating f_j twice and using integration by parts that

$$f_j''+(\lambda_0-\tilde{q})f_j=2(\tilde{z}_j y_j)' y_0. \qquad (15)$$

If the expressions for f_j, f_j', f_j'' are

$$f_j=\tilde{z}_j\int_0^x y_j y_0 dt$$

$$f_j' = \tilde{z}_j y_j y_0 + \tilde{z}_j' \int_0^x y_j y_0 dt$$

$$f_j' = 2\, \tilde{z}_j' y_j y_0 + \tilde{z}_j y_j' y_0 + \tilde{z}_j y_j y_0' + \tilde{z}_j'' \int_0^x y_j y_0 dt.$$

If you write instead of the above expressions to find the left side of equation (15)

$$f_j'' + (\lambda_0 - \tilde{q}) f_j = 2\left(\tilde{z}_j' y_j\right)' y_0 + (\lambda_0 - \lambda j)\tilde{z}_j \int_0^x y_j y_0 + \left[y_j y_0' - y_j' y_0 \right] \tilde{z}_j dt$$

If we consider the integral part

$$I = (\lambda_0 - \lambda j) \int_0^x y_j y_0 dt$$

$$= \int_0^x y_j \lambda_0 y_0 dt - \int_0^x \lambda_j y_j y_0 dt$$

$$= \int_0^x y_j [\tilde{q} y_0 - y_0''] dt - \int_0^x [\tilde{q} y_0 - y_0''] y_0 dt$$

$$= \int_0^x \left(y_j'' y_0 - y_0'' y_j \right) dt$$

$$= \int_0^x \left(y_j' y_0 - y_0' y_j \right)' dt$$

We can now derive (7) and (8). By differentiating (14) and using (15) we obtain

$$y_0' \text{-} \tilde{y}_0' = \frac{1}{2} \sum_j f_j'$$

$$y_0'' \text{-} \tilde{y}_0'' = (\tilde{q} - \lambda_0)\left(y_0 \text{-} \tilde{y}_0\right) + \sum (\tilde{z}_j y_j)' y_0.$$

Thus eq. (7) follows by setting $x = 0$ in the first equation. To derive (8) from the second equation we use that $y_0'' = (q - \lambda_0) y_0$ and $\tilde{y}_0'' = (\tilde{q} - \lambda_0)\tilde{y}_0$ and note that the eigenfunction y_0 is positive in the whole interval.

We will show that $\sqrt{\lambda_j}\tilde{z}_j$, and $\tilde{z}_j$ are $O(\lambda_j - \tilde{\lambda}_j)$. Let z_j, be the eigenfunction of (1.2) corresponding to the eigenvalue $\tilde{\lambda}_j$. We will compare $\tilde{y}_j$ with z_j and let $w_j = (\tilde{y}_j - u_j)/\ (\lambda_j - \tilde{\lambda}_j)$.

Thus $w = w_j$ satisfies the differential equation

$$w'' + (\lambda - \tilde{p})w = -u_j$$

$$w(0) = w'(0) = 0$$

with $\lambda = \lambda_j$. Let φ_1 and φ_2 be solutions of the homogeneous equation, i.e. (3), with initial conditions $\varphi_{i+1}^{(j)} = \delta_{ij}$ at $x = 0$. The solution of the inhomogeneous equation is then given by

$$w(x) = \int_0^x [\varphi_1(x)\varphi_2(z) - \varphi_2(x)\varphi_1(z)]u_j(z)dz \tag{16}$$

To estimate w we must estimate φ_1, φ_2 and z_j. We first observe that φ_1 is the solution

of the Volterra integral equation (9) with $h = 0$ and p replaced by $\tilde{p}$ and $\lambda = \lambda_j$

Let $\|\tilde{p}\|_1 = \int_0^1 |\tilde{p}|dx$ Since the potential $\tilde{p}$ is symmetric we find that $u_j(1-u) = k_j u_j(x)$ and $\tilde{v}_j(x) = \tilde{y}_j(1-x)$.It therefore follows from the definition of w, and formula (6) that

$$\tilde{z}_j(x) = 2\frac{\lambda_j - \tilde{\lambda}_j}{\omega'(\lambda_j)}[w_j(1-x) - k_j w_j(x)] \tag{17}$$

To complete our study of $\tilde{z}_j$ we must find a lower bound for $|\omega'(\lambda_j)| = \|y_j\|_2$

Here y_j is the eigenfunction of (1) normalized such that $y_j(0) = 1$.

We can now prove the validity of the formal arguments leading to (7) and (8). From the solution of equation (15) follows that

$$\frac{1}{2}\sum f_j - \frac{1}{2}\sum \tilde{z}_j(0)\varphi_2(x) - \int_0^x G(x,y)\sum(\tilde{z}_j y_j)' y_0 dz \tag{18}$$

Here $G(x,y) = \varphi_1(x)\varphi_2(z) - \varphi_2(x)\varphi_1(z)$, where φ_1 and φ_2 are solutions of the homogeneous (3) with $\lambda = \lambda_j$ and satisfy the initial conditions $\varphi_{i+1}^{(j)} = \delta_{ij}$ at $x = 0$. To interchange the order of integration and summation we have used that $\sum(\tilde{z}_j y_j)'$ converges in L^∞, On the other hand, $y = y_0 - \tilde{y}_0$ satisfies the differential equation $y'' + (\lambda_0 - \tilde{p})y = (p - \tilde{p})y_0$ with the initial conditions $y(0) = 0$ and $y'(0) = h - \tilde{h}$ and consequently

$$y_0 - y_0 = (h - \tilde{h})\varphi_2(x) - \int_0^x G(x,y)(p - \tilde{p})y_0 dz.$$

The theorem now follows from (14) by comparing the last two equations and using the uniqueness theorem for ordinary differential equations with summable coefficients, Neumark [18]. This completes the proof.

2. Uniqueness results

The explicit formula (1.8) for the difference between two potentials is well suited for deriving some well-known uniqueness results. We will show that the potential and the boundary conditions are uniquely determined by the full spectrum. Moreover, we will prove that if the boundary conditions are known then the potential is uniquely determined by the reduced spectrum. Here the reduced spectrum is the full spectrum with the lowest eigenvalue omitted. Finally, we will show that the assumption $\sum|\lambda_j - \tilde{\lambda}_j|$ in Theorem 1 is quite restrictive. In particular, it implies that if the comparison spectrum $\{\tilde{\lambda}_j\}$ corresponds to a Sturm-Liouville problem with constant coefficients, then the Fourier series for the potential $p(x)$ will be absolutely convergent.

Corollary 2.1. Consider the eigenvalue problem (1) where p is integrable in $[0,1]$. If $p(x) = p(1-x)$ almost everywhere in $0 < x < 1$ then $p(x)$ and h are uniquely determined by the spectrum.

Proof. We have two Sturm-Liouville problems, let's have the eigenvalues $\lambda_j = \tilde{\lambda}_j$. From equations (3) and (4) follows that $\tilde{y}_j$ is an eigenfunction, and since the potential $\tilde{p}$ is symmetric we conclude that $v_j = k_j\tilde{y}_j$. This shows that all $\tilde{z}_j$ vanish identically and the right hand sides of equations (7) and (8) are zero. This completes the proof.
The symmetric property of the potential indicates that the eigenfunctions have single or double $\frac{1}{2}$ functions. For this reason, we can divide the eigenvalue problem over $[0,1]$ into two problems with the boundary conditions of Dirichlet or Neumann at $x = 1/2$. It has been shown by [12,15,16], that the potential and the boundary conditions are uniquely determined by two spectra, and can be reconstructed from this data. The result is sometimes credited to [5,15]. However, Borg proved a different, but equally precise result, namely that if the boundary conditions are given then two-possible reduced-spectra determine the potential uniquely.

Corollary 2.2. Consider the eigenvalue problem (1) where p is integrable on $[0,1]$ and satisfies $p(x) = p(1-x)$ a.e. If $p(x)$ is replaced by another symmetric potential $\tilde{p}(x)$ and the two problem8 have the same reduced spectrum then $p = \tilde{p}$.

Proof. Let λ_j and $\tilde{\lambda}_j$ be the eigenvalues corresponding to p and $\tilde{p}$. If $h = \tilde{h}$ and $\lambda_j = \tilde{\lambda}_j$ for $j = 1,2,..$ then we from Theorem-1 that $\tilde{z}_0(0) = 0$ and

$p-\tilde{p}=(\tilde{z}_0 y_0)'$. We will Show that $\tilde{z}_0$ vanishes identically. From (6) follows that $\tilde{z}_0(1-x)=\tilde{z}_0(x)$ and thus $\tilde{z}_0$ vanishes at $x=0,\frac{1}{2}$ and 1.Since $\lambda_0 < \lambda_1 = \tilde{\lambda}_1$ we conclude from Sturm comparison theorem that if $\tilde{z}_0 \neq 0$ then the eigenfunction $\tilde{y}(x,\lambda_1)$ will have at least two zeros in $(0,1)$. This is contrary to Sturm's theory of oscillation and the result follows contraception.
We can think that the problems discussed in the 1st Chamber and the 2 nd Criminal can emerge in practice. Thus, the inverse eigen problem of a cylinder can be reduced
To solve the two inverse Sturm-Liouville problem, in this case, the boundary conditions in the second eigenproblem, the first property spectrum
The eigenvalue problem using Corollary I. In Corollary 2, the lowest eigenvalue plays a special role.

References

1.V. Ambarzumian, Uber eine Frage der Eigenwerttheorie. Z. Physik., **53**, 690-695 (1929)
2.V. Barcilon, Iterative solution of the inverse Sturm-Liouville problem. J. Mathematical Phys., **15,** 287-298 (1974)
3.G.Borg, Eine Umkehrung der Sturm-Liouvillesehen Eigenwertaufgabe. Acta Math., **78,** 1-96 (1946)
4.E.Coddinoton, N.Levinson, Theory of ordinary differential equations. McGraw-Hill, New York, (1955)
5. G.M.Gasymov, B.M.Levitan, On Sturm-Liouville differential operators with discrete spectra. Amer. Math. Soc. Transl. Series 2, **68,** 21-33 (1968).
6. I. M. Gel'fand , B.M.Levitan, On the determination of a differential equation from its spectral function. Amer. Math. Soe. Transl. Series 2, **1**, 253-304 (1955)
7. O.H.Hald, Inverse eigenvalue problems for layered media. Comm. Pure Appl. Math., **30**, 69-94 (1977)
8.H. Hochstadt, The inverse Sturm-Liouville problem. Comm. Pure Appl. Math., **26**, 715-729 (1973.
9. F.W.J. Olver, Introdiction to Asymptotics and Spesial Functions,New-York and London,Academic Pres.(1974)
10. C.J.Chysan, J.Henderson, Positive solutions for singular higher order nonlinear equations,Differential Equations Dynam.Systems 2 ,153-160 (1994)
11.K. Jörgens, Spectral theory of second-order ordinary differential operators, Lectmyes delivered at Aarhus Universitet, 1962/63, Aarhus, (1964)
12. M. G. Krein, Solution of the inverse Sturm-Liouville problem. Dokl. Akad. Vaulc SSSR (N.S.), **76**, 21-24 (1941)
13. On the transfer function of a one-dimensional boundary problem of second order. Dokl. Akat. 1Vaulc SSSR (N.S.), **88**, 405-408 (1953)
14.N. Levinson, The inverse Sturm-Liouville problem. Mat. Tidsskr. B., 25-30 (1949)
15. B.M.Levitan, On the determination of a Sturm-Liouville equation by two spectra. Amer. Math. Soc. Transl. Series 2, **68**, 1-20 (1968)
16. V.A. Marcenko, Concerning the theory of a differential operator of the second order. Dokl. Akad. Nauk SSSR (N.S.), **72**, 457-460 (1950)
17. C. Willis, Inverse Sturm- liouville problems with two discontinuites. Inverse problems 2, 111-130 (1986)
18. M. A Neumark, Lineare Differential Operatoren. Akademie-Verlag, Berlin, (1963)

19. E. C.Titchmarsh, The theory of functions. Oxford University Press, London, (1939)
20. M.Koboyashi, A uniqueness proof for discontinuous inverse Sturm- liouville problems with symmetric potentials, 767-781(1989)
21.V.V.Zrkov, On inverse Sturm-Liouville problems on a finite segment. Math. USSR--Izv. 1 923-934 (1967)
22. A.Zygmund, Trigonometric series,vol.I,2nd ed.Cambridge University Press,London, (1959)

On Some Spectral Problems for Diffusion Operator

Mine Babaoglu[1,*], *Etibar S. Panakhov*[2]

[1]Kahramanmaras Sutcu Imam University, Faculty of Education, Kahramanmaras, Turkey
[2]Baku State University, Institute of Applied Mathematics, Baku, Azerbaijan

Abstract. In this study, we attain several spectral results for Diffusion operator. In particular, the solution functions belong to Paley-Wiener space:

$$PW_\pi = \left\{ f \text{ entire}, \ |f(\mu)| \le Ce^{\pi|\operatorname{Im}\mu|}, \ \int_R |f(\mu)|^2 d\mu < \infty \right\}.$$

so that required theorems are proved.

1 Introduction

The aim of this paper is to solve the following diffusion equation

$$-y'' + \left[2\lambda p(x) + q(x)\right] y = \lambda^2 y, \quad x \in [0,\pi] \tag{1}$$

where the functions $p(x)$ and $q(x)$ are real-valued and $p(x) \in W_2^{m+1}[0,\pi]$, $q(x) \in W_2^m[0,\pi]$ for $m \ge 1$. Some spectral problems were extensively solved for the diffusion operator in references [1-4].

Consider the problem

$$-y'' + \left[2\lambda p(x) + q(x)\right] y = \lambda^2 y, \tag{2}$$

$$y(0) = 1, \quad y'(0) = h, \tag{3}$$

where h is a finite number. Let us denote by $\varphi(x,\lambda)$ the solution of (2) satisfying the initial conditions (3). Let [2]

$$\varphi(x,\lambda) = \cos\left[\lambda x - \alpha(x)\right] + \int_0^x A(x,t)\cos\lambda t \, dt + \int_0^x B(x,t)\sin\lambda t \, dt, \tag{4}$$

where

$$\alpha(x) = x.p(0) + 2\int_0^x \left\{ A(\xi,\xi)\sin\alpha(\xi) - B(\xi,\xi)\cos\alpha(\xi) \right\} d\xi$$

* Corresponding author: mnbabaoglu@gmail.com

$$
\begin{aligned}
&q(x)=-p^2(x)+2\frac{d}{dx}\{A(x,x)\cos\alpha(x)+B(x,x)\sin\alpha(x)\} \\
&A(0,0)=h, \quad B(x,0)=0, \quad \left.\frac{\partial A(x,t)}{\partial t}\right|_{t=0}=0, \quad \alpha(x)=\int_0^x p(t)dt
\end{aligned} \tag{5}
$$

and

$$
\lambda_n = n + c_0 + \frac{c_1}{n} + \frac{c_{1,n}}{n}, \tag{6}
$$

be the nth eigenvalue where

$$
\begin{aligned}
&c_0 = \frac{1}{\pi}\int_0^\pi p(x)dx, \quad \sum_n \left|c_{1,n}\right|^2 < \infty, \\
&c_1 = \frac{1}{\pi}\left(h + H + \frac{1}{2}\int_0^\pi \left[q(x)+p^2(x)\right]dx\right),
\end{aligned} \tag{7}
$$

and H is a finite number.

Consider the Diffusion equation with more general separable boundary conditions

$$
\begin{aligned}
&-y'' + \left[2\lambda p(x)+q(x)\right]y = \lambda^2 y, \quad x\in[0,\pi], \\
&a_{11}y(0,\lambda) - a_{12}y'(0,\lambda) = 0, \\
&a_{21}y(\pi,\lambda) + a_{22}y'(\pi,\lambda) = 0.
\end{aligned} \tag{8}
$$

where $a_{11}^2 + a_{12}^2 \neq 0$, $a_{21}^2 + a_{22}^2 \neq 0$. So, let $\lambda = \mu^2$ and $y(x,\mu^2)$ denote the solution of the initial value problem

$$
\begin{aligned}
&-y'' + \left[2\mu^2 p(x)+q(x)\right]y = \mu^4 y, \\
&y(0,\mu^2) = a_{12}, \quad y'(0,\mu^2) = a_{11}.
\end{aligned}
$$

The eigenvalues of (8) are the square of the zeroes of the boundary function $B(\mu)$,

$$
B(\mu) := a_{21}y(\pi,\mu^2) + a_{22}y'(\pi,\mu^2).
$$

In the Dirichlet case, this boundary function is an entire function of μ of order 1 and type π and is square integrable on the real line. Therefore, it belongs to the Paley-Wiener space.

2 Main Results

Let

$$
\begin{aligned}
v_1(x,\mu) &= y(x,\mu^2) - \cos\left[\mu^2 x - \alpha(x)\right] \\
&= \int_0^x A(x,t)\cos(\mu^2 t)dt + \int_0^x B(x,t)\sin(\mu^2 t)dt
\end{aligned} \tag{1}
$$

In the following, we shall make use of the estimates [5],

$$
|\cos u| \le e^{|\mathrm{Im}\,u|}, \quad |\sin u| \le c_0 e^{|\mathrm{Im}\,u|}, \tag{2}
$$

where c_0 is some constant (we may take $c_0 = 1.72$ for numerical purposes).

Define the constants

$$c_1 = \int_0^\pi \max_{0\le x\le\pi} \left|A(x,t)\right| dt, \quad c_2 = \int_0^\pi \max_{0\le x\le\pi} \left|B(x,t)\right| dt, \tag{3}$$

$$c_3 = \left(c_1 + c_0c_2\right)\left|a_{21}\right| + c_0c_2\left|a_{22}\right|.$$

we claim the subsequent results.

Theorem 1 $v_1(x,\mu) \in PW_x$ is a function of μ for each x and the following estimate hold:

$$\left|v_1(x,\mu)\right| \le \left(c_1 + c_0c_2\right) e^{x\left|\operatorname{Im}\mu^2\right|} \tag{4}$$

Proof In the first instance

$$\left|v_1(x,\mu)\right| \le \int_0^x \left|A(x,t)\right|\left|\cos\left(\mu^2 t\right)\right| dt + \int_0^x \left|B(x,t)\right|\left|\sin\left(\mu^2 t\right)\right| dt \tag{5}$$

Thus,

$$\left|v_1(x,\mu)\right| \le \int_0^x \left|A(x,t)\right| e^{\left|\operatorname{Im}\left(\mu^2 t\right)\right|} dt + c_0 \int_0^x \left|B(x,t)\right| e^{\left|\operatorname{Im}\left(\mu^2 t\right)\right|} dt$$

$$\le e^{x\left|\operatorname{Im}\mu^2\right|}\left(\int_0^x \left|A(x,t)\right| dt + c_0 \int_0^x \left|B(x,t)\right| dt\right)$$

$$\le e^{x\left|\operatorname{Im}\mu^2\right|}\left(\int_0^\pi \max_{0\le x\le\pi}\left(\left|A(x,t)\right|\right) dt + c_0 \int_0^\pi \max_{0\le x\le\pi}\left(\left|B(x,t)\right|\right) dt\right)$$

from which we obtain

$$\left|v_1(x,\mu)\right| \le e^{x\left|\operatorname{Im}\mu^2\right|}\left(c_1 + c_0c_2\right)$$

So, we have proved the estimate. Therefore, $v_1(x,\mu)$ is entirely of type x order 1 and square integrable on the real line as a function of μ for each x.

Theorem 2 $v_2(x,\mu) \in PW_x$ is a function of μ for each x and the following estimate hold:

$$\left|v_2(x,\mu)\right| \le c_0c_2 e^{x\left|\operatorname{Im}\mu^2\right|} \tag{6}$$

Proof

$$v_2(x,\mu) = \int_0^x \left\{A(x,t)\cos\left(\mu^2 t\right) dt + B(x,t)\sin\left(\mu^2 t\right)\right\} dt - \int_0^x \left\{A(x,t)\cos\left(\mu^2 t\right)\right\} dt$$

$$\left|v_2(x,\mu)\right| \le \int_0^x \left|B(x,t)\right|\left|\sin\left(\mu^2 t\right)\right| dt$$

$$\le \int_0^x c_0\left|B(x,t)\right| e^{\left|\operatorname{Im}\left(\mu^2 t\right)\right|} dt$$

$$\le c_0 e^{x\left|\operatorname{Im}\mu^2\right|}\left(\int_0^\pi \max_{0\le x\le\pi}\left|B(x,t)\right| dt\right)$$

from which we get

$$\left|v_2(x,\mu)\right| \le (c_0 c_2) e^{x\left|\operatorname{Im}\mu^2\right|}$$

Hence, function $v_2(x,\mu)$ is entirely of type x order 1 and square integrable on the real line as a function of μ for each x.

The boundary function(characteristic equation) $B(\mu)$ is not necessarily in PW_π as in the Dirichlet-Dirichlet case. However, we have the following theorem.

Theorem 3 $\tilde{B}(\pi,\mu) = a_{21} v_1(\pi,\mu) + a_{22} v_2(\pi,\mu) \in PW_\pi$ is a function of μ and the following estimate holds:

$$\left|\tilde{B}(\pi,\mu)\right| \le c_3 e^{\pi\left|\operatorname{Im}\mu^2\right|} \tag{7}$$

Proof Primarily, we have

$$\left|\tilde{B}(x,\mu)\right| \le |a_{21}|\left|v_1(x,\mu)\right| + |a_{22}|\left|v_2(x,\mu)\right|$$

from which we acquire

$$\begin{aligned}\left|\tilde{B}(x,\mu)\right| &\le |a_{21}|(c_1 + c_0 c_2) e^{x\left|\operatorname{Im}\mu^2\right|} + |a_{22}| c_0 c_2 e^{x\left|\operatorname{Im}\mu^2\right|} \\ &\le e^{x\left|\operatorname{Im}\mu^2\right|}\left((c_1 + c_0 c_2)|a_{21}| + c_0 c_2 |a_{22}|\right)\end{aligned}$$

so

$$\left|\tilde{B}(\pi,\mu)\right| \le c_3 e^{\pi\left|\operatorname{Im}\mu^2\right|}$$

$\left|\tilde{B}(\pi,\mu)\right|$ is easily seen from above mentioned inequality. Hence, the following theorem is applicable.

Theorem 4 Let $f \in PW_\pi$, then

$$f(\mu) = \sum_{k=-\infty}^{\infty} f(k) \frac{\sin \pi(\mu - k)}{\pi(\mu - k)}, \tag{8}$$

where the series converges uniformly on compact set of $\mathbb{R}$ and also in $L^2_{d\mu}$ [6].

Let $\tilde{B}_N(\pi,\mu)$ denote the truncation of $\tilde{B}(\pi,\mu)$

$$\tilde{B}_N(\pi,\mu) = \sum_{k=-N}^{N} \tilde{B}(\pi,k) \frac{\sin \pi(\mu - k)}{\pi(\mu - k)}, \tag{9}$$

and $B_N(\pi,\mu)$ the corresponding approximation to $B(\pi,\mu)$.

Conclusion

In this paper, we examined Diffusion operator and succeeded in performing our approach for Diffusion equation. The approach is based on the well established technique: Shannon's sampling theorem. Thus, we obtained satisfactory results by using the Paley-Wiener spaces.

References

1. E. Bairamov, Ö. Çakar, and A. O. Çelebi, Quadratic pencil of Schrödinger operators with spectral singularities, discrete spectrum and principal functions, J. Math. Anal. Appl. **216** (1997), 303-320.
2. M. G. Gasymov, and G. Sh. Guseinov, Determination of diffusion operator on spectral data, Dokl. Akad. Nauk Azerb. SSSR **37**, 2 (1981), 19-23.
3. H. Koyunbakan, and E. S. Panakhov, Half inverse problem for diffusion operators on the finite interval, J. Math. Anal. Appl. **326** (2007), 1024-1030.
4. B. Chanane, Sturm-Liouville problems with parameter dependent potential and boundary conditions, Journal of Computational and Applied Mathematics **212** (2008), 282-290.
5. K. Chadan, and P.C. Sabatier, Inverse Problems in Quantum Scattering Theory, Second edition, Springer-Verlag (1989).
6. A. I. Zayed, Advances in Shannon's Sampling Theory, CRC Press, Boca Raton (1993).
7. B. M. Levitan, and I. S. Sargsjan, Introduction to Spectral Theory: Selfadjoint Ordinary Differential Operators, American Mathematical Society, Providence, Rhode Island (1975).
8. B. Chanane, Computing eigenvalues of regular Sturm-Liouville problems, Appl. Math. Lett. **12** (1999), 119-125.
9. A. Boumenir, and B. Chanane, Computing eigenvalues of Sturm-Liouville systems of Bessel type, Proc. Edinburgh Math. Soc. **42** (1999), 257-265.
10. B. Chanane, Computing the eigenvalues of singular Sturm-Liouville problems using the regularized sampling method, Appl. Math. Com. **184** (2007), 972-978.
11. A. Boumenir, and B. Chanane, Eigenvalues of Sturm-Liouville systems using sampling theory, Applicable Analysis **62** (1996), 323-334.
12. B. Chanane, High Order Approximations of the Eigenvalues of Regular Sturm-Liouville Problems, J. Math. Anal. Appl. **226** (1998), 121-129.

AUTHOR INDEX

Aabidi M.H. 232
Akbas Kübra Elif 207
Aksan Emine Nesligül 139
Altin Yavuz 53
Altinok Hifsi 53
Ammar Hany 232
Antonov Valery 264
Asmouh Ilham 144
Atas Sibel Sehriban 149, 157
Ay Karakus Betül 244
Aydin Galip 244
Babaoglu Mine 281
Baidada Chafik 232
Bal-Incebacak Belgin 14, 39, 108
Baltürk Yücel 255
Bas Erdal 58
Baskonus Haci Mehmet 144, 157, 165, 225
Bildik Necdet 1
Bulut Hasan 139, 149, 157, 165, 171, 192, 229
Çalik Sinan 196
Çetin Ismail 97
Çilingir Süngü Inci 212
Davidov Roman 264
Degirmenci Ozge Irmak 192
Demir Duygu Dönmez 73
Demir Hüseyin 255
Demirdag Betul 165
Deniz Sinan 1
El Mahi Bouziane 232
Erdogan Fevzi 144
Ersoy Esen 14, 39, 97, 108
Esen Alaattin 250
Et Mikail 53, 68
Gençoglu Muharrem Tuncay 84, 89
Giri Debasis 178
Gökdere Gökhan 200
Gül Ozan 260
Güral Yunus 196
Gürcan Mehmet 196
Hallaç Ibrahim Riza 244
Isik Erdem 149
Isik Hilal Arslanoglu 171
Isik Mahmut 207
Isik Ozlem 192
Jakimi Abdeslam 232
Kalinin Nikolay 264
Karaagac Berat 250
Karay Mehmet 126
Kayhan Miraç 139
Koçak Zeynep Fidan 64
Lisitsyna Irina 229
Maitra Tanmoy 178
Makinist Semiha 244
Ozarslan Ramazan 58
Ozkul Arif 144
Panakhov Etibar S. 281
Polismakova Maria 239
Pronin Vitaliy 225
Ryakhovskiy Alexey 264
Sandulyak Alexander 239
Sandulyak Anna 225, 229, 239
Sandulyak Darya 239
Sendurur Emine 97
Shkatov Petr 225, 229
Sulaiman Tukur Abdulkadir 171
Tan Nusret 260
Tuz Münevver 272
Yagmurlu N. Murat 250
Yel Gülnur 64
Zeybek Aylin 73